SEISMIC AMBIENT NOISE

The seismic ambient field allows us to study interactions between the atmosphere, the oceans, and the solid Earth. The theoretical understanding of seismic ambient noise has improved substantially over recent decades, and the number of its applications has increased dramatically. With chapters written by eminent scientists from the field, this book covers a range of topics including ambient noise observations, generation models of their physical origins, numerical modeling, and processing methods. The later chapters focus on applications in imaging and monitoring the internal structure of the Earth, including interferometry for time-dependent imaging and tomography. This volume thus provides a comprehensive overview of this cutting-edge discipline for graduate students studying geophysics and for scientists working in seismology and other imaging sciences.

NORI NAKATA is a principal research scientist in geophysics at the Massachusetts Institute of Technology. He received the Mendenhall Prize from the Colorado School of Mines in 2013 and the Young Scientist Award from the Seismological Society of Japan in 2017. His research interests include crustal and global seismology, exploration geophysics, volcanology, and civil engineering.

LUCIA GUALTIERI is a postdoctoral research associate at Princeton University, mainly interested in studying the coupling between the solid Earth and the other Earth systems, and in using seismic signals to image the Earth's structure. She received Young Scientist Awards from the Italian Physical Society in 2016, the American Geophysical Union (AGU) in 2017, and the New York Academy of Sciences in 2018.

ANDREAS FICHTNER is a professor leading the Seismology and Wave Physics Group at ETH Zürich. His research interests include inverse theory and tomography, numerical wave propagation, effective medium theory, and seismic interferometry. He received early career awards from the AGU and the IUGG, and is the recipient of an ERC Starting Grant. He serves as a consultant in the development of Salvus (a suite of full waveform modeling and inversion software) with a focus on seismic and seismological applications.

SEISMIC AMBIENT NOISE

Edited by

NORI NAKATA
Massachusetts Institute of Technology, USA

LUCIA GUALTIERI
Princeton University, USA

and

ANDREAS FICHTNER
ETH Zürich, Switzerland

CAMBRIDGE
UNIVERSITY PRESS

CAMBRIDGE
UNIVERSITY PRESS

Shaftesbury Road, Cambridge CB2 8EA, United Kingdom

One Liberty Plaza, 20th Floor, New York, NY 10006, USA

477 Williamstown Road, Port Melbourne, VIC 3207, Australia

314–321, 3rd Floor, Plot 3, Splendor Forum, Jasola District Centre, New Delhi – 110025, India

103 Penang Road, #05–06/07, Visioncrest Commercial, Singapore 238467

Cambridge University Press is part of Cambridge University Press & Assessment, a department of the University of Cambridge.

We share the University's mission to contribute to society through the pursuit of education, learning and research at the highest international levels of excellence.

www.cambridge.org
Information on this title: www.cambridge.org/9781108417082

DOI:10.1017/9781108264808

First published 2019

A catalogue record for this publication is available from the British Library

Library of Congress Cataloging-in-Publication data
Names: Nakata, Nori, 1984– editor. | Gualtieri, Lucia, 1986– editor. |
Fichtner, Andreas, 1979– editor.
Title: Seismic ambient noise / edited by Nori Nakata, Lucia Gualtieri, and Andreas Fichtner.
Description: Cambridge ; New York, NY : Cambridge University Press, 2019. |
Includes bibliographical references and index.
Identifiers: LCCN 2018041954 | ISBN 9781108417082 (alk. paper)
Subjects: LCSH: Seismic waves. | Seismometry. | Earthquake sounds. |
Seismology – Research – Methodology.
Classification: LCC QE538.5 .S383 2019 | DDC 551.22–dc23
LC record available at https://lccn.loc.gov/2018041954

ISBN   978-1-108-41708-2   Hardback

Contents

R. SNIEDER, A. DURAN, AND A. OBERMANN

Color plate section to be found between pages 164 and 165

Contributors

Fabrice Ardhuin *Laboratoire d'Océanographie Physique et Spatiale, Université de Bretagne Occidentale, CNRS, Ifremer, IRD, Plouzané, France*

Richard I. Boaz *Software Architect and Scientific Programmer, Boaz Consultancy*

Florent Brenguier *ISTerre, Université Grenoble-Alpes, Grenoble, France*

Michel Campillo *ISTerre, Université Grenoble-Alpes, Grenoble, France*

Alejandro Duran *Swiss Seismological Service, ETH Zürich, Zürich, Switzerland*

Lili Feng *Department of Physics, University of Colorado at Boulder, Boulder, CO, USA*

Andreas Fichtner *Department of Earth Sciences, ETH Zürich, Zürich, Switzerland*

Martin Gal *School of Physical Sciences (Earth Sciences), University of Tasmania, Hobart, Australia*

Lucia Gualtieri *Department of Geosciences, Princeton University, Princeton, NJ, USA*

Koichi Hayashi *OYO Corporation / Geometrics, Inc., San Jose, CA, USA*

Daniel E. McNamara *USGS National Earthquake Information Center, Golden, CO, USA*

Nori Nakata *Department of Earth, Atmospheric and Planetary Sciences, Massachusetts Institute of Technology, Cambridge, MA, USA*

Kiwamu Nishida *Earthquake Research Institute, University of Tokyo, Tokyo, Japan*

Anne Obermann *Swiss Seismological Service, ETH Zürich, Zürich, Switzerland*

Anya M. Reading *School of Physical Sciences (Earth Sciences), University of Tasmania, Hobart, Australia*

Michael H. Ritzwoller   *Department of Physics, University of Colorado at Boulder, Boulder, CO, USA*

Christoph Sens-Schönfelder   *GFZ German Research Centre for Geosciences, Potsdam, Germany*

Nikolai M. Shapiro   *Institut de Physique du Globe de Paris, CNRS-UMR7154, Université Paris Diderot, Paris, France*

Roel Snieder   *Center for Wave Phenomena, Colorado School of Mines, Golden, CO, USA*

Eléonore Stutzmann   *Institut de Physique du Globe de Paris, Paris, France*

Victor C. Tsai   *Seismological Laboratory, California Institute of Technology, Pasadena, CA, USA*

Foreword

This volume presents a series of contributions that provide an overview of the latest developments in the recent and rapidly developing field of noise-based seismology. While earthquake records have already revealed most of the features of the deep Earth, seismic sensors are now sensitive enough to resolve the permanent vibrations of the ground that continue between earthquake shakings. Often referred to as "noise," these weak permanent vibrations are the physical signals that are produced by seismic waves, although their sources are often poorly understood, and they change their nature and location with the frequency band considered. This natural noise is particularly strong in the microseism band (0.04–0.2 Hz), which is also of interest for imaging purposes at the lithospheric and global scales. The relationships between meteorological activity, oceanic swell and microseisms were noted in the earliest days of the study of seismograms. The issue of using noise to study Earth structure has attracted the attention of seismologists for a long time, with notable contributions such as those of Aki in 1957, Claerbout in 1968, and Nogoshi and Igarashi in 1971.[1]

Back in 1980, in his presidential address to the Seismological Society of America, Aki[2] proposed to use the microseisms generated under the oceans:

We know now it is possible to simultaneously determine the locations and origin times of local earthquakes and the structure of the earth in which the earthquakes are taking place, if we have a sufficient number of stations. In principle, this method of simultaneously determining the source and structure parameters can be extended to the microseisms, because the source of microseisms can probably be described by a finite number of parameters.

[1] Aki, K. (1957), Space and time spectra of stationary stochastic waves with special reference to microtremors, *Bull. Earthq. Res. Inst.*, **35**, 415–456.
Claerbout, J. (1968), Synthesis of a layered medium from its acoustic transmission response: *Geophysics*, **33**, 264–269.
Nogoshi, M. and Igarashi, T. (1971), On the amplitude characteristics of microtremor (part 2). (*in Japanese with English abstract*). *J. Seismol. Soc. Japan*, **24**, 26–40.

[2] The full text is available from: Aki, K. (1980) Presidential Address: Possibilities of seismology in the 1980's, *Bull. Seismol. Soc. Am.*, **70**(5), 1969–1976.

Interestingly, he also indicated:

The advantage of microseisms is that they exist 24 hours a day, all year around. A signal processing method may be developed to extract the body wave part of microseisms, in order to use them for deep structure studies.

These sentences look particularly pertinent today, although so far they have not been followed by many direct applications of the principles that they state. Indeed the sources of noise have proved to be multiple, and they result from complex processes and are difficult to reduce from the seismological measurements to a small number of parameters.

The potential to exploit the continuous noise records has recently become a critical issue. This comes with the advent of large networks that can provide huge amounts of continuous digital data that can be easily accessible from public databases and are now ready for massive processing with our improved computing capabilities. To take advantage of the wealth of noise records, an approach slightly different from the one foreseen by Aki can be proposed. This consists of seeking information on the medium in the propagation properties of the waves themselves, with limited reference to their sources. This was done 15 years ago with the use of long-term averages of the correlations between the noise records of distant stations, to retrieve the propagation properties between the two stations.[3] Correlations of coda waves and ambient seismic vibrations are effectively and widely used now to reconstruct impulse responses between two distant passive receivers, as if a source was placed at one of them. We have to acknowledge Aki for his 1957 study where he initially proposed a way to study the local structure with noise records, without knowing the source of the noise. This was an important conceptual advance. Assuming a single surface wave mode, Aki used a specific local array geometry to remove the effect of directivity of the noise by azimuthal averaging. The modern version relies on the spatio-temporal variability of the noise source, or on scattering, to produce the required averaging. In practice, the central part of the processing of noise records is the estimation of a time-average cross-correlation that can be identified as the impulse seismic response of the Earth between two sensors. Note that under restrictive conditions on the nature of the noise, firm theoretical ground exists to support this approach without inter-station distance limitations or hypotheses as to the structure of the medium. Although these conditions on the noise properties are rarely fully satisfied, the magic of waves is operating in the form of time reversibility and spatial reciprocity that leads to partial reconstruction of the impulse responses between two points from the correlations of

---

[3] A short report of the emergence of noise correlations is given in Courtland, R. (2008), Harnessing the hum, *Nature*, **453**, 146–148.

passive records. Noise correlations can be understood as a realization of an experiment of time-reversal self-focusing.[4] This simple physical interpretation explains the robustness of the method and the positive role of scattering on the quality of the retrieval of the impulse responses that are observed when dealing with the data. Importantly for imaging applications, the precision of the measurements of travel times deduced from noise records can be quantified, and it has been shown to be sufficient in practice for most tomographic applications.

On these grounds, passive imaging has grown rapidly, first with surface-wave tomography, as surface waves are the most easily retrieved components of the seismic field. With a large array of N three-component station arrays, a set of $N \cdot (N-1)/2$ inter-station paths can be built for each nine cross-component correlation to be computed for positive and negative time. This means that large sets of local measurements can be produced that allow in practice not only to repeat what can be done with earthquake data, but also to reach unprecedented resolution. This approach has been applied worldwide at scales ranging from shallow layers for engineering purposes to lithospheric imaging. The reconstruction of body waves is more problematic with sources distributed at the surface. High-frequency body-wave retrieval appears to be possible through the significant scattering that occurs at high frequencies. Long-period body waves are weakly sensitive to scattering, and specific analyses have to be done before apparent travel times can be used in correlations for inferring Earth structures. This is a typical example where detailed knowledge of the structure of the wavefield is required.

The characteristics of the noise are well studied nowadays. The ambient noise wavefield is investigated from the available large sensor arrays, with specific processing and beamforming techniques. With the same class of techniques, it is possible to locate the sources of noise and to validate the physical models of generation based on oceanographic data. Together with refined models of microseism sources constructed from seismological analysis and from independent meteorological and oceanographic observations, the initial strategy of Aki in the 1980s to invert for sources and structures can be reconsidered with up-to-date inversion procedures.

An exciting opportunity offered by using the ambient noise is to repeat the measurements at different dates, and to move forward to time-dependent imaging of the Earth. It has been shown that very small changes in the elastic properties of rock at depth can be observed through measures of the temporal changes of seismic velocities evaluated with the scattered parts of the retrieved impulse responses.

---

[4] A heuristic presentation of the correlation methods in a general framework of wave physics can be found in Derode, A., Larose, E., Campillo, M., and Fink, M. (2003), How to estimate the Green's function of a heterogeneous medium between two passive sensors? Application to acoustic waves, *Appl. Phys. Lett.*, **83**(15), 3054–3056.

This is a perspective for various applications, including the monitoring of underground industrial activities, or the detection of the small precursory changes before instabilities related to natural hazards, such as volcanic eruptions, landslides, and earthquakes. For the study of deep processes, the difficulty is that the Earth's crust is also changing through various external forces (e.g., rainfall, snow load, tidal and thermal effects), and that these effects have to be removed from the actual temporal observations to isolate the changes related to internal processes of natural or industrial origins. At the same time, this shows that seismology can contribute to a vast domain of environmental sciences, including hydrology, man-induced hazards, and monitoring of climate-related processes.

Ambient noise seismology is a promising field that is in rapid evolution, and that has turned out to be one of the main components of geophysical imaging. In this time of development of observations with new sensors (e.g., large nodal arrays, optical fibers, rotation sensors), the possibilities are numerous, and wide avenues are open for new applications of passive seismology. The perspectives merged in this volume will help the reader to understand the present-day challenges. As solutions are proposed, new questions appear, more and more scientists are involved, and novel data analyses lead to discoveries; there can be no doubt that the years to come will be fascinating. Ambient noise still holds a lot of the mysteries of the Earth to reveal.

Michel Campillo
ISTerre, Université Grenoble-Alpes, Grenoble, France

Acknowledgments

We would like to acknowledge colleagues and friends who helped us with this project during the past one and a half years. First and foremost, we thank all the authors who contributed a chapter to this book. We know very well that a book chapter is sometimes (and we think incorrectly) not considered a top-level scientific contribution, which makes it even harder to reserve time to write. We nevertheless hope that the prospect of reaching many students and colleagues will be sufficiently rewarding. Without the authors' deep knowledge, a book on such a diverse and rapidly growing topic like seismic ambient noise would have been impossible.

Critical readings of each chapter were undertaken by (in alphabetical order) Michael Afanasiev, Florent Brenguier, Evan Delaney, Laura Ermert, Koichi Hayashi, Dirk-Phillip van Herwaarden, Naiara Korta, Lion Krischer, Kiwamu Nishida, Anne Obermann, Patrick Paitz, Anya Reading, Michael Ritzwoller, Korbinian Sager, Christoph Sens-Schönfelder, Leonard Seydoux, Yixiao Sheng, Roel Snieder, and Victor Tsai. We appreciate their valuable comments, advice, criticism, and understanding of the concept of the educational aspects.

We are particularly grateful to Cambridge University Press. Susan Francis supported us in establishing the concept of this book and kept encouraging us to complete it. Zoë Pruce and Sarah Lambert always promptly and patiently replied to countless emails, making it as easy as possible for us to finalize this book. We are grateful for the opportunity to edit and write this book, and for the pleasure and little bit of struggle during this task.

Nori, Lucia & Andreas

Introduction

ANDREAS FICHTNER, LUCIA GUALTIERI,
AND NORI NAKATA

In the late 1860s and early 1870s, the Italian priest Timoteo Bertelli (1826–1905) mounted a pendulum on the wall of the college where he taught natural sciences. With the help of a lens, he observed the phenomenon that he had traced in historic records back to the year 1643: small, spontaneous movements that did not have any obvious explanation. Following improvements of his instrument and thousands of experiments, he was able to exclude passing vehicles, wind, and temperature variations as possible sources (Davison, 1927). In 1872 he concluded that some of his microseismic observations coincided with distant earthquakes but also with barometric depressions (Bertelli, 1872). Furthermore, they seemed to be stronger in winter than in summer. He had made some of the first reliable observations of the ambient seismic field, providing a first indication of its possible sources.

I.1 The Ambient Seismic Field

The surface of the solid Earth is subject to continuously acting forces caused by the whole bandwidth of human activities, and by a wide range of natural phenomena. The ambient field generated by these forces constitutes the seismic background radiation of the Earth, historically referred to as *microseisms*. Observations in the early days of instrumental seismology already suggested a close relation between microseisms and meteorological conditions (e.g., Klotz, 1910; Burbank, 1912; Banerji, 1925), long before the first physical theories for microseism generation were proposed (e.g., Miche, 1944; Longuet-Higgins, 1950; Hasselmann, 1963).

The ambient field is omnipresent, and its amplitude varies with position, time, and frequency. The quasi-random nature of the ambient field, a small snapshot of which is shown in Figure I.1, disables the detection of distinct arrivals, well-known from the analysis of earthquakes, explosions, or other sources of short duration that

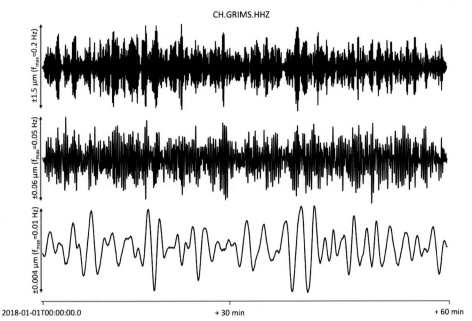

CH.GRIMS.HHZ

2018-01-01T00:00:00.0   + 30 min   + 60 min

Figure I.1. One hour of ambient seismic field recording on the vertical component of station GRIMS, located at 1746 m altitude inside a tunnel system in the Swiss Alps. Shown are three traces, low-pass filtered at 0.2 Hz (top), 0.05 Hz (middle), and 0.01 Hz (bottom). Note the different vertical scales.

have traditionally been the subject of seismological investigations. The absence of an easily identifiable, deterministic signal has led to the common classification of the ambient field as *seismic ambient noise*. Despite being quasi-random, the ambient field obeys the laws of physics, which imprint a coherent structure into the apparent disorder. Most fundamentally, ambient noise is a seismic wavefield travelling between two points $\mathbf{x}_A$ and $\mathbf{x}_B$ at an apparent velocity v controlled by the elastic properties of the Earth's interior — just as any other wavefield. A wave packet $u(\mathbf{x}, t)$ passing by position $\mathbf{x}_A$ will reach position $\mathbf{x}_B$ after some time $\Delta t = |\mathbf{x}_A - \mathbf{x}_B|/v$. Therefore, we expect the wavefield recordings $u(\mathbf{x}_A, t)$ and $u(\mathbf{x}_B, t - \Delta t)$ to be statistically similar, or correlated.

Though being conceptually simple, this line of arguments suggests that the hidden coherence of the ambient field might be extracted through the computation of correlations between pairs of stations. Indeed, when averaging over long enough times, the interstation correlation

$$C(\mathbf{x}_A, \mathbf{x}_B, t) = \int u(\mathbf{x}_A, \tau)\, u(\mathbf{x}_B, t + \tau)\, d\tau \,, \qquad \text{(i)}$$

approximates the wavefield that would be recorded at receiver position $\mathbf{x}_B$ if a source had acted at receiver position $\mathbf{x}_A$. Equation (i) effectively turns stations A

and *B* into a large interferometer that emphasizes those parts of the ambient field that travel coherently between them, while suppressing incoherent components. The interferogram $C(\mathbf{x}_A, \mathbf{x}_B, t)$ has two outstanding properties: It can be repeatedly computed at any time because noise is always present, and the virtual source *A* can be positioned anywhere, independent of any real wavefield sources.

Expressions that relate correlations of noise to a deterministic wave travelling between two points in space have been known for decades (e.g., Aki, 1957; Claerbout, 1968). As illustrated in Figure I.2, knowledge that coherent signals may be extracted largely precedes the recent boom of ambient noise seismology and of ambient noise tomography in particular (e.g., Sabra et al., 2005; Shapiro et al., 2005). This suggests that the apparent simplicity of equation (i) might be deceiving. Indeed, it hides the presence of instrumental noise and of transient signals, for

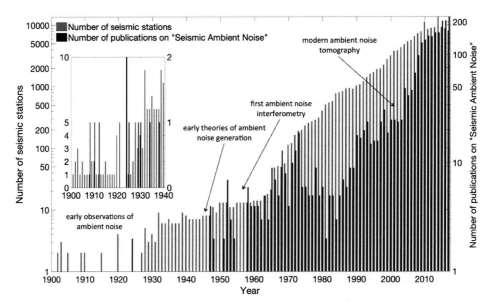

Figure I.2. Histograms showing the number of seismic stations in the *IRIS* data archive (www.iris.edu) and the number of publications in the *Web of Science* since the beginning of the 20th century, containing the words "seismic ambient noise" or "microseisms." Early observations mostly related ambient noise recordings to meteorological phenomena (e.g., Klotz, 1910; Burbank, 1912; Banerji, 1925). Pioneering theories for ambient noise generation and interferometry appeared in the 1950s and 1960s (e.g., Miche, 1944; Longuet-Higgins, 1950; Aki, 1957; Hasselmann, 1963; Claerbout, 1968; Haubrich and McCamy, 1969). The number of publications experienced a rapid growth after the first applications of seismic interferometry to image the Earth structure (e.g., Sabra et al., 2005; Shapiro et al., 2005). The present book is intended to respond to the needs of an increasing number of scientists who are approaching the field of ambient noise seismology.

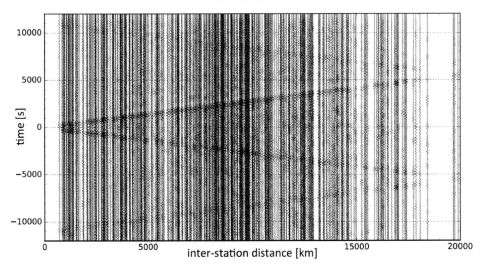

Figure I.3. Ambient noise correlations of 420 globally distributed station pairs, averaged over 1 year, from 1 January 2014 to 1 January 2015. The frequency range is 3–5 mHz. Coherent arrivals, corresponding to vertical-component Rayleigh waves, are clearly visible. (Figure prepared by Laura Ermert and Alexey Gokhberg.)

instance from earthquakes, that may, by far, overwhelm the low-amplitude ambient signal. As a consequence, the emergence of noise interferometry as a widely used technique had to await improvements of seismic instruments (with lower instrumental noise compared to ambient noise), the deployment of seismic arrays that enable the detection of weak coherent signals, and the development of processing techniques to suppress large-amplitude transients.

Today, the seismic ambient field is known to be coherent over length scales exceeding the circumference of the Earth (e.g., Nawa et al., 1998; Nishida et al., 2009), as illustrated in Figure I.3. It is used constructively in applications ranging from volcano and reservoir monitoring to global seismic tomography. Though the usefulness of the ambient field is meanwhile undisputed, its classification as noise stubbornly persists – probably as a deliberate antagonism between the past when most seismologists tried to suppress the ambient field, and the modern era where noise has indeed become signal. In choosing the title of this book, we could not resist following this trend.

I.2 The Scope of This Book

Writing a book about a topic that is as dynamic as ambient noise seismology is not a trivial task. On the one hand, we are convinced that such a book is needed, not only for our academic colleagues, but also for many young students working in this field. On the other hand, we are well aware of the fact that the very latest developments may not be included, and that the half-life of some of the book's content may be rather short.

Despite being a comparatively young branch of geophysics, ambient noise seismology is already too diverse to be covered with sufficient competence by a single author. This led us to compile a sequence of chapters written by well-known experts in their fields. While this strategy promotes depth and completeness, it also comes with the risk of slight divergences in opinion, notation, and presentation style. We hope that the benefits of this diversity overcompensate for its disadvantages.

First and foremost, this book is written for students and professionals with no or little experience in ambient noise seismology. Its character is largely educational. Therefore, the chapters are more or less self-contained, starting at the level of mathematics and physics of a beginning PhD student with some background in natural sciences.

We tried to cover a wide range of topics, including the observation and processing of ambient noise signals, the physical origin and numerical modeling of ambient noise, the theoretical foundations of noise and coda-wave interferometry, as well as applications from local to global scales. With limited space available, we had to make decisions concerning topics and the level of detail. We very much hope that colleagues whose topics we were unable to cover will understand this decision.

I.3 Outline

This book is roughly organized to proceed from the phenomenology and observations of the ambient noise field and its sources, to theory and methods for random wavefield interferometry, and finally to applications at various scales.

Chapter 1 by McNamara and Boaz (2018) describes the spectral properties of the ambient field, from the high frequencies that result from human activity, to low frequencies caused largely by ocean waves. The chapter introduces the power-spectral density as one of the most useful tools to characterize the noise level experienced by a seismic network. Knowing the noise power-spectral density is crucial for applications like earthquake monitoring and detection where small but interesting events may easily drown within the noise.

The wave type and propagation direction of the ambient field are the topics of Chapter 2 by Gal and Reading (2018), who introduce beamforming and polarization analysis. A special focus is on single-component plane-wave beamforming and on data-adaptive beamforming. These techniques are widely used to infer, for instance, the location and strength of ambient field sources in space and time.

The effect of seismic array geometry on the frequency-dependent resolution of a beamforming technique is discussed in detail.

Beamforming and polarization analysis hint at the actual physics of ambient noise generation, covered in Chapter 3 by Ardhuin et al. (2018). At frequencies below 1 Hz, where anthropogenic noise excitation is weak, ocean waves are the dominant source. Though seismic waves have wavelengths much longer than ocean waves, for a given frequency, two mechanisms are shown to provide efficient coupling to the solid Earth: the direct interference of ocean waves with shallow ocean bottom topography, and the interference of pairs of ocean waves that exert small pressure variations also at greater depths.

The extraction of coherent, deterministic signals from quasi-random noise recordings is the subject of Chapter 4 by Fichtner and Tsai (2018). Assuming wavefield equipartitioning or homogeneously distributed noise sources, correlations of noise as in equation (i) approximate interstation Green's functions, an important result that forms the basis of most ambient noise studies of the Earth's (time-dependent) internal structure. More general, emerging, approaches to noise interferometry make no assumptions on the nature of the ambient field, but also require the joint consideration of noise sources and Earth structure.

To the frustration of ambitious ambient field seismologists, the precious noise tends to be superimposed by transient signals with much higher amplitude, for instance, from earthquakes. The challenge of emphasizing noise so as to promote the emergence of correlation functions that reliably approximate a Green's function is the subject of Chapter 5 by Ritzwoller and Feng (2018). The authors describe many of the techniques that have been developed during the past decade, and that have proven to be effective in a wide range of settings.

Quasi-random wavefields may arise from quasi-random sources, such as ocean waves, but also from multiple scattering in the heterogeneous Earth. In this regard, scattering-generated coda waves and the ambient field are close relatives. As shown by Snieder et al. (2018) in Chapter 6, subtle changes of coda waves are valuable indicators of changes in Earth structure. Despite being diffuse in nature, coda waves may in fact be used to locate temporal velocity variations.

Chapter 7 by Shapiro (2018) is the first in a series of chapters focused on applications. It illustrates the use of surface waves in ambient noise to image Earth structure, at scales ranging from the top hundred meters to the upper hundred kilometers. Surface wave observables, such as frequency-dependent traveltimes and amplitudes, may be extracted from interstation correlations. When dense arrays with an interstation spacing smaller than the wavelength are available, ambient noise surface wave phase velocities can be extracted directly and used for subsurface velocity and anisotropy imaging.

While ambient noise correlations are often dominated by surface waves, coherent body wave arrivals tend to be more difficult to extract. In Chapter 8, Nakata and Nishida (2018) show that advanced processing techniques, partly developed for dense arrays, can promote the emergence of body waves in noise correlations. The extracted signals can be used in applications ranging from the imaging of subsurface reservoirs to the mapping of discontinuities at several hundred kilometers depth.

In contrast to transient signals from earthquakes or explosions, ambient noise is omnipresent, which enables repeated measurements and the monitoring of subsurface changes. In Chapter 9, Sens-Schönfelder and Brenguier (2018) explain how coda waves extracted from ambient noise correlations can be used in practice to detect and locate minute velocity changes in the Earth. Their survey of recent applications includes observations of time-variable Earth structure related to volcanic processes, earthquake-related stress adjustments, and environmental factors, such as local hydrology. Often, observed velocity changes are larger than expected for ideal rock-forming minerals, suggesting that micro-structures and their associated nonlinear effects play a significant role.

Finally, Chapter 10 by Hayashi (2018) provides an introduction to the use of ambient noise methods for near-surface engineering of the upper tens to hundreds of meters. This includes a review of the spatial auto-correlation (SPAC) method, which provides robust inferences of near-surface shear velocity structure, and applications for ground motion predictions based on the estimated 2D shear wave velocity structure.

A book intended to serve as an introduction to a dynamic and expanding field can never be complete. We nevertheless hope to have found a reasonable balance between observations, theory, and applications that whets the reader's appetite for more.

References

Aki, K. 1957. Space and time spectra of stationary stochastic waves, with special reference to microtremors. *Bull. Earthq. Res. Inst., Univ. Tokyo*, **35**, 415–457.

Ardhuin, F., Gualtieri, L., and Stutzmann, E. 2018. Physics of ambient noise generation by ocean waves. Pages 69–108 of: Nakata, N., Gualtieri, L., and Fichtner, A. (eds.), *Seismic Ambient Noise*. Cambridge University Press, Cambridge, UK.

Banerji, S. K. 1925. Microseisms and the Indian monsoon. *Nature*, **116**, 866–866.

Bertelli, T. 1872. Osservazioni sui piccoli movimienti dei pendoli. *Bullettino Meteorologico dell'Osservatorio*, **XI(11)**.

Burbank, J. E. 1912. Microseisms caused by frost action. *Am. J. Sci.*, **33**, 474–475.

Claerbout, J. F. 1968. Synthesis of a layered medium from its acoustic transmission response. *Geophysics*, **33**, 264–269.

Davison, C. 1927. *The Founders of Seismology*. Cambridge University Press, Cambridge, UK.

Fichtner, A., and Tsai, V. 2018. Theoretical foundations of noise interferometry. Pages – of: Nakata, N., Gualtieri, L., and Fichtner, A. (eds.), *Seismic Ambient Noise*. Cambridge University Press, Cambridge, UK.

Gal, M., and Reading, A. M. 2018. Beamforming and polarization analysis. Pages – of: Nakata, N., Gualtieri, L., and Fichtner, A. (eds.), *Seismic Ambient Noise*. Cambridge University Press, Cambridge, UK.

Hasselmann, K. 1963. A statistical analysis of the generation of microseisms. *Rev. Geophys.*, **1**(2), 177–210.

Haubrich, R. A., and McCamy, K. 1969. Microseisms: coastal and pelagic sources. *Bull. Seis. Soc. Am.*, **7**(3), 539–571.

Hayashi, K. 2018. Near-surface engineering. Pages – of: Nakata, N., Gualtieri, L., and Fichtner, A. (eds.), *Seismic Ambient Noise*. Cambridge University Press, Cambridge, UK.

Klotz, O. 1910. Microseisms. *Science*, **32**, 252–254.

Longuet-Higgins, M. S. 1950. A theory of the origin of microseisms. *Philosophical Transactions of the Royal Society of London. Series A, Mathematical and Physical Sciences*, **243**(857), 1–35.

McNamara, D., and Boaz, R. 2018. Visualization of the Ambient Seismic Noise Spectrum. Pages – of: Nakata, N., Gualtieri, L., and Fichtner, A. (eds.), *Seismic Ambient Noise*. Cambridge University Press, Cambridge, UK.

Miche, M. 1944. Mouvements ondulatoires de la mer en profondeur constante ou décroissante. *Annales des Ponts et Chaussées*, **114**, 42–78.

Nakata, N., and Nishida, K. 2018. Body wave exploration. Pages – of: Nakata, N., Gualtieri, L., and Fichtner, A. (eds.), *Seismic Ambient Noise*. Cambridge University Press, Cambridge, UK.

Nawa, K., Suda, N., Fukao, Y., Sato, T., Aoyama, Y., and Shibuya, K. 1998. Incessant excitation of the Earth's free oscillations. *Earth Planets Space*, **50**, 3–8.

Nishida, K., Montagner, J.-P., and Kawakatsu, H. 2009. Global surface wave tomography using seismic hum. *Science*, **326**, 5949.

Ritzwoller, M. H., and Feng, L. 2018. Overview of pre- and post-processing of ambient noise correlations. Pages – of: Nakata, N., Gualtieri, L., and Fichtner, A. (eds.), *Seismic Ambient Noise*. Cambridge University Press, Cambridge, UK.

Sabra, K. G., Gerstoft, P., Roux, P., and Kuperman, W. A. 2005. Surface wave tomography from microseisms in Southern California. *Geophys. Res. Lett.*, **32**, doi:10.1029/2005GL023155.

Sens-Schönfelder, C., and Brenguier, F. 2018. Noise-based monitoring. Pages – of: Nakata, N., Gualtieri, L., and Fichtner, A. (eds.), *Seismic Ambient Noise*. Cambridge University Press, Cambridge, UK.

Shapiro, N. 2018. Applications with surface waves extracted from ambient seismic noise. Pages – of: Nakata, N., Gualtieri, L., and Fichtner, A. (eds.), *Seismic Ambient Noise*. Cambridge University Press, Cambridge, UK.

Shapiro, N. M., Campillo, M., Stehly, L., and Ritzwoller, M. 2005. High resolution surface wave tomography from ambient seismic noise. *Science*, **307**, 1615–1618.

Snieder, R., Duran, A., and Obermann, A. 2018. Locating velocity changes in elastic media with coda wave interferometry. Pages – of: Nakata, N., Gualtieri, L., and Fichtner, A. (eds.), *Seismic Ambient Noise*. Cambridge University Press, Cambridge, UK.

1

Visualization of the Seismic Ambient Noise Spectrum

DANIEL E. MCNAMARA AND RICHARD I. BOAZ

Abstract

In this chapter we visualize and review the major components of the Earth's ambient seismic noise spectrum. Short-period noise is dominated by human activity, while mid- and long-period noise is dominated by effects from ocean waves. For earthquake-monitoring and detection purposes it is important to know the seismic ambient noise levels experienced by a network of seismic stations. To make this assessment we describe the probability density function of power spectral density (PSDPDF) method in full. We discuss both operational and research applications of ambient seismic noise analysis, concluding with software and data resources available to students and researchers interested in studying the spectral characteristics of ambient seismic noise.

1.1 Introduction to Ambient Seismic Noise

For many seismologists, the study of seismic ambient noise aims to improve the fidelity of signals generated from events of interest such as earthquakes and nuclear explosions. In fact, the study of seismic noise illuminates a variety of signals which can reveal characteristics of a broad range of sources that include ocean waves (e.g., Gutenberg, 1936; Bromirski, 2009; Díaz, 2016) (see Chapter 3), climate (Aster et al., 2008, 2010; Ebeling, 2012), glaciers (O'Neel et al., 2007; Walter et al., 2010; Ekström et al., 2003, 2006), ship wakes, wind waves (McNamara et al., 2011), urban areas (e.g., McNamara and Buland, 2004), and even rugby matches (e.g., Boese et al., 2015). For operational seismology, the study of seismic noise can provide information about the performance of seismic instrumentation, recording station construction designs, and the quality of metadata (Hutt et al., 2017; McNamara et al., 2005, 2009; Ringler et al., 2010, 2015).

With advances in modern broadband instrumentation over the past few decades, theoretical understanding of seismic ambient noise has rapidly improved, and the number of its applications has increased dramatically. As you will see throughout the chapters of this book, seismic ambient noise/signals have been used to study many natural climate and environmental processes and improve the resolution of earth velocity and attenuation models (e.g., Bensen et al., 2008) (see Chapter 5). In this chapter we will use the standard seismology nomenclature and refer to any non-earthquake signal as noise. However, as we will demonstrate, noise is often a very useful signal.

1.1.1 Motivation to Visualize the Ambient Seismic Noise Spectrum

For earthquake-monitoring and detection purposes it is important to know the seismic ambient noise levels experienced by a network of seismic stations. Detailed frequency (1/period) dependent observations can contribute to improved seismic station and network design, which will in turn improve the detection of earthquakes and other seismic sources of interest, such as nuclear explosions. Beginning with the publication of high and low seismic background displacement curves of Brune and Oliver (1959) and Frantti et al. (1962), Earth noise models that have been used as baselines for evaluating seismic station site characteristics and construction methods, instrument quality, and environmental seismic noise sources. Later studies compared seismic noise models for different tectonic regions such as island arcs versus stable continental landmasses using more modern digital seismic instrumentation (Peterson, 1993; Stutzmann et al., 2000; Berger et al., 2004; Butler et al., 2004; Wilson et al., 2002; Bahavar and North, 2002; Reif et al., 2002). Using long-term broadband records from stations within the GSN, the standard Peterson (1993) new low-noise model (NLNM) (Figure 1.1a) was constructed by removing earthquakes and other transient signals from the seismic records to identify the minimum noise levels representative of quiet periods at continental interior seismic stations distributed around the world. The new high-noise model (NHNM) (Figure 1.1a) was constructed with the same processing methods, though using mostly island-based stations within the GSN (Peterson, 1993).

While it may be scientifically interesting to determine the absolute quietest noise levels achieved by a network, more recent studies find that these low noise levels are generally very low probability occurrences (< 10%) and do not closely track the significantly higher probability power levels. In fact, statistical median (50%) noise levels are as much as 25 dB above short-period (0.1–1 s) minimums (Figure 1.1a). For earthquake monitoring, it is important to know the distribution of ambient seismic noise in a probabilistic sense since noise significantly impacts magnitude detection threshold (e.g., McNamara et al., 2016).

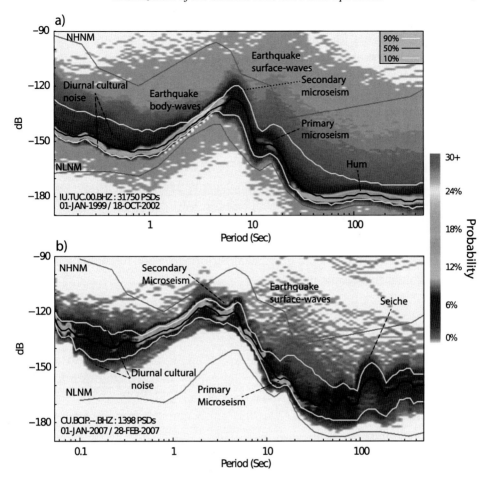

Figure 1.1. Major components of the seismic ambient noise spectrum. Red to green colors (in the color version; lighter grey colors in the greyscale version) indicate the highest probability seismic ambient noise power levels at each period. Baseline statistics shown as solid lines (10th% lower white line, 50th% black line and 90th% upper white line). New low-noise model (NLNM) and new high-noise model (NHNM) shown as gray lines (Peterson, 1993). (a) PSDPDF computed using data from a GSN station in Tucson, Arizona (IU.TUC.00.BHZ, 31,750 PSDs, 01-JAN-99-18-OCT-02). (b) PSDPDF computed using data from a GSN station on Isla Barro Colorado, Panama (CU.BCIP.00.BHZ, 1,398 PSDs, 01-JAN-07-28-FEB-07). (A black-and-white version of this figure appears in some formats. For the color version, please refer to the plate section.)

For example, many seismic stations are surrounded by urban areas and have considerably higher median noise levels at periods important to earthquake detection (0.1–1 s), and thus will significantly impact earthquake body and surface wave-wave detection capabilities (Figure 1.1). This is the principal reason that the noise levels of the NLNM have such low probabilities of occurrence for stations worldwide (Figure 1.1). For many stations such low levels of noise are unattainable, suggesting that for routine earthquake-monitoring purposes a global noise threshold based on higher probability noise levels is more useful.

There is no need to screen for system transients, earthquakes or general data artifacts since they map into a background, low-probability level of occurrence. Examination of artifacts related to station operation and episodic seismic noise, which would have been removed using past methods, allows us to estimate both the overall station quality and a baseline level that is specific to each site. This renders the results of this analysis useful for characterizing the performance of a seismic sensor, detecting operational problems within the recording system, and evaluating the overall quality of data produced by a particular station. The main advantage of the probability density function of power spectral density (PSDPDF) is the probabilistic view across a broad spectrum of periods. This allows for long-term assessment of the seismic instrument performance and research on the major sources of ambient seismic noise.

1.1.2 Sources of the Ambient Seismic Noise Spectrum

The broadband seismic ambient noise spectrum is multimodal with distinctly different physical mechanisms transferring energy into the solid Earth as seismic waves (e.g., Figure 1.1a). The broadband PSDPDF of ambient noise at a seismic station will thus describe noise coming from several different sources predominantly from machinery, water waves, and atmospheric oscillations.

At short periods (0.1–1 s) ambient noise power levels are generally dominated by human-generated (i.e., "cultural") seismic energy radiated from the electrical grid, cars, trains, and machinery within a few kilometers of the recording station. Intermediate periods (1–30 s) are dominated by microseisms, that can be many orders of magnitude higher in power than other parts of the seismic spectrum. Long-period (30–500 s) signals are generally caused by ocean infragravity waves generated by storm-forced, shoreward-directed winds, commonly referred to as "Hum." These major noise source characteristics are discussed in detail in Section 1.3.

This chapter provides the reader with the following: First, we review the PSDPDF method to visualize the spectral characteristics of long-term seismic ambient noise (Section 1.2 and Appendix 1.A). Second, we visualize and describe numerous sources of ambient seismic noise (Section 1.3). Third, we discuss operational and research applications of the PSDPDF (Section 1.4). And finally, we conclude with software and data resources available to students and researchers (Appendix 1.C).

1.2 Visualization Methods Overview

In this section we describe the PSDPDF method using two months of seismic data recorded at station CU.BCIP at the Smithsonian Tropical Research Station on Barro Colorado Island in the Panama Canal (Figure 1.1b). The example demonstrates how the PSDPDF improves the visualization of spectrum details by mapping varying and overlapping PSDs into a probability distribution. See Appendix 1.A for a detailed description of the PSD and PDF methods. See Appendix 1.B, Table 1.1 and Figure 1.7 for station location details used for the figures throughout this chapter.

The process begins with an instrument-corrected vertical-component of motion displacement seismic time series, in this case using seismic data from channel CU.BCIP.--.BHZ provided in Figure 1.2a. (See Appendix 1.C for seismic station and data channel naming conventions.) A Fourier Transform (Cooley and Tukey, 1965) is applied to convert the data from the time domain into the frequency domain, the resultant PSD computed as the normalized square of the displacement spectrum (Peterson, 1993), as seen in Figure 1.2b. Figure 1.2c shows two consecutive two-hour PSDs (with a 50% overlap). Minor variations in power levels are already apparent at short (< 1 s) and long (> 10 s) periods. Figure 1.2d shows 22 PSDs computed for a complete day, January 1, 2007. Significant power variation is further observed (20–30 dB) at the short and long periods over this single day. Figure 1.2e shows 158 hourly PSDs spanning the first week of January 2007, again showing significant variation and overlap of individual PSDs at short and long periods, this overlap potentially obscuring useful detail. Figure 1.2f shows 319 hourly PSDs that span the first two weeks of January 2007. Further significant variation of PSDs at short and long periods is observed with considerable overlap of the PSDs even further obscuring detailed information. In addition, surface waves in the 10–100 s period band are observed over several hours of decaying PSD power.

Grid cells shown in Figure 1.2f represent the discretization used to construct the PSDPDF. PSDs are gathered into 1/8 octave period bins and 1 dB power bins. The PDF is computed by counting the number of PSD intersections per grid cell and then dividing each grid cell sum by the total number of PSDs (319) to obtain a percentage for each period/power bin combination (McNamara and Buland, 2004). The PSDPDF example for seismic data channel CU.BCIP.--.BHZ (Figure 1.1b) contains significant detail at the short and long periods that is obscured by numerous overlapping individual PSDs (Figure 1.2f) but readily visualized in the PSDPDF. Also, at long periods (> 100 s) there is an obvious bimodal distribution of power (Figure 1.1b) not observed in the overlaid PSDs (Figure 1.2f). Section 1.4 provides additional details on the source mechanism of this bimodal distribution of long-period power.

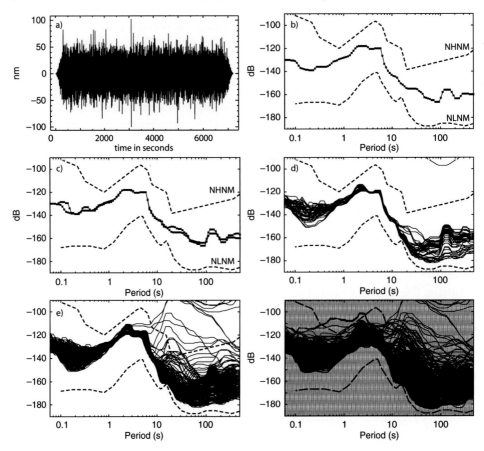

Figure 1.2. Example of the PSDPDF method. (a) CU.BCIP.-.BHZ two-hour (7200 s) seismogram beginning on 2007-01-01 00:00:00. (b) PSD calculated for the two-hour segment in Figure 1.2a and smoothed by averaging powers over full octaves in 1/8 octave intervals, center points of averages shown. The dashed line shows the NLNM and NHNM from Peterson (1993). (c) Two continuous PSD segments. (d) PSDs for one day (22 segments). (e) PSDs for 1 week (158 segments). (f) PSDs for 2 weeks (319 segments) gridded for the PDF calculation.

1.3 PSDPDF Visualization and Applications

Long-term variation of seismic power is well visualized in a single PSDPDF plot including instrumentation problems, geographic location differences, and diurnal and seasonal variations. This can be useful for observing time-varying (i.e., diurnal and seasonal) changes in one location as well as for making comparisons between locations. Advances in instrumentation have reduced noise levels in long-period bands (> 20 s), while local noise sources which vary from station to station, such as roads and population density, can increase short-period (< 1 s) absolute power levels and diurnal variations. Ambient noise in the mid-period band (1–20 s) is strongly affected by geographic location (mainly due to proximity of the station to coastlines) as well as the general effect of seasonal variation (McNamara and Buland, 2004).

Problems such as telemetry issues and incorrect instrument response calibrations are well visualized. For example, problems with overall sensor gain as well as transfer function shape, defined by poles and zeros, can be diagnosed by comparing the power levels versus standard baselines computed from PSDPDFs (McNamara et al., 2009). In addition, due to the probability estimate, it is possible to determine how often system problems occur, such as gaps, sensor mass re-centering, spiking, and clipping.

Optimal seismic station siting and good vault design are both important for reducing noise levels at short and long periods (McMillan, 2002; Hutt et al., 2017; Busby et al., 2013). For example, because sensors are often placed underground on concrete pads, tilt of the pad due to thermal contraction and expansion is readily observed on long-period PSDPDFs. This can usually be mitigated with a simple application of insulation, though in extreme cases deeper burial may be required to achieve low noise levels at longer periods.

In addition to seismic station and instrumentation quality control for operational applications, information can be obtained on the source characteristics of seismic ambient noise. PSDPDFs are particularly useful for the visualization of diurnal and seasonal variations of both cultural and environmental noise sources. For example, the cultural noise from cars, machinery, and even ships in the Panama Canal generally increases during the day-time hours. Seismic noise due to ocean waves has strong seasonal variation in power and has been effectively used for studies of storms and long-term decadal ocean oscillations, important for climate change monitoring (Aster et al., 2008, 2010). In the next sections we provide details on the use of PSDPDFs for studying sources of seismic ambient noise and for use in instrumentation and data quality control.

1.3.1 A Review of the Dominant Sources of Seismic Ambient Noise

Here we describe numerous examples of PSDPDFs from stations distributed throughout the various tectonic regions of the Earth (Appendix 1.B Figure 1.7, and Table 1.1). We separate noise sources into three period bands and discuss several examples of the dominant sources of seismic ambient noise within each of these bands. In general, short-period noise is dominated by human activity while mid- and long-period noise is dominated by ocean waves, but not always.

Short-Period Seismic Noise

Cultural Noise. The most common source of short-period seismic noise is from the actions of human beings at or near the surface of the Earth. This is often referred to as "cultural noise" and originates primarily from the coupling of automobile traffic and machinery energy into the solid Earth. Cultural noise propagates mainly as short-period (0.1–1 s) surface waves that attenuate within several kilometers in distance and depth. Cultural noise is often bimodal due to diurnal variations in human activity, as reflected by the fact that power levels are higher during the daytime hours and lower during the night, as well as on weekends and holidays. Figure 1.3a shows a PSDPDF from a seismic station built in a neighborhood in

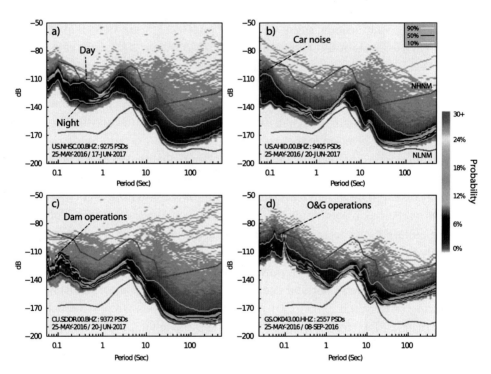

Figure 1.3. Short-period cultural noise PSDPDF examples. (a) PSDPDF of station US.NHSC.00.BHZ, constructed using 9,275 PSDs, showing bimodal, cultural, short-period noise with diurnal variation. (b) PSDPDF of station US.AHID.00.BHZ, constructed using 9,405 PSDs, showing short-period noise from car traffic on a nearby road. (c) PSDPDF of station CU.SDDR.00.BHZ, constructed using 9,372 PSDs showing short-period noise from operations at the Sabaneta Dam in the Dominican Republic. (d) PSDPDF of station GS.OK043.00.HHZ, constructed using 2,557 PSDs, showing a distinct short-period noise band from activity at oil and gas production and wastewater injection sites. (A black-and-white version of this figure appears in some formats. For the color version, please refer to the plate section.)

New Hope, South Carolina, with a clear bimodal distribution of power levels at short periods (0.1–1 s). The higher levels occur during the daytime when people are actively driving cars and operating machinery. For example, automobile traffic along a dirt road only 20 m from station US.AHID, in Auburn Hills, Idaho, creates a 30–35 dB increase in power in the 0.1 s period range and is observable in the PSDPDF (Figure 1.3b) as a region of low probability (−130 dB) at short periods (0.1–1 s).

Machinery such as diesel generators and water pumps can also have a strong influence on seismic power levels. Figure 1.3c shows station CU.SDDR, located in the high central mountains of the Dominican Republic on the grounds of the Sabaneta Dam (McNamara et al., 2006). Strong variations in short-period (0.1– 0.5 s) power are clearly observed with the highest probability roughly 10 dB greater than the lower power levels. Investigation of the low power levels indicate that most occur between 1 and 3 pm local time, indicating regular periods of operational shutdown.

Often, industrial machinery will display a very sharp and distinct characteristic period band. Station GS.OK043 (Figure 1.3d) is located within one km of multiple oil and gas production and wastewater disposal sites with continuously operating pumps and generators. The seismic energy generated from this activity, near 0.1 s period, is clearly observed in the PSDPDF and suggests that there is very little downtime in the activity since low power levels in this period band are very low probability ($\sim$ 1%).

Glacier Calving. Not all short-period seismic noise is due to human activity. Station YM.BBB was located near the terminus of the Columbia Glacier on the Great Nunatak in the Prince William Sound of Alaska (O'Neel et al., 2007). Comparing visual observations of calving events and seismic energy recorded at YM.BBB, calving at the terminus was determined to generate energy in the $\sim$ 0.5 s period band (Figure 1.4a). As the glacier retreated several kilometers from 2005 to 2008, the terminus became ungrounded (floating) causing calving energy to dissipate into the water rather than couple into the earth (Walter et al., 2010). In section 1.1 we introduced the secondary microseism, the prominent peak between 2 and 10 s period, that is generated by ocean wave action at nearby coastlines or in the deep ocean, clearly observed at all global seismic stations. The larger peak of energy between periods 0.2 and 1 s lies within the calving seismicity band (O'Neel et al., 2007) and exists only in the 2004–2005 PSDPDF. In 2008–2009, however, this energy is absent (Figure 1.4b), strongly suggesting a significant change in the calving dynamics. The 0.2–1 s period energy peak is characteristic of the tidewater glacier environment, and its disappearance indicates a decrease in seismic energy related to calving during the second deployment when the terminus was floating

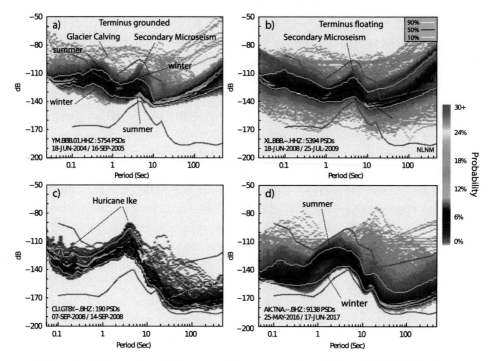

Figure 1.4. Examples of the spectral content and variation of PSDPDFs from the interaction of glaciers, storms, and sea ice with ocean wave energy. (a) PSDPDF of calving seismicity for the time period (June 18, 2004, to September 16, 2005). Station YM.BBB.01.HHZ was located at the terminus of the Columbia Glacier in the Prince William Sound of Alaska (Appendix 1.B Figure 1.7) and displays a clear power peak near 0.5 s period from calving ice. The peak near 5 s period is the secondary microseism and displays the opposite seasonal variation as the calving peak. (b) Station XL.BBB.–.HHZ occupied the same location as YM.BBB (June 18, 2008, to July 25, 2009). The 0.5 s period power peak from calving ice disappeared from the PSDPDF after the glacier terminus went from grounded to floating. (c) Hurricane Ike recorded at Guantanamo Bay seismic station (CU.GTBY.–.BHZ) as it passed by on September 7–8, 2008. Significant increases in microseism and short-period noise occurred. (d) Alaska earthquake center seismic station located on the western tip of the Seward Peninsula (AK.TNA.–.BHZ) shows the seasonal variation of Bering sea ice influence on the power level and bandwidth of the secondary microseism. (A black-and-white version of this figure appears in some formats. For the color version, please refer to the plate section.)

(Walter et al., 2010). Also, interestingly, the calving peak has the opposite seasonal variation trend than for the secondary microseism. Microseism energy in Alaska is greatest in the northern hemisphere winter months when large Pacific storms are frequent, generating large ocean wave activity, while the calving power is highest in the summer months when the temperatures are warm and the glacier is actively calving (Figure 1.4a).

Wind. Another common source of short-period seismic noise is observed from the interaction of wind with the surface of the earth. Wind can cause vibration of trees and structures as well as movement of debris, coupling energy into the solid earth. Withers et al. (1996) showed a strong correlation between wind speed greater than about 3 m/s and short-period (< 1 s) seismic noise at a remote seismic station in New Mexico. Figure 1.4c shows an example of wind noise at station CU.GTBY located at the Guantanamo Bay Navy Base in Cuba (Appendix 1.B, Figure 1.7). High-power, short-period (< 0.1 s) PSDPDF levels occurred as hurricane Ike passed over on September 7 and 8, 2007. In addition, noise in the mid-period band (1–10 s) increased due to ocean wave action.

Mid-Period Seismic Noise

Microseisms. The seismic ambient noise spectrum of the Earth is dominated by broad and dominant peaks near the periods of 7 and 14 seconds, principally due to seismic surface waves. This excitation, the microseism, is observed globally, even in deep continental interiors (Figure 1.1) though strongest at coastal and island stations (Figure 1.5), and has long been recognized as a proxy for ocean wave and storm intensity (Gutenberg, 1947; Grevemeyer et al., 2000; Bromirski, 2009; Aster et al., 2008, 2010; Ebeling and Stein, 2011; Ebeling, 2012; Gualtieri et al., 2018). Intense cyclonic low-pressure storm systems have steep pressure gradients that cause strong surface winds, causing a transfer of atmospheric energy into ocean swell (Kedar et al., 2008; Pierson and Moskowitz, 1964; Ebeling and Stein, 2011), subsequently becoming microseism energy. The seasonal distribution of ocean wave energy and its long-term variability are influenced by broad-scale climate factors (Bromirski et al., 2005a,b) that are also reflected in microseism level variations (Aster et al., 2008, 2010).

The primary, or single-frequency, and secondary, or double-frequency microseism PSD peaks, are largely unaffected by common anthropogenic noise sources. Primary microseisms are generated at the causative ocean swell period via direct pressure fluctuations at the ocean bottom in shallow water resulting from breaking and/or shoaling waves (Hasselmann, 1963). The amplitude of the much higher-power secondary microseism is proportional to the product of standing wave components of the ocean wave field (Longuet-Higgins, 1950), where incoming swells interact with coastal reflections (Bromirski, 2009; Bromirski et al., 1999; Kedar et al., 2008; Ardhuin et al., 2011; Gualtieri et al., 2013, 2014) (see Chapter 3). Whether this mechanism occurs mostly in deep or shallow water, or in both environments is still a matter of debate. For studies of Rayleigh wave sources in the deep ocean see Webb and Constable (1986); Cessaro (1994); Stehly et al. (2006); Kedar et al. (2008); Obrebski et al. (2012). For studies on body-wave sources in the deep ocean see Gerstoft and Tanimoto (2007); Koper et al. (2009, 2010); Landès

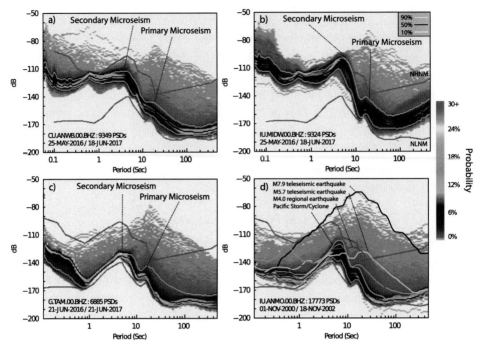

Figure 1.5. Examples of the variation in microseism power levels and bandwidth at island-based and continental interior seismic stations. (a) CU.ANWB, Barbuda Island, Caribbean Sea. (b) IU.MIDW, Midway Island, South Pacific Ocean. (c) G.TAM, located in remote southern Algeria thousands of kilometers from the nearest coast. (d) IU.ANMO, located in Albuquerque, New Mexico thousands of kilometers from the nearest coast. IU.ANMO PSD comparison between a Pacific storm and earthquake observations. Shown is a vertical acceleration PSDPDF constructed from continuous 20 Hz data from station IU.ANMO near Albuquerque, New Mexico from November 1, 2000, to November 19, 2002, displaying 17,773 2-hour, 50% overlapping PSD estimates. Gray lines show the NLNM and NHNM low- and high-noise models (Petersen, 1993). Earthquake signals generate strong long-period (>10 s) surface wave energy relative to Earth's background spectrum. The black line shows the PSD for an Alaska M 7.9 teleseismic earthquake at origin time 2002-11-03 22:12:41 UTC. The thick white line shows the PSD for a mid-Atlantic M5.7 teleseismic earthquake at origin time 2001-07-13 19:48:19 UTC. The dark grey line shows the PSD for a Southern Colorado M 4.0 regional earthquake at origin time 2001-09-04 12:45:53 UTC, located 293 km from IU.ANMO.00.BHZ. The light grey line shows a prominent Pacific storm-associated PSD during this time, demonstrating the spectral characteristics of an individual extreme storm event (12/14/00). (A black-and-white version of this figure appears in some formats. For the color version, please refer to the plate section.)

et al. (2010). For observations of both shallow and deep sources read Haubrich and McCamy (1969); Friedrich et al. (1998); Chevrot et al. (2007b,a); Bromirski and Duennebier (2002).

As previously mentioned, the microseism peaks are observed globally at broadband seismic stations and are orders of magnitude higher in power than short- and long-period seismic ambient noise levels (Figure 1.5). Stations located near the coast and on islands will have the highest and broadest band microseism power levels, often near the NHNM (Figures 1.5a and 1.5b). Figures 1.5a and 1.5b demonstrate, for example, that median microseism power levels were higher in the Pacific at Midway Island (IU.MIDW) than in the North Atlantic on the island of Barbuda (CU.ANWB) from May 2016 to June 2017.

Island-based stations often show higher-power broadband microseism peaks that are generated by surface waves associated with ocean wave action (Figures 1.5a and 1.5b). This can be a problem for earthquake monitoring systems since high noise levels at short periods can adversely impact P-wave detection. However, even in the presence of high power in the microseism band, if power levels remain low at longer periods (>20 s) when the station is well-constructed, the data will still be useful for earthquake source products such as moment tensors and finite faults (Figures 1.5a and 1.5b).

Seismic surface waves generated by the microseism can propagate hundreds to thousands of kilometers within stable continental interiors. Figure 1.5c shows a clear and prominent microseism peak for roughly the same time period (June 2016– June 2017) at a seismic station far from the nearest coastline in southern Algeria (G.TAM). IU.ANMO is also located far from coastlines in central New Mexico and has low microseism power levels near the NLHM (Figure 1.5d).

Extra-Tropical Storms/Tropical Cyclones. Figure 1.4c demonstrates that Hurricane Ike significantly increased PSDPDF power over a broad range of periods. Recall, short-period (< 1 s) power is due to wind and debris; however, mid-period power also increases due to ocean wave activity created by the hurricane. Additional examples of the spectral characteristics of individual storms are shown in Figure 1.5d. An individual PSD for a prominent Pacific storm is compared to a PSDPDF for seismic station IU.ANMO.00.BHZ for the same time period. The Pacific storm increases power in the mid-period and short-period bands, while long-period power remains relatively low (Ebeling and Stein, 2011; Sufri et al., 2014; Gualtieri et al., 2018). In comparison, earthquake signals generate high long-period (> 30 s) surface wave power relative to Earth's background spectrum (Figure 1.5d).

Sea Ice. Microseism power is potentially affected by any process that alters ocean wave intensity. Because thick sea ice prevents large ocean waves from forming, this ice significantly affects microseism amplitudes (Stutzmann et al., 2009; Grob et al., 2011; Anthony et al., 2014). By examining several island and coastal seismic stations along the Bering Sea, Tsai and McNamara (2011) demonstrated that sea ice can impact the bandwidth and absolute power levels of the secondary microseism.

Take, for example, the Seward Peninsula (Alaska). It is surrounded by sea ice between roughly December and May of each year. Sea ice begins forming in the northern Bering Sea in December and moves steadily southward. The ice then begins to recede in the southern Bering Sea in May, this recession progressing northward through the summer months. Broadband seismic station AK.TNA is located on the western end of the Seward Peninsula adjacent to the Bering Strait at the northern extent of the Bering Sea and was used by Tsai and McNamara (2011) in their study (Figure 1.4d). The PSDPDF for AK.TNA.--.BHZ displays significant variations in the bandwidth and power of the secondary microseism (Figure 1.4d) due to the sea ice damping the ocean wave energy. During the summer months ocean waves generate a broadband (0.2–10 s) microseism. In winter, when sea ice dampens ocean wave heights, short-period (0.1–1.2 s) microseism power is significantly reduced (10–20 dB).

Long-Period Sources of Seismic Ambient Noise

In general, power in long-period bands greater than 20 s is of interest to the seismology community for earthquake surface waves (Figure 1.1) and long-period resonances in the solid Earth. A less common but useful signal in this band are surface waves generated by large calving glaciers in Greenland (Ekström et al., 2003, 2006). Often, optimal noise levels in the long period are difficult to record since it requires both a high-quality broadband sensor and a well-built seismic station.

Hum. Figure 1.1a shows that when seismic stations are built well, as is the case for IU.TUC in Tucson Arizona, the long-period power levels are very low, near the NLNM, and are able to resolve the Earth's "hum." Hum is a term referring to bell-like ringing associated with the fundamental long-period resonant spheroidal oscillations of Earth (Rhie and Romanowicz, 2004; Webb, 2007; Uchiyama and McWilliams, 2008; Bromirski, 2009; Bromirski and Gerstoft, 2009; Ardhuin et al., 2015).

Hum excitation is due to the portion of ocean swell that reaches coastlines and is transformed into much longer periods (> 50 s) (Webb, 2007; Ardhuin et al., 2015). For a detailed review see Chapter 3 of this book. Similar to the microseisms, hum amplitudes depend on ocean swell that impacts coasts, which also depend on climate factors affecting storm track and storm intensity (Aster et al., 2008,

2010). Consequently, hum power levels are climate-related and vary on seasonal and decadal cycles (Bromirski, 2009).

Seiche. A unique and interesting long-period ($100-200$ s) signal is apparent in the PSDPDF for station CU.BCIP.--.BHZ and is related to ship traffic in the Panama Canal (Figure 1.1b). The signal is in the same period band as the hum but does not have the seasonal distribution as expected by climate-related seismic signals. Instead, it has a diurnal distribution with the highest power during daytime hours and essentially absent in the early morning (1 and 6 AM local time). While the canal is always open, the anomalous long-period signal is dominant when major shipping operations occur (McNamara et al., 2011). CU.BCIP.--.BHZ is built in a relatively deep vault, at a depth of roughly 3 m, that is affected by tilt induced from movement of the surrounding tropical rain forest soils. Long-period noise results from a Seiche wave that forms in the canal channel from the constructive interference of the wakes of passing ships and wind-forced waves reflecting off the nearby shorelines. These seiche waves crash onto the shore and cause movement of the soils, resulting in tilt of the seismic vault at a period of ~ 100 s (Figure 1.1b).

1.4 Application to Earthquake Monitoring Observatories

Often, body and surface waves from earthquakes, or system transients and instrumental glitches such as data gaps, clipping, spikes, mass re-centers, or calibration pulses are removed from seismic noise studies. These signals, and all others, are included as part of the PSDPDF method since they are low-probability occurrences that do not contaminate the high-probability seismic ambient noise observed in the final PSDPDF. In actual fact, transient signals become quite useful for evaluating seismic station performance. In addition to visualizing the spectral source characteristics of earthquakes and environmental and cultural seismic noise, PSDPDFs have proven useful in operational seismology for monitoring seismic data and metadata quality, improving seismic station construction practices and optimizing the distribution and design of networks for earthquake monitoring and tsunami warning. In this section we summarize several practical applications of PSDPDFs in an operational environment.

1.4.1 Seismic Station Design and Construction

Short-period cultural noise propagates mainly as short-period surface waves (0.1–1 s) that attenuate within several kilometers in distance and depth. For this reason cultural noise will generally be significantly reduced in boreholes, deep caves, and

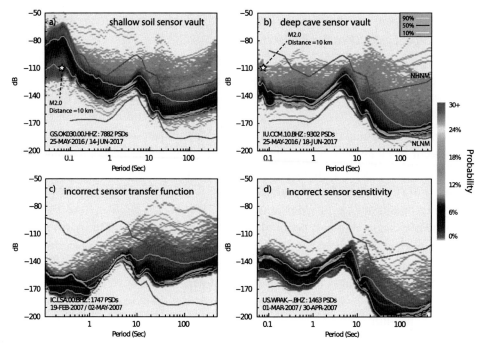

Figure 1.6. PSDPDF comparison of seismic station construction practices and instrument response problems. (a) Station GS.OK030 shows high short-period cultural noise due to a shallow surface vault near a busy road which services the Cushing, OK oil tank storage facility, but with a low power microseism power due to its large distance from the coasts. The white star is the estimated power (dB) of a P-wave from a Mw 2.0 earthquake at 10 km distance from the station. (b) Station IU.CCM, located deep in a cave in Missouri, shows low noise levels at both short and long periods. (c) An incorrect instrument response applied to IC.LSA.00.BHZ results in unrealistically low powers, near and below the NLNM, at short periods. (d) When an incorrect sensitivity of several orders of magnitude too large is applied to US.WRAK.–.BHZ, this results in an overcorrection of the data and PSDs are unrealistically low, near and below the NLNM. (A black-and-white version of this figure appears in some formats. For the color version, please refer to the plate section.)

tunnels relative to shallow surface seismic vaults (Hutt et al., 2017). Figure 1.6 shows examples of seismic sensors deployed in shallow surface vaults (Figure 1.6a) versus sensors installed deep in a cave (Figure 1.6b). Station GS.OK030.-.HHZ was deployed in Oklahoma to record small-magnitude induced earthquakes (McNamara et al., 2015). The white stars on Figures 1.6a and 1.6b indicate the estimated power (dB) of a P-wave from a Mw 2.0 earthquake at 10 km distance from the station (Brune, 1970; Kanamori, 1977; McNamara et al., 2016). This station then is only able to detect a Mw 2.0 at 10 km distance approximately 10% of the time, clearly unable to achieve its originally stated goal. For comparison, station IU.CCM.00.BHZ is located deep in a cave in the U.S. state of Missouri and can always detect a Mw 2.0 at 10 km distance. IU.ANMO.00.BHZ (Figure 1.5d) is located 300 m below the surface in a deep bore hole and displays similar low-power ambient noise levels at short periods. Small earthquake magnitude detection capability is significantly improved when the seismic sensor is located well below the surface.

By examining seismic sensors in shallow vaults, Wolin et al. (2015) demonstrated that the timing and amplitude of long-period noise power variations correlate with variations in atmospheric pressure (Ewing and Press, 1953). Based on this and other studies, PSDPDFs have been used to evaluate seismic network performance for the specific purpose of improving sensor vault construction practices (Rastin et al., 2012; Ringler and Hutt, 2010; McNamara et al., 2005). For additional studies on the noise characteristics of different vault designs see Aderhold et al. (2015); Bormann (2012); Spriggs et al. (2014).

1.4.2 Quality Assessment of Seismic Data and Metadata

As the number and availability of real-time seismic stations have increased over the past few decades, it became important for earthquake monitoring systems to develop tools to ensure that only the highest quality data are used in rapid earthquake monitoring and products. In addition to seismic station construction practices, PSDPDFs contribute to improving other components of the system such as communications, metadata quality, real-time data processing procedures, and the optimization of seismic monitoring networks (McNamara et al., 2016).

At the USGS NEIC thousands of real-time seismic stations stream into the earthquake detection system. At any given moment as many as 20% of the channels can have missing data, problems with instrumentation, or metadata inaccuracies. Because it is very important that automated earthquake detection and source characterization products use only the highest quality data, the NEIC needed to develop automated systems to optimize its data streams. Below we describe one system based on PSDPDFs that automatically detects bad instrument responses and other undesirable system-generated transient signals.

The basic concept is that short-term deviations, due to system problems, are easily detected relative to long-term baselines (McNamara et al., 2009). For the USGS NEIC example, the NHNM and NLNM represent a global standard baseline to which individual stations and time periods can be compared. The NHNM and NLNM are used as a standard operating range for seismic ambient noise. Outlier noise conditions, such as instrument response changes or systems transients, are detected by comparing a short-term station's PSDPDF percentile statistics to the NHNM and NLNM baselines.

Figure 1.6 demonstrates examples of how individual station PSDPDFS can help identify problems with instrument response shape and total sensitivity. The baseline outlier detection method is very sensitive to instrument response problems such as incorrect input units (resulting in an incorrect transfer function) and incorrect sensitivities.

Figure 1.6c is an example using station IC.LSA.00.BHZ. In order to demonstrate the sensitivity of this method to an error in instrument response units, an incorrect instrument response was applied to 3493 hours of data, during the days of the year 2007 from day 050 (2007:050) to 2007:122. If units of acceleration are mistakenly output from the transfer function instead of velocity by adding an extra zero to the instrument response definition, the error is clearly observed and easily detectable as low power at short periods and high power at long periods, relative to the long-term baseline noise model. In contrast, applying output units of displacement instead of velocity, due to a missing zero in the response file, results in high power at short periods and low power at long periods, relative to the long-term baseline model (Peterson, 1993).

Figure 1.6d is an example using ANSS backbone station US.WRAK.--.BHZ. In this case an incorrect sensitivity is applied to 2927 hours of data from days of the year 2007:060 to 2007:120. This demonstrates the effect of incorrectly applying the Streckeisen STS-2 high gain (20,000 volts/(meter/s)) when the standard gain (1500 volts/(meter/s)) should be used. The result is low power across the entire spectrum relative to the long-term station baseline model. In all cases discussed here, an incorrect instrument response causes anomalous amplitudes that can transfer to earthquake products which rely on accurate estimates of ground motion.

1.5 Conclusion

In this chapter we have described the motivations for the study of seismic ambient noise, the PSDPDF statistical methods developed to provide long-term probabilistic visualization of seismic ambient noise, and research, operational, and practical applications of the PSDPDF method. In Appendix 1.C we provide additional information on data and software resources that will enable readers to generate and work with PSDPDFs for their own seismic data quality assurance and research applications, all intended to further the understanding of seismic instrumentation, data and metadata QC/QA, and the detailed source characteristics of seismic ambient noise.

Acknowledgments

The authors thank the following colleagues for their contributions and guidance as we developed the methods and conducted the research described in this chapter: R. Buland, P. Earle, E. Wolin, H. Benz, L. Gee, T. Ahern, B. Weertman, B. Beaudoin, K. Anderson, S. Rastin, F. Vernon, P. Davis, R. Aster, P. Bromirski, C. Hutt, H. Bolton, J. Mayer, A. Ringler, D. Wilson, R. Herrmann, J. McCarthy, D. Mason, S. ONeel, V. Tsai, and P. Launy. The authors also thank the editors (Nori Nakata, Lucia Gualtieri, Andreas Fichtner) and reviewers (Anya M. Reading, Lion Krischer) of this chapter whose critical and helpful comments improved the presentation of the concepts. Figures were made using SQLX, SAC (Goldstein and Snoke, 2005), GMT (Wessel and Smith, 1991), and MATLAB.

References

Aderhold, K., Anderson, K. E., Reusch, A. M., Pfeifer, M. C., Aster, R. C., and Parker, T. 2015. Data quality of collocated portable broadband seismometers using direct burial and vault emplacement. *Bulletin of the Seismological Society of America*, **105**(5), 2420–2432.

Ahern, T., Casey, R., Barnes, D., Benson, R., Knight, T., and Trabant, C. 2007. SEED Reference Manual, version 2.4. *IRIS (www.iris.edu/software/pqlx/)*.

Anthony, R. E., Aster, R. C., Wiens, D., Nyblade, A., Anandakrishnan, S., Huerta, A., Winberry, J. P., Wilson, T., and Rowe, C. 2014. The seismic noise environment of Antarctica. *Seismological Research Letters*, **86**(1), 89–100.

Ardhuin, F., Gualtieri, L., and Stutzmann, E. 2015. How ocean waves rock the Earth: Two mechanisms explain microseisms with periods 3 to 300 s. *Geophysical Research Letters*, **42**(3), 765–772.

Ardhuin, F., Stutzmann, E., Schimmel, M., and Mangeney, A. 2011. Ocean wave sources of seismic noise. *Journal of Geophysical Research: Oceans*, **116**(C9), 765–772.

Aster, R., McNamara, D., and Bromirski, P. 2008. Multidecadal climate-induced oceanic microseism variability. *Seismological Research Letters*, **79**(2), 194–202.

———. 2010. Global trends in extremal microseism intensity. *Geophysical Research Letters*, **37**(14), 194–202.

Bahavar, M., and North, R. 2002. Estimation of background noise for international monitoring system seismic stations. Pages 911–944 of: *Monitoring the Comprehensive Nuclear-Test-Ban Treaty: Data Processing and Infrasound*. Springer.

Bendat, J. S., and Piersol, A. G. 1971. *Random data: analysis and measurement procedures*. Vol. 729. John Wiley & Sons.

Bensen, G., Ritzwoller, M., and Shapiro, N. M. 2008. Broadband ambient noise surface wave tomography across the United States. *Journal of Geophysical Research: Solid Earth*, **113**(B05306).

Berger, J., Davis, P., and Ekström, G. 2004. Ambient earth noise: a survey of the global seismographic network. *Journal of Geophysical Research: Solid Earth*, **109**(11307).

Boaz, R., and McNamara, D. 2008. PQLX: A data quality control system, uses and applications. *Orfeus Newsletter*, **8**(1).

Boese, C., Wotherspoon, L., Alvarez, M., and Malin, P. 2015. Analysis of anthropogenic and natural noise from multilevel borehole seismometers in an urban environment,

Auckland, New Zealand. *Bulletin of the Seismological Society of America*, **105**(1), 285–299.

Bormann, P. 2012. *New manual of seismological observatory practice (NMSOP-2)*. IASPEI, GFZ German Research Centre for Geosciences, Potsdam.

Bromirski, P. D. 2009. Earth vibrations. *Science*, **324**(5930), 1026–1027.

Bromirski, P. D., Cayan, D. R., and Flick, R. E. 2005b. Wave spectral energy variability in the northeast Pacific. *Journal of Geophysical Research: Oceans*, **110**(C03005).

Bromirski, P. D., and Duennebier, F. K. 2002. The near-coastal microseism spectrum: Spatial and temporal wave climate relationships. *Journal of Geophysical Research: Solid Earth*, **107**(B8), 2166.

Bromirski, P. D., Duennebier, F. K., and Stephen, R. A. 2005a. Mid-ocean microseisms. *Geochemistry, Geophysics, Geosystems*, **6**(4), Q04009.

Bromirski, P. D., Flick, R. E., and Graham, N. 1999. Ocean wave height determined from inland seismometer data: Implications for investigating wave climate changes in the NE Pacific. *Journal of Geophysical Research: Oceans*, **104**(C9), 20753–20766.

Bromirski, P. D., and Gerstoft, P. 2009. Dominant source regions of the Earth's "hum" are coastal. *Geophysical Research Letters*, **36**(13).

Brune, J. N. 1970. Tectonic stress and the spectra of seismic shear waves from earthquakes. *Journal of Geophysical Research*, **75**(26), 4997–5009.

Brune, J. N., and Oliver, J. 1959. The seismic noise of the earth's surface. *Bulletin of the Seismological Society of America*, **49**(4), 349–353.

Busby, R., Frassetto, A., Hafner, K., Woodward, R., and Sauter, A. 2013. Sensor Emplacement Techniques and Seismic Noise Analysis for USArray Transportable Array Seismic Stations. Page 13298 of: *EGU General Assembly Conference Abstracts*, vol. 15.

Butler, R., Lay, T., Creager, K., Earl, P., Fischer, K., Gaherty, J., Laske, G., Leith, B., Park, J., Ritzwolle, M., et al. 2004. The Global Seismographic Network surpasses its design goal. *Eos, Transactions American Geophysical Union*, **85**(23), 225–229.

Cessaro, R. K. 1994. Sources of primary and secondary microseisms. *Bulletin of the Seismological Society of America*, **84**(1), 142–148.

Chevrot, S., Sylvander, M., Benahmed, S., Ponsolles, C., Lefevre, J., and Paradis, D. 2007a. Correction to "Source locations of secondary microseisms in western Europe: Evidence for both coastal and pelagic sources." *Journal of Geophysical Research: Solid Earth*, **112**(B12).

———. 2007b. Source locations of secondary microseisms in western Europe: Evidence for both coastal and pelagic sources. *Journal of Geophysical Research: Solid Earth*, **112**(B11).

Cooley, J., and Tukey, J. 1965. An algorithm for machine calculation of complex Fourier series. Mathematics of computing, reprinted 1972. *Digital signal processing*. IEEE Press, New York, NY, 223–227.

Díaz, J. 2016. On the origin of the signals observed across the seismic spectrum. *Earth-Science Reviews*, **161**, 224–232.

Ebeling, C. W. 2012. Inferring ocean storm characteristics from ambient seismic noise: A historical perspective. Pages 1–33 of *Advances in Geophysics*, vol. 53. Elsevier.

Ebeling, C. W., and Stein, S. 2011. Seismological identification and characterization of a large hurricane. *Bulletin of the Seismological Society of America*, **101**(1), 399–403.

Ekström, G., Nettles, M., and Abers, G. A. 2003. Glacial earthquakes. *Science*, **302**(5645), 622–624.

Ekström, G., Nettles, M., and Tsai, V. C. 2006. Seasonality and increasing frequency of Greenland glacial earthquakes. *Science*, **311**(5768), 1756–1758.

Ewing, M., and Press, F. 1953. Further study of atmospheric pressure fluctuations recorded on seismographs. *Eos, Transactions American Geophysical Union*, **34**(1), 95–100.

Frantti, G., Willis, D., and Wilson, J. T. 1962. The spectrum of seismic noise. *Bulletin of the Seismological Society of America*, **52**(1), 113–121.

Friedrich, A., Krueger, F., and Klinge, K. 1998. Ocean-generated microseismic noise located with the Gräfenberg array. *Journal of Seismology*, **2**(1), 47–64.

Gerstoft, P., and Tanimoto, T. 2007. A year of microseisms in southern California. *Geophysical Research Letters*, **34**(20).

Goldstein, P., and Snoke, A. 2005. SAC availability for the IRIS community. *Incorporated Research Institutions for Seismology Newsletter*, **7**(UCRL-JRNL-211140).

Grevemeyer, I., Herber, R., and Essen, H.-H. 2000. Microseismological evidence for a changing wave climate in the northeast Atlantic Ocean. *Nature*, **408**(6810), 349.

Grob, M., Maggi, A., and Stutzmann, E. 2011. Observations of the seasonality of the Antarctic microseismic signal, and its association to sea ice variability. *Geophysical Research Letters*, **38**(11).

Gualtieri, L., Camargo, S. J., Pascale, S., Pons, F. M., and Ekström, G. 2018. The persistent signature of tropical cyclones in ambient seismic noise. *Earth and Planetary Science Letters*, **484**, 287–294.

Gualtieri, L., Stutzmann, E., Capdeville, Y., Ardhuin, F., Schimmel, M., Mangeney, A., and Morelli, A. 2013. Modeling secondary microseismic noise by normal mode summation. *Geophysical Journal International*, **193**(3), 1732–1745.

Gualtieri, L., Stutzmann, É., Farra, V., Capdeville, Y., Schimmel, M., Ardhuin, F., and Morelli, A. 2014. Modeling the ocean site effect on seismic noise body waves. *Geophysical Journal International*, **197**(2), 1096–1106.

Gutenberg, B. 1936. On microseisms. *Bulletin of the Seismological Society of America*, **26**(2), 111–117.

———. 1947. Microseisms and weather forecasting. *Journal of Meteorology*, **4**(1), 21–28.

Hasselmann, K. 1963. A statistical analysis of the generation of microseisms. *Reviews of Geophysics*, **1**(2), 177–210.

Haubrich, R. A., and McCamy, K. 1969. Microseisms: Coastal and pelagic sources. *Reviews of Geophysics*, **7**(3), 539–571.

Hutt, C. R., Ringler, A. T., and Gee, L. S. 2017. Broadband seismic noise attenuation versus depth at the Albuquerque Seismological Laboratory. *Bulletin of the Seismological Society of America*, **107**(3), 1402–1412.

Kanamori, H. 1977. The energy release in great earthquakes. *Journal of Geophysical Research*, **82**(20), 2981–2987.

Kedar, S., Longuet-Higgins, M., Webb, F., Graham, N., Clayton, R., and Jones, C. 2008. The origin of deep ocean microseisms in the North Atlantic Ocean. Pages 777–793 of *Proceedings of the Royal Society of London A: Mathematical, Physical and Engineering Sciences*, vol. 464. The Royal Society.

Koper, K. D., De Foy, B., and Benz, H. 2009. Composition and variation of noise recorded at the Yellowknife Seismic Array, 1991–2007. *Journal of Geophysical Research: Solid Earth*, **114**(B10), B10310.

Koper, K. D., Seats, K., and Benz, H. 2010. On the composition of Earth's short-period seismic noise field. *Bulletin of the Seismological Society of America*, **100**(2), 606–617.

Landès, M., Hubans, F., Shapiro, N. M., Paul, A., and Campillo, M. 2010. Origin of deep ocean microseisms by using teleseismic body waves. *Journal of Geophysical Research: Solid Earth*, **115**(B5), B05302.

Longuet-Higgins, M. S. 1950. A theory of the origin of microseisms. *Phil. Trans. R. Soc. Lond. A*, **243**(857), 1–35.

McMillan, J. R. 2002. *Methods of installing United States national seismographic network (USNSN) stations: A construction manual.* US Department of the Interior, US Geological Survey.

McNamara, D., Benz, H. M., and Leith, W. 2005. An assessment of seismic noise levels for the ANSS backbone and selected regional broadband stations. *USGS Open-File Report*, **1077**, 19.

McNamara, D., Hutt, C., Gee, L., Benz, H. M., and Buland, R. 2009. A method to establish seismic noise baselines for automated station assessment. *Seismological Research Letters*, **80**(4), 628–637.

McNamara, D., McCarthy, J., and Benz, H. 2006. Improving earthquake and tsunami warnings for the Caribbean Sea, the Gulf of Mexico, and the Atlantic coast. *U.S. Geol. Surv. Fact Sheet*, **3012**, 4.

McNamara, D., Ringler, A., Hutt, C., and Gee, L. 2011. Seismically observed seiching in the Panama Canal. *Journal of Geophysical Research: Solid Earth*, **116**(B4).

McNamara, D. E., and Boaz, R. 2005. Seismic noise analysis system, power spectral density probability density function: stand-alone software package. *United States Geological Survey Open File Report*, **1438**, 30.

McNamara, D. E., and Boaz, R. I. 2011. *PQLX: A seismic data quality control system description, applications, and users manual.* Tech. rept. US Geological Survey.

McNamara, D. E., and Buland, R. P. 2004. Ambient noise levels in the continental United States. *Bulletin of the Seismological Society of America*, **94**(4), 1517–1527.

McNamara, D. E., Rubinstein, J. L., Myers, E., Smoczyk, G., Benz, H. M., Williams, R., Hayes, G., Wilson, D., Herrmann, R., McMahon, N. D., et al. 2015. Efforts to monitor and characterize the recent increasing seismicity in central Oklahoma. *The Leading Edge*, **34**(6), 628–639.

McNamara, D. E., von Hillebrandt-Andrade, C., Saurel, J.-M., Huerfano, V., and Lynch, L. 2016. Quantifying 10 years of improved earthquake-monitoring performance in the Caribbean region. *Seismological Research Letters*, **87**(1), 26–36.

Obrebski, M., Ardhuin, F., Stutzmann, E., and Schimmel, M. 2012. How moderate sea states can generate loud seismic noise in the deep ocean. *Geophysical Research Letters*, **39**(11), L11601.

O'Neel, S., Marshall, H. P., McNamara, D. E., and Pfeffer, W. T. 2007. Seismic detection and analysis of icequakes at Columbia Glacier, Alaska. *Journal of Geophysical Research: Earth Surface*, **112**(F3).

Peterson, J. 1993. Observations and modeling of seismic background noise. *U.S. Geol. Surv. Tech. Rept.*, **93-322**, 1–95.

Pierson, W. J., and Moskowitz, L. 1964. A proposed spectral form for fully developed wind seas based on the similarity theory of SA Kitaigorodskii. *Journal of Geophysical Research*, **69**(24), 5181–5190.

Rastin, S., Unsworth, C., Gledhill, K., and McNamara, D. 2012. A detailed noise characterization and sensor evaluation of the North Island of New Zealand using the PQLX data quality control system. *Bulletin of the Seismological Society of America*, **102**(1), 98–113.

Reif, C., Shearer, P. M., and Astiz, L. 2002. Evaluating the performance of global seismic stations. *Seismological Research Letters*, **73**(1), 46–56.

Rhie, J., and Romanowicz, B. 2004. Excitation of Earth's continuous free oscillations by atmosphere-ocean-seafloor coupling. *Nature*, **431**(7008), 552.

Ringler, A., and Hutt, C. 2010. Self-noise models of seismic instruments. *Seismological Research Letters*, **81**(6), 972–983.

Ringler, A., Gee, L., Hutt, C., and McNamara, D. 2010. Temporal variations in global seismic station ambient noise power levels. *Seismological Research Letters*, **81**(4), 605–613.

Ringler, A., Steim, J., van Zandt, T., Hutt, C. R., Wilson, D., and Storm, T. 2015. Potential improvements in horizontal very broadband seismic data in the IRIS/USGS component of the Global Seismic Network. *Seismological Research Letters*, **87**(1), 81–89.

Spriggs, N., Bainbridge, G., and Greig, W. 2014. Comparison study between vault seismometers and posthole seismometers. In: *EGU General Assembly Conference Abstracts*, vol. 16.

Stehly, L., Campillo, M., and Shapiro, N. 2006. A study of the seismic noise from its long-range correlation properties. *Journal of Geophysical Research: Solid Earth*, **111**(B10), B10306.

Stutzmann, E., Roult, G., and Astiz, L. 2000. GEOSCOPE station noise levels. *Bulletin of the Seismological Society of America*, **90**(3), 690–701.

Stutzmann, E., Schimmel, M., Patau, G., and Maggi, A. 2009. Global climate imprint on seismic noise. *Geochemistry, Geophysics, Geosystems*, **10**(11).

Sufri, O., Koper, K. D., Burlacu, R., and de Foy, B. 2014. Microseisms from superstorm Sandy. *Earth and Planetary Science Letters*, **402**, 324–336.

Tsai, V. C., and McNamara, D. E. 2011. Quantifying the influence of sea ice on ocean microseism using observations from the Bering Sea, Alaska. *Geophysical Research Letters*, **38**(22), L22502.

Uchiyama, Y., and McWilliams, J. C. 2008. Infragravity waves in the deep ocean: Generation, propagation, and seismic hum excitation. *Journal of Geophysical Research: Oceans*, **113**(C7).

Walter, F., O'Neel, S., McNamara, D., Pfeffer, W., Bassis, J. N., and Fricker, H. A. 2010. Iceberg calving during transition from grounded to floating ice: Columbia Glacier, Alaska. *Geophysical Research Letters*, **37**(15), L15501.

Webb, S. C. 2007. The Earth's "hum" is driven by ocean waves over the continental shelves. *Nature*, **445**(7129), 754.

Webb, S. C., and Constable, S. C. 1986. Microseism propagation between two sites on the deep seafloor. *Bulletin of the Seismological Society of America*, **76**(5), 1433–1445.

Wessel, P., and Smith, W. H. 1991. Free software helps map and display data. *Eos, Transactions American Geophysical Union*, **72**(41), 441–446.

Wilson, D., Leon, J., Aster, R., Ni, J., Schlue, J., Grand, S., Semken, S., Baldridge, S., and Gao, W. 2002. Broadband seismic background noise at temporary seismic stations observed on a regional scale in the southwestern United States. *Bulletin of the Seismological Society of America*, **92**(8), 3335–3342.

Withers, M. M., Aster, R. C., Young, C. J., and Chael, E. P. 1996. High-frequency analysis of seismic background noise as a function of wind speed and shallow depth. *Bulletin of the Seismological Society of America*, **86**(5), 1507–1515.

Wolin, E., van der Lee, S., Bollmann, T. A., Wiens, D. A., Revenaugh, J., Darbyshire, F. A., Frederiksen, A. W., Stein, S., and Wysession, M. E. 2015. Seasonal and diurnal variations in long-period noise at SPREE stations: The influence of soil characteristics on shallow stations' performance. *Bulletin of the Seismological Society of America*, **105**(5), 2433–2452.

Appendix 1.A PSDPDF Method

Using two months of continuous broadband seismic data sampled at 40 Hz from station CU.BCIP, we describe in detail the processing procedures applied to the seismic data in order to generate a PSDPDF. This appendix details, by section: (1.A.1) how data are prepared for the FFT calculation, (1.A.2) the FFT calculation itself and the procedures to compute the PSD from the FFT, and (1.A.3) how the PSDs are accumulated to generate a PDF.

1.A.1 Data Processing and Preparation

Record Length. We begin with a finite-length, evenly sampled, seismic time series (Figure 1.2a). The seismic data record length is chosen on the basis of the longest period of interest to the user of this method. In general, the record length is chosen such that it is between ten and twenty times the longest period of interest. In our example, we parse a continuous 14-day time series, sampled at 40 Hz, into two-hour (7200 s) time series segments (so as to resolve to 500 seconds), over-lapping by 50%; overlapping time series segments are commonly used to reduce variance in the PSD estimate (Bendat and Piersol, 1971).

Pre-Processing. Preparing each time series segment for PSD computation con-sists of four significant operations. First, in order to significantly improve the FFT speed ratio by reducing the number of operations, the number of samples in the time series is truncated to the next lowest power of two. Second, in order to further reduce the variance of the final PSD estimate, each modified time series segment is divided into 13 subsegments, overlapping by 75% (e.g., Peterson, 1993). Third, in order to minimize long-period contamination, the time series is corrected to a zero-mean value and any long-period linear trend is removed by the average slope method. Long-period trend is defined as any frequency component whose period is longer than the record length. If trends are not eliminated in the data, large distor-tions can occur in spectral processing by nullifying the estimation of low-frequency spectral quantities. Fourth, in order to suppress side-lobe leakage in the resulting FFT, a 10% cosine taper is applied to the ends of the time series. Tapering the time series removes the effect of the discontinuity between the beginning and end of the time series. The time series variance reduction is quantified by the ratio of the total power in the raw FFT to the total power in the smoothed filter (Bendat and Piersol, 1971), and is used to correct absolute power in the final spectrum.

1.A.2 Power Spectral Density Method

The fundamental operation required to convert time series data from the time domain to the frequency domain is the Fast Fourier Transform (FFT), known as the "Cooley-Tukey Method" (Cooley and Tukey, 1965). For direct comparison to the Peterson new high- and low-noise models, the algorithm used to develop the NHNM and NLNM (Peterson, 1993) is also used to calculate the PSDs for all plots in this chapter.

The standard method for quantifying seismic ambient background noise is to calculate the PSD (Peterson, 1993; Reif et al., 2002). The most common method for estimating a PSD for stationary random seismic data makes use of the discrete Fourier transform. This method computes a PSD using a finite-range FFT of the original data and is advantageous for its computational efficiency (Cooley and Tukey, 1965).

The finite-range Fourier transform of a periodic time series y(t) is given by:

$$Y(f, T_r) = \int_0^{T_r} y(t)e^{-i2\pi f t}dt \tag{1.1}$$

where T_r is the length of the time series segment, and f is frequency. For implementation on a computer the discrete FFT is required, where Fourier frequency components, f_k, are defined as:

$$Y(f_k) = \frac{Y(f_k, T_r)}{\Delta t} \tag{1.2}$$

for $f_k = k/N\Delta t$ when $k = 1, 2, \ldots, N-1$, where Δt is the sample interval and N is the number of samples in each time series segment, $N = T_r/\Delta t$. Using the Fourier components as defined by Cooley and Tukey (1965), the total PSD estimate (Bendat and Piersol (1971)) is defined as the normalized square of each component where:

$$P_k = \frac{2\Delta t}{N}|Y(f_k)|^2 \tag{1.3}$$

This PSD process is then repeated for each of the 13 separate time series segments described in the pre-processing section. After all segment PSD estimates are complete, PSD estimates are then averaged for the 13 separate time segments. As is apparent from equation (1.3), the total power, P_k, is the square of the amplitude spectrum with a normalization factor of $2\Delta t/N$. The normalization factor is common practice in seismology to arrive at physical units of acceleration. Next, the PSD estimate, P_k, is corrected to account for the 10% cosine taper applied earlier in the pre-processing section.

Instrument Response Removal. Finally, the seismometer instrument response is removed through division of the PSD by the instrument transfer function to physical units of acceleration in the frequency domain, this being equivalent to deconvolution in the time domain. This procedure is necessary since the uncorrected time series data also contains the response of the instrument itself, which differs from instrument type to instrument type. The instrument response transfer function contains a frequency-dependent amplification to the time series due to the sensitivity of the sensor mass and data processing electronics in the system.

Standard seismological software and instrument transfer function formats exist to perform this instrument deconvolution (SAC [Goldstein and Snoke 2005] and EVALRESP). For additional details on instrument response transfer functions and deconvolution procedures see Ahern et al. (2007) and the IRIS DMC MUSTANG website (http://service.iris.edu/mustang/metrics/docs/1/desc/transfer _function/). Also see the IRIS DMC metadata aggregator for individual seismic station instrument response transfer function metadata (http://ds.iris.edu/mda).

Conversion to Units of Decibels. For direct comparison to the original standard NLNM baseline (Peterson, 1993), the PSD estimate, P_k, is converted to units of decibels (dB) with respect to acceleration $(m/s^2)^2/Hz$ with the equation:

$$P_{dB} = 10 \log_{10}(P_k) \tag{1.4}$$

1.A.3 Probability Density Function Method

The PSD method described above is useful for evaluating the amplitude-frequency distribution of a finite-length time series. Seismologists use these PSDs in numerous applications estimating the spectral content of ground motion for specific transient earthquakes and to study the characteristics of non-earthquake seismic energy. Below we describe the method used to visualize the spectral content of ground motions over longer periods of time using a probability density function (PDF) of the PSDs. To estimate the long-term variation of seismic ambient noise at a given station, we generate a PDF from potentially thousands of PSDs processed using the method discussed in the previous section. Next, we describe how to combine these numerous PSDs into a single long-term PSDPDF visualization.

First, in order to reduce the data storage requirements of the PSDs and simplify the visualization of the data, full-octave geometric means are taken in 1/8 octave intervals (McNamara and Buland, 2004). While this smoothing step reduces computation time and data storage requirements, it does have the disadvantage of removing some detail, such as smoothing out the fine structure of the microseisms. For our example of 40 Hz data, this procedure reduces the number of frequencies by a factor of 169 from nfft = 16,385 to 97. These geometric means are then evenly sampled in log space. The average power for each octave is stored as the power value for the center period of the octave to be used for future analysis. This smoothing step is repeated for every PSD estimate, resulting in potentially thousands of decimated PSD estimates for each station/component pair.

The PDF for a given center period, T_c, can be estimated as:

$$P(T_c) = N_{PTc}/N_{ctc} \tag{1.5}$$

where N_{PTc} is the number of spectral estimates that fall into a 1-dB power bin, P, having a center period T_c. N_{ctc} is the total number of spectral estimates over all powers having center period. A wealth of seismic noise information may be obtained from the resulting statistical view of these collected PSDs. Numerous examples are discussed in detail in the sections 1.3 and 1.4 of this chapter.

1.B Seismic Station Information

Figure 1.7 shows the locations of the seismic sensors used.

1.C Software and Data Resources

The PSDPDF method, presented in Section 1.2 and Appendix 1.A, was originally developed to determine global ambient noise levels using worldwide broadband stations within the Global Seismographic Network (GSN) and the Advanced National Seismic System (ANSS). The algorithm and initial software were first developed at the United States Geological Survey (USGS) as a part of the data and network quality control (QC) system for the Advanced National Seismic System (ANSS) (McNamara and Buland, 2004). Further development, supported by the

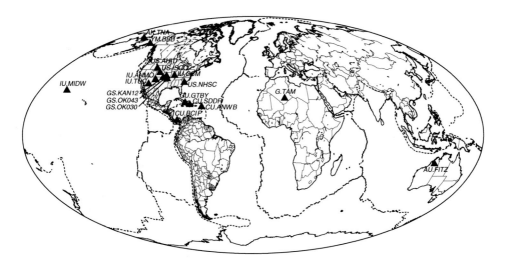

Figure 1.7. Global map showing locations of seismic stations (black triangles) used in this chapter as PSDPDF examples. Dashed black lines are tectonic plate boundaries. See Table 1.1 for station coordinates. See Appendix 1.C for a description of seismic station naming conventions.

Table 1.1 *Stations used in PSDPDF*
examples and mapped in Figure 1.7.

Net.Sta	Lat	Lon
IU.TUC	32.310	−110.785
CU.BCIP	9.167	−79.837
US.NHSC	33.107	−80.178
US.AHID	42.765	−111.100
CU.SDDR	18.982	−71.288
GS.OK043	36.433	−98.746
YM.BBB	61.122	−147.048
CU.ANWB	17.669	−61.786
IU.MIDW	28.216	−177.370
G.TAM	22.792	5.528
IU.ANMO	34.946	−106.457
CU.GTBY	19.927	−75.111
AK.TNA	65.560	−167.926
GS.KAN12	37.297	−97.998
GS.OK030	35.928	−96.784
IU.CCM	38.056	−91.245
US.ISCO	39.800	−105.613
AU.FITZ	−18.098	125.640

Incorporated Institutions in Seismology (IRIS) (https://iris.edu), allowed for the implementation of the algorithm to be applied to the large global data holdings at the IRIS Data Management Center (DMC) (https://ds.iris.edu) and is available to researchers and seismic network operators.

Since the utility of the PSDPDF method was initially demonstrated in a study of the USGS ANSS and GSN networks, several professional software upgrades to the original PSDPDF research codes have been developed. Here we provide a list of previous and current implementations available to the seismology community.

First, the PDF Stand-Alone version was developed for implementation at the IRIS DMC as part of the QUACK Data QC system (McNamara and Boaz, 2005). Both Quack and the PDF Stand-Alone systems are now obsolete. QUACK was replaced by the Mustang Data QA system, and the PDF Stand-Alone software package replaced by PQLX. The new IRIS DMC Mustang Data QA system can be found here: https://ds.iris.edu/ds/nodes/dmc/quality-assurance/mustang.

PQLX is an open-source implementation integrating the PDF Stand-Alone package with IRIS Passcal Quick Look (PQL), providing an interactive GUI and formal database for storage of analysis results (McNamara and Boaz, 2011; Boaz and McNamara, 2008). Written primarily in C, it is no longer actively developed nor supported, but remains available for download and compilation from here: https://ds.iris.edu/ds/nodes/dmc/software/downloads/pqlx.

SQLX is a closed-source implementation further expanding on PQLX functionality, e.g., adding many new types of plots, a frequency domain picker, normalization of PSD period bins across all sample rates, configurable PSD calculation parameters, and more. It is available for Windows, MAC, and Unix-like operating systems, is actively maintained, and is available for download from here: https://sqlx.science.

The ObsPy project provides open-source Python code for the PSDPDF calculation, including documentation and a tutorial, and is available for download from here (search for PPSD): https://obspy.org

Seismic Station and Data Channel Naming Convention and Availability. All seismic times series data processed in this chapter were downloaded from the Incorporated Institutions in Seismology (IRIS) (https://iris.edu) Data Management Center (DMC) (https://ds.iris.edu). For all data used in this chapter we conform to the standard international Federation of Digital Seismograph Networks standard exchange of earthquake data (SEED) seismic station data channel naming convention defined in Ahern et al. (2007). For the example shown in Figure 1.1b, CU is the network code abbreviation of the Caribbean USGS regional portion of the GSN, BCIP is the station code abbreviation of Barro Colorado Island Panama, "--" is the location code of the seismic sensor at the seismic station (used to distinguish a single station having potentially multiple instruments), and BHZ is the channel code abbreviation for broadband data with a vertical component of motion.

2

Beamforming and Polarization Analysis

MARTIN GAL AND ANYA M. READING

Abstract

Beamforming and polarization analysis are two approaches to seismic data analysis which aim to determine the wave type and source location of the incoming signal. Beamforming is viable where seismic signals are recorded by an array of one- or three-component stations, while polarization analysis techniques exist for both three-component arrays and single three-component stations. The majority of permanent seismic arrays were deployed as part of the International Monitoring System to improve the detection of nuclear explosions. Such arrays are optimized for events occurring one at a time, with an impulsive onset, tightly constrained location, and short duration. In contrast, ambient seismic noise occurs at many locations simultaneously with relatively low amplitude, has no sharp onset and may have moving sources. The range of methods that may be used to estimate the slowness and back azimuth of incoming energy arriving at the array are reviewed and discussed as well as developments of such methods that improve the observation of ambient noise, its location and character.

Single-component plane wave beamforming in the form of conventional frequency-wavenumber (f-k) analysis and data adaptive beamforming (MVDR /Capon) is introduced. Almost all array observations of seismic noise, e.g., source localization of Rayleigh, L_g, P-waves, and power measurements, can be attributed to these two techniques, which have solidified the understanding between theory, numerical modeling, and observations. The choice of array design impacts on the resolution for a given frequency range for the intended task. The theory for the underlying beamforming techniques is related to the array design and further strategies are discussed to optimize beamforming for the analysis of microseisms.

Methods for investigating the polarization of incoming ambient noise signals are presented for seismic arrays and three-component stations. Multiple ways are reviewed to extend the beamformers described previously for an analysis of the three-component wavefield, which allows for the observation of shear surface and body waves and improves overall performance. For single station three-component sensors, a review of the most commonly used polarization methods in the context of microseism analysis is given.

2.1 Introduction

Seismologists often aim to determine the type of seismic waves present in an incoming signal and also to determine the back azimuth, which gives information on the source direction of the signal. Beamforming and polarization analysis may both be used to analyze signals from seismic events with defined onset times such as earthquakes and explosions. Ambient seismic noise, the background signal ubiquitous on all seismic records, may come from many sources simultaneously and heightened levels of seismic noise typically have no sharp onset, thus presenting additional challenges for its analysis. In this chapter, the observation and estimation of the properties of the ambient seismic wavefield are discussed. For such seismic noise investigations, beamforming analysis is the preferred approach if seismic one- or three-component array data are available and the polarization of incoming signals may be investigated using three-component arrays. If only single-station three-component data exist for a particular receiver location, then polarization analysis of these records may yield some of the desired wave properties.

2.2 Beamforming

The general idea of beamforming in seismology is to estimate the coherent portion of seismic wave energy that propagates over the seismic array and to determine its propagation characteristics (e.g., Krim and Viberg, 1996; Schweitzer et al., 2002; Rost and Thomas, 2002). This is achieved by simultaneously utilizing all sensor information of the seismic array and stacking the seismic traces according to some wave propagation model. This can be visualized with the help of a schematic example of the delay-and-sum (DAS) beamforming process (e.g., Kelly, 1967) shown in Figure 2.1. In this example, five time-shifted sensor outputs $x_i(t)$ are shown and the general form of the beam output is given as

$$y(t) = \frac{1}{M} \sum_i w_i x_i(t - \tau_i), \qquad (2.1)$$

where $y(t)$ is the beam trace, M is the number of sensors, w_i are the sensor weights that differ for each type of beamformer, and τ_i are the time delays. A time shift and

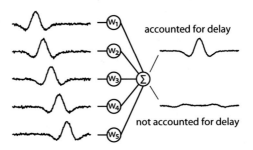

Figure 2.1. Schematic example of a delay-and-sum process on 5 synthetic traces. If the phase delay between the seismic traces is known, the summation will preserve the true signal while down weighting the instrument noise. A summation with incorrect phase delays will lead to a reduction of the true signal.

weight is applied to each sensor before the traces are combined to calculate the seismic array beam trace $y(t)$. If the true time shifts are known, one can combine all traces constructively to increase the signal-to-noise ratio and suppress individual sensor noise. The coherent summation of all sensor traces leads to the maximization of the beam power P

$$P = \int_{-\infty}^{\infty} y^2(t)dt = \int_{-\infty}^{\infty} |Y(f)|^2 \, df, \qquad (2.2)$$

while an incoherent summation, i.e., the sensor traces are stacked with incorrect phase delays, leads to a degradation of the beam power. The right-hand side was obtained with the help of Parseval's Theorem, where $Y(f)$ is the Fourier transform of $y(t)$. The equivalent discretized equation for a beam trace of length L is given as

$$P = \frac{1}{L} \sum_{t=1}^{L} y^2(t) = \frac{1}{L^2} \sum_{f} |Y(f)|^2. \qquad (2.3)$$

The normalization factor of the inverse length is important to give a measure of beam power which is independent of the seismic trace length.

The phase delays τ_i are in general unknown and have to be estimated. A common approximation is to allow only phase delays that are induced by plane waves travelling over the array in the far-field approximation, i.e., the wavefront approaches the array as a straight line without any curvature. The time delays τ_i are then derived from the knowledge of the spatial sensor locations and a wave propagation model which predicts travel times between all sensors given a slowness vector $\mathbf{s}$. As $\mathbf{s}$ is unknown, one performs a grid search over all possible slownesses to estimate the beam power P. The angle θ of the slowness vector $\mathbf{s}$

$$\mathbf{s} = s \begin{bmatrix} \sin\theta \\ \cos\theta \end{bmatrix} \qquad (2.4)$$

which maximizes P is commonly referred to as direction of arrival (back azimuth) and s is the estimated slowness of the signal.

2.2.1 Frequency-Wavenumber Beamforming

In frequency-wavenumber (f-k) beamforming, the delay-and-sum approach is implemented in the frequency domain. It assumes plane wave propagation over the area of the seismic array and therefore assumes the seismic sources to be in the far field. From the definition of a plane wave with frequency f, imaginary number i, wavenumber vector $\mathbf{k}$, and spatial coordinate vector $\mathbf{r}$

$$x(\mathbf{k}) = e^{-i\mathbf{kr}} = e^{-2\pi i f s r} \quad \text{with} \quad \mathbf{k} = \omega \mathbf{s} = 2\pi f \mathbf{s} \qquad (2.5)$$

one can calculate the phase delay between all station pairs defined as the steering vector

$$\mathbf{a}(f, \mathbf{s}) = e^{-2\pi i f s(\mathbf{r}_n - \mathbf{r}_0)}, \qquad (2.6)$$

where $\mathbf{s}$ is the slowness vector of the plane wave and $\mathbf{r}_n - \mathbf{r}_0$ the distance vector between sensor n and some reference sensor at $\mathbf{r}_0$. The steering vector $\mathbf{a}(f, \mathbf{s})$ defines the phase delays induced by a plane wave propagating over the array with a slowness $\mathbf{s}$ at frequency f. These synthetic phase delays are compared to the delays induced by the observed seismic wavefield via a grid search by iterating over slowness and back azimuth of the steering vector to estimate power contributions from a variety of plane waves.

The beam trace in the frequency domain for M stations is given as

$$y(t) \xrightarrow{\mathcal{F}} Y(f) = \frac{1}{M} \sum_i w_i^* X_i(f) = \frac{1}{M} \mathbf{w}^H \mathbf{X}(f), \qquad (2.7)$$

where ^{H} denotes the conjugate transpose, the superscript $*$ is the complex conjugate, $\mathcal{F}$ represents the Fourier transform of the beam trace $y(t)$ to its spectral representation $Y(f)$, and $\mathbf{X}(f)$ are the Fourier transformed seismic traces. The resulting beam power is given as

$$P(f) = \frac{1}{L^2} |Y(f)|^2 = \frac{1}{L^2 M^2} \mathbf{w}^H \mathbf{X}(f) \mathbf{X}^H(f) \mathbf{w} = \frac{1}{M^2} \mathbf{w}^H \mathbf{C}(f) \mathbf{w}. \qquad (2.8)$$

The middle term

$$\mathbf{C}(f) = \frac{1}{L^2} \mathbf{X}(f) \mathbf{X}^H(f) \qquad (2.9)$$

is the cross spectral density matrix (CSDM) of $M \times M$ form. For this expression the following Fourier transform convention is used

$$X(f) = \int_{-\infty}^{\infty} x(t) e^{-i2\pi f t} dt. \qquad (2.10)$$

The CSDM is a Hermitian matrix that governs all auto and cross spectra information between all sensors and the diagonal terms C_{nn} form the power spectral densities $\int_{-\infty}^{\infty} C_{nn}(f)df$ of the corresponding stations (see McNamara and Boaz, 2018).

The f-k approach is formulated to maximize the beam power $\max\{P(f)\}$. Hence one wants to find the weights which maximize the last term in equation (2.8). To find the weights we reformulate the expression to

$$\mathbf{w}^H \mathbf{C}(f)\mathbf{w} = \mathbf{w}^H \mathbf{X}(f)\mathbf{X}^H(f)\mathbf{w} = \left|\mathbf{w}^H \mathbf{X}(f)\right|^2, \qquad (2.11)$$

and the term simplifies to an inner product between two complex vectors (normalization is omitted in the above equation). The expression is maximized when $\mathbf{w} = \mathbf{X}(f)e^{i\phi}$, where ϕ is an arbitrary phase, given that each multiplication of the inner product results in the same complex angle (coherent phase stacking) and is summed before the calculation of the absolute value. Since one assumes plane wave propagation, the weights are set as a plane wave

$$\mathbf{w}_{fk} = \mathbf{a}(f, \mathbf{s}) \qquad (2.12)$$

defined in equation (2.6). The weights are data independent and solely bound by the array geometry and the propagation characteristics of a plane wave. Hence the f-k method falls under the category of data-independent beamforming techniques. The beam power then follows from equation (2.8) as

$$P_{fk}(f, \mathbf{s}) = \frac{\mathbf{a}^H(f, \mathbf{s})\mathbf{C}(f)\mathbf{a}(f, \mathbf{s})}{M^2}. \qquad (2.13)$$

The denominator normalizes the array beam power to reflect the mean power of the signal recorded at one station and hence can differ from other normalizations which express the combined beam power (e.g., Krim and Viberg, 1996). Rewriting equation (2.13) shows the summation in detail as

$$P_{fk}(f, \mathbf{s}) = \frac{1}{M^2} \sum_n \sum_m C_{nm}(f)e^{-2\pi i f \mathbf{s}(\mathbf{r}_m - \mathbf{r}_n)}. \qquad (2.14)$$

Thus, the beam power is calculated as a comparison between the cross spectrum between stations n m and the phase delay of a synthetic plane wave governed by $f\mathbf{s}(\mathbf{r}_m - \mathbf{r}_n)$. The f-k approach is also known as Bartlett beamforming or the Bartlett processor. Identical to the time domain case, when the synthetic phase delays match the delays from the CSDM, the beam power is maximized and, in the case of a delay mismatch, the beam power degrades.

The above derivation of the beam power was carried out under the assumption of planar wave propagation, i.e., all seismic stations do not vary in elevation. For large differences in elevation and shorter signal wavelengths one is required to account for the difference, which can be accomplished by taking into account the crustal velocity to estimate the phase delay in the vertical direction.

Alternatively, instead of correlating the observed wavefield with a synthetic plane wave, one can extend the correlation to an arbitrary wavefield configuration which leads to the class of matched field processing (Bucker, 1976; Baggeroer et al., 1993) and allows to extend beamforming to a more general form which among others can account for near field sources, heterogeneous propagation media, wavefront bending, and the sphericity of the Earth.

2.2.2 Imprint of the Array

The spatial geometry of the seismic array has a direct influence on the resolution capabilities of beamforming. A way of classifying the performance of each array is to study the beam power under the influence of a single monochromatic wave propagating over the array. The resulting power spectrum is also known as the array response function (ARF) and is defined as

$$P_{ARF}(f, \mathbf{s}) = \frac{1}{M^2} \mathbf{a}^H(f, \mathbf{s})\mathbf{a}(f, \mathbf{s}')\mathbf{a}^H(f, \mathbf{s}')\mathbf{a}(f, \mathbf{s}), \qquad (2.15)$$

where $\mathbf{C}(f) = \mathbf{a}(f, \mathbf{s}')\mathbf{a}^H(f, \mathbf{s}')$ is the CSDM of a plane wave propagating with slowness $\mathbf{s}'$. For the special case of $\mathbf{s}' = 0$, all entries in the CSDM are equal to 1, which is a popular way to display the ARF. In detail, the equation can be written as

$$P_{ARF}(f, \mathbf{s}) = \frac{1}{M^2} \sum_n \sum_m e^{-2\pi i f(\mathbf{s}-\mathbf{s}')(\mathbf{r}_m - \mathbf{r}_n)}. \qquad (2.16)$$

Three ARFs are shown in Figure 2.2 for publicly available seismic arrays, i.e., (a) Gräfenberg Array, (b) Warramunga Array, and (c) Southern California Network Array. In this case $\mathbf{s}' = 0$, which represents a wave exciting all sensors simultaneously (vertical arrival). Hence the highest beam power is located at $s_x, s_y = 0$ in the power spectrum. All three arrays differ in station count, spatial configuration, and aperture, which leads to unique features in their respective ARFs.

Even though the array records only a single monochromatic plane wave, the power spectrum shows power contributions at slowness values which differ from the true slowness $\mathbf{s}'$. This is a direct result of the sparse spatial sampling (array station configuration) of the wavefield which leads to power contributions for other slownesses than $\mathbf{s}'$.

The power spectrum of an ARF can be classified into two separate areas: the beam main lobe, which is the power distribution at and in close vicinity of $\mathbf{s}'$, i.e., the true arrival direction of the signal, and the beam side lobes, which are energy contributions at other slownesses. In general, it is desired to decrease the width of

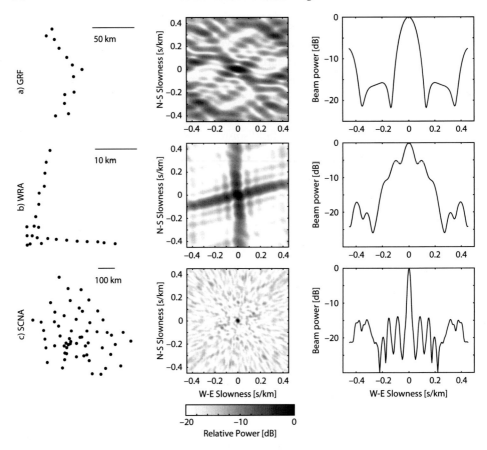

Figure 2.2. In the first column, the station configurations are displayed for (a) the Gräfenberg Array, (b) the Warramunga Array, and (c) the Southern California Network Array (not all stations). The second column displays the corresponding ARFs at $f = 0.2$, $f = 0.6$, and $f = 0.1$ Hz for GRF, WRA, and SCNA. These values are closely related to frequencies at which ambient noise studies have been performed at these arrays. The right column displays a W-E cross section of the ARFs for each array. The unique station configuration and array aperture have a direct influence on the form of the ARF.

the beam main lobe (higher resolution) and decrease the power contribution from beam side lobes. The extent of the main lobe is controlled by the geometry and aperture of the array and the frequency at which beamforming is performed.

The ARF can be seen as an elementary function of the respective array as it is the response to a plane wave. It has the unique property of shift invariance, i.e., the ARF power spectrum retains its form but is shifted in the slowness plane so that the beam main lobe is at position $\mathbf{s}'$. If multiple plane waves with differing $\mathbf{s}'$ propagate over the array, the resulting power spectrum is a superposition of ARFs centered at the corresponding slowness vectors. Alternatively, it can be seen as a convolution between point sources at $\mathbf{s}'_i$ and the ARF. In the presence of multiple sources, beam side lobes sum constructively and can generate local maxima in the power spectrum that can be misinterpreted as real source, hence, an understanding of the ARF is important. It should be noted that, in the case of non-planar wave propagation, the form of the ARF changes.

Most of the permanent seismic arrays available globally were deployed for the purpose of nuclear signal monitoring, and hence were designed to have a high resolution for body waves above 1 Hz. Furthermore, some of these arrays were deployed to have increased resolution toward particular azimuths (e.g., the Warramunga Array) while sacrificing resolution toward others.

For the analysis of ambient noise, it is common to observe multiple arrivals of surface and body waves from a variety of azimuths, which pose differing requirements for the array geometry. It is desired to have omnidirectional resolution with minimal side lobes for an accurate estimation of multiple wave arrivals. To obtain a sharper/narrower main lobe, it is required to extend the aperture of the array, and to suppress side lobes it is required to deploy many stations which cover a variety of wave vectors $\mathbf{k}$ between all inter-station pairs. Hence, with a limited number of stations, one faces a trade-off between main lobe resolution and side lobe suppression.

2.2.3 Data Adaptive Beamforming (MVDR/Capon)

Data adaptive beamforming describes a class of beamformers that take into account the data structure to improve the power spectrum (e.g., Krim and Viberg, 1996). Unlike the data-independent f-k beamformer, whose resolution solely depends on the array configuration, the weights of data adaptive beamformers are derived from the input data.

Among the many approaches in adaptive beamforming, the minimum variance distortionless response (MVRD) or Capon beamformer (Capon, 1969) was successfully applied to ambient noise and showed superior resolution in comparison to the f-k approach (Capon, 1973). The idea is to choose the weights in a way to keep the gain constant in the array "look" direction (direction in which the array beam is focused) while suppressing contributions from all other wavenumbers. This results in a reduction of side lobes dependent on the input data. The weights of the Capon beamformer can be derived through the following minimization problem

$$\min \mathbf{w}^H \mathbf{C} \mathbf{w}$$
$$\text{s.t.} \quad \mathbf{w}^H \mathbf{a}(\mathbf{s}, f) = 1. \tag{2.17}$$

Hence one tries to minimize the output power of the beamformer under the constraint of maintaining the true power into the look direction. The optimization problem can be solved with the Lagrangian multiplier method

$$\mathcal{L} = \mathbf{w}^H \mathbf{C}\mathbf{w} + \Lambda(\mathbf{w}^H \mathbf{a} - 1) = \mathbf{w}^H \mathbf{C}\mathbf{w} + \Lambda(\mathbf{a}^H \mathbf{w} - 1). \qquad (2.18)$$

Solving

$$\frac{\partial \mathcal{L}}{\partial \mathbf{w}} = 0, \qquad (2.19)$$

leads to

$$\mathbf{w}^H \mathbf{C} = -\Lambda \mathbf{a}^H$$
$$\mathbf{C}^H \mathbf{w} = -\Lambda^* \mathbf{a} \qquad (2.20)$$
$$\mathbf{w} = -\mathbf{C}^{-1} \mathbf{a} \Lambda^*.$$

Inserting the condition (2.17) and reordering yields the expression for the Lagrangian parameter

$$\Lambda^* = -\frac{1}{\mathbf{a}^H \mathbf{C}^{-1} \mathbf{a}}. \qquad (2.21)$$

The weight vector follows from equation (2.20) using equation (2.21)

$$\mathbf{w}_C = \frac{\mathbf{C}^{-1} \mathbf{a}}{\mathbf{a}^H \mathbf{C}^{-1} \mathbf{a}}, \qquad (2.22)$$

and leads to the Capon power spectrum

$$P_C = \mathbf{w}_C^H \mathbf{C} \mathbf{w}_C = \frac{1}{\mathbf{a}^H \mathbf{C}^{-1} \mathbf{a}}. \qquad (2.23)$$

In the data adaptive case, the response of the array to a plane wave differs from the conventional f-k case. The shape of the ARF is now data dependent and shift invariance is not assured. The suppression of side lobes is best for a monochromatic plane wave and zero noise (i.e., synthetic case). Increasing the number of signals and/or noise will decrease suppression capabilities of the Capon beamformer and can, in the case of steering vector errors, result in a worse power estimation than the f-k case (e.g., Seligson, 1970). In general, the required noise levels to degrade the performance below the resolution of the f-k approach are not encountered during the analysis of ocean-induced microseisms.

The increased resolution comes at the cost of robustness (the term robustness in the beamforming jargon refers to the stability of the approach). Unlike the f-k case, the weight vector in the Capon approach is calculated via the inverse of the CSDM, which can be rank deficient in the case of limited data. Rank-deficient matrices, which necessarily have determinant equal to zero, do not have an inverse and one is forced to modify the CSDM in order to obtain an estimate of their inverse. Successful modifications include, but are not limited to, the use of the Moore-Penrose pseudo inverse, temporal and frequency averaging, and diagonal loading (Featherstone et al., 1997).

Another robustness issue is the sensitivity to departures from the plane wave model. The Capon beamformer is subject to strong performance degradation in the presence of a propagation model mismatch (Seligson, 1970). While this is also the case for f-k, the Capon beamformer is more sensitive toward a propagation mismatch and can lead to a worse performance than the f-k approach.

2.3 Optimized Beamforming for the Analysis of Ambient Noise

The nature of the continuous background oscillations of ambient noise can be exploited in order to optimize performance and robustness of the beamforming framework. Unlike in earthquake seismology, where the source information is constrained to a relatively small time window, the signal information from ambient noise is spread over long durations and can be exploited to better estimate the resulting phase delays between all sensor pairs.

2.3.1 Pre-Processing

Quality Control – Excluding Outliers

To guarantee accurate functionality and power estimation of beamforming, a quality control of the seismic data is important. The majority of unwanted signals are either earthquakes or incoherent noise due to local sources or malfunctioned receivers. The unwanted signals become a problem once their amplitude is comparable to the amplitude of the signal. Such incoherent noise will have an imprint on the frequency spectrum and interfere with the beamforming analysis, causing the beampower to degrade. To identify seismic traces/snapshots that are perturbed by earthquake signals, a common approach is to use an event catalogue (e.g., ISC Bulletin) to identify time frames where strong earthquakes occurred and exclude them from the beamforming process. Alternatively, if accurate amplitude information is not important, one can substitute the CSDM with the spectral coherence (e.g., Capon, 1969), which normalizes the frequency spectrum (spectral whitening) and retains phase information only (discussed in more detail in *Modification of the CSDM* in this section). This is especially useful in combination with temporal averaging (Gerstoft et al., 2008) as the influence of the transient earthquake signal is averaged out. If amplitude information is of interest (e.g., Gualtieri, 2014; Farra et al., 2016), one can either identify snapshots with earthquake contributions given their differing spectral characteristics in comparison to ambient noise (Aster et al., 2010), or one can perform a type of temporal power estimation and exclude snapshots which show power above a set threshold (e.g., Meschede et al., 2017).

Receivers that display a high, continuous, incoherent noise need to be excluded from the beamforming process.

Temporal Averaging

Given the relatively stationary nature of seismic ambient noise (within a few hours), an often deployed method is temporal averaging (e.g., Gerstoft et al., 2008). The idea is to take an array time series $\mathbf{x}(t)$ and split it into N shorter time durations of equal temporal length $\mathbf{x}_i$ with $i \in 1, \ldots, N$. The shorter duration time series, also known as snapshots or temporal sub-windows, are used to construct N CSDM's which are averaged into one final CSDM

$$\mathbf{C}(f) = \frac{1}{L^2 N} \sum_i^N \mathbf{X}_i(f) \mathbf{X}_i^H(f). \tag{2.24}$$

The advantage of this averaging is that phase delays that persist over the duration of the averaging are constructively summed, while phase delays from incoherent signals are averaged out. This also means that if non-stationary signals are present, e.g., moving sources, their power will not be accurately estimated but is reduced due to the averaging procedure. Another advantage is the increase in the rank of the CSDM $\mathbf{C}$, which is especially important in the case of the Capon beamformer.

The time length, which is divided into snapshots, can reach from minutes/hours (e.g., Friedrich et al., 1998; Gerstoft et al., 2008) to multiple weeks (e.g., Euler et al., 2014) depending on the task. While temporal averaging in general increases the robustness of the beamformer, it can have a different effect on data adaptive and data-independent (f-k) beamforming. Since the Capon beamformer estimates its weight from the CSDM, its side lobe suppression capabilities are governed by the signal distribution propagating over the array. If one averages over a time duration where the wavefield is non-stationary, a variety of wavenumbers $\mathbf{k}$ populate the CSDM and reduce the suppression capabilities of the Capon beamformer.

The optimal snapshot length is dependent on the frequency of interest, the observed wavefield, and the aperture of the array. If the array aperture is large (several 100 km) and one is interested in surface waves, the snapshot length must be chosen with care given that the surface waves require a certain time to travel between all stations. This is especially important for wavefields that strongly vary with time (extreme case is surface waves from an earthquake).

In an example the importance of the snapshot length is further highlighted. Let there be a signal of interest at 0.2 Hz and our sensor has a sampling rate of 1 Hz. If one selects a snapshot length of 50 s, i.e., 10 periods of the signal are present in one snapshot, the Fourier transformed signal consists of 26 positive valued frequency bins (including 0), which are $f_n = 0.02n$ with $n \in 0, 1, \ldots, 25$, where $n = 25$ equals the Nyquist frequency. If the CSDM at 0.2 Hz is computed, one can see that additional frequencies are present, i.e., $f = 0.2 \pm 0.01$ Hz. Furthermore, one faces the issue of spectral leakage (e.g., Thomson, 1982), which is more pronounced for short snapshots. Increasing the snapshot length reduces the frequency bin width and reduces the effect of spectral leakage, and hence reduces the presence of neighboring frequencies. However, increasing the snapshot length too far (multiple hundreds to thousands of periods in one snapshot) results in a noise amplification in the phases of $\mathbf{X}(f)$, which translates into a degradation of the beamformer's performance.

With the above information one can tailor the snapshot length to a desired task, i.e., where a narrow frequency band of the wavefield is of interest one will choose a longer snapshot length, while in cases where the signal-to-noise ratio is low one might choose a shorter snapshot length.

Pre-Processing of Snapshots

Once the desired snapshot length is chosen, standard time series processing should be applied. For each snapshot the trend, mean, and sensor response should be removed and optionally a taper function should be applied. Taper functions are used to reduce finite window effects which occur during the Fourier transformation. Since each snapshot is considered as a periodic signal when the Fourier transform is applied, discontinuities at the edges are likely to appear which induce unwanted spectral information. Taper functions modify each snapshot in order to reduce certain characteristics during the Fourier transform. Commonly used taper functions include the Cosine and Hann window function to reduce spectral leakage and are more important when the snapshot length is chosen on the shorter side. One should keep in mind that taper functions directly influence the strength of signals and hence this needs to be taken into account if the signal power is of interest.

Modification of the CSDM

Overlapping snapshots is a popular method to increase the rank and robustness of the CSDM. This method is particularly useful if little data are available and one obtains a better estimated CSDM. If taper functions are used, overlapping snapshots is a great method to include the information that has been tapered off in the previous snapshot.

Another popular modification is the sensor array normalization (e.g., Capon, 1969; Haubrich and McCamy, 1969) also known as coherence. The coherence matrix is defined as

$$\hat{C}_{nm}(f) = \frac{X_n(f)X_m^H(f)}{\sqrt{|X_n(f)|^2 |X_m|^2}}. \tag{2.25}$$

This normalization retains only phase information and removes improper sensor equalization, which can occur due to path propagation effects, geographical location, faulty sensors, and gain fluctuations. The normalization becomes increasingly beneficial for seismic arrays with large apertures (e.g., Haubrich and McCamy, 1969; Gerstoft et al., 2008; Euler et al., 2014) and many stations where path propagation effects and faulty sensors due to the increased number are likely. Furthermore, it reduces the influence of earthquakes if combined with temporal averaging as their information is averaged out (e.g., Gerstoft et al., 2008).

Additionally to the above modifications it can prove beneficial to remove the diagonal elements of the CSDM, which are the auto-spectra. In the case of strong incoherent noise, the auto-spectra may result in a distorted beamforming result (e.g., Westwood, 1992; Ruigrok et al., 2017).

2.3.2 Extension to Broadband Analysis

Both the f-k and Capon beamformer as laid out in the above sections are narrowband techniques. The signal propagation characteristics are examined at one frequency f to yield the power spectrum $P(f, \mathbf{s})$. In the case of ocean-induced microseisms, where the sources are distributed in a broad frequency range, one observes only a fraction of the information present in the wavefield (see McNamara and Boaz, 2018 and Ardhuin et al., 2018). The extension to estimate the power spectrum in a broad frequency range can be accomplished through a variety of approaches (e.g., Wax et al., 1984; Wang and Kaveh, 1985; Kværna and Doornbos, 1985; Kværna and Ringdal, 1986; Westwood, 1992; Chiou and Bolt, 1993; Soares and Jesus, 2003) which can be divided into two groups. The first group focuses on combining the estimated narrowband spectra over a given frequency range in order to obtain a broadband representation. It is also known under incoherent averaged signal (IAS) method, where the incoherent averaging relates to the summation after the beam power is estimated. The second group focuses on modifying the CSDM in order to obtain the broadband representation and is also known as coherent averaged signal (CAS) method, where coherent averaging refers to methods which combine the broadband signal before the beam power is estimated.

The simplest way is to estimate narrowband spectra at multiple narrowband frequencies, and their combination results in the broadband representation

$$P_{IAS}(\mathbf{s}) = \sum_i P(f_i, \mathbf{s}). \tag{2.26}$$

Alternatively, the summation can be replaced by the median in order to suppress sharp artifacts from individual narrowband spectra. The simplicity of this method allows it to be applied over a broad frequency range. This has the additional advantage that the ARF changes with frequency, and the contribution of side lobes can average out. Another advantage is that sources present at differing frequencies do not mix with each other, which allows adaptive beamforming methods to yield better resolution. Potential disadvantages are the increased computational burden as multiple narrowband spectra need to be calculated, and stability issues in scenarios with low signal-to-noise ratio (Yoon et al., 2006).

Coherent approaches have received more attention mainly for the reason that they exploit the broadband characteristics for increased performance (e.g., Chiou and Bolt, 1993). Although most of the work has focused on the construction of a single broadband CSDM, mainly for the purpose of subspace-based beamforming (another type of adaptive beamforming), e.g., MUSIC (Schmidt, 1986), approaches that are useful for the analysis of microseisms, where multiple signals are present, are discussed.

A straightforward coherent approach is to average the CSDM $\mathbf{C}(f)$ over neighboring frequencies

$$\mathbf{C}_{CAS} = \sum_i \mathbf{C}(f_i). \tag{2.27}$$

This procedure increases the rank of the CSDM and better estimates the average phase delay between stations. A disadvantage of this method arises due to the mix of multiple frequencies in $\mathbf{C}_{CAS}$ and the subsequent projection on to a monochromatic plane wave. Hence the beamformer correlates a plane wave with a broadband signal. As the beamformer assumes that information in the CSDM is monochromatic with frequency f_p, signals that differ from f_p are shifted in slowness since $\mathbf{k} = 2\pi f \mathbf{s}$. From this linear relationship it follows that a signal that differs by $\pm 10\%$ from f_p will be estimated with a modified slowness by $\pm 10\%$ (additional information can be found in Gal et al. (2014)). Furthermore, the mixing of a frequency range is responsible for the decrease in signal coherence (Menon et al., 2014). For negligible slowness errors, this method can be used in order to increase the robustness of beamforming. The approach is similar to the short snapshot length case where wide frequency bins harbor a range of frequencies. Hence using short snapshots can lead to biased slowness estimates in the same way.

The above shortcomings can be bypassed by explicitly accounting for phase mismatch between f_n and f_p. Let there be a frequency f_n which differs from the projection frequency f_p; one can formulate the transformation matrix $\mathbf{T}$ which transforms between two steering vectors at differing frequencies

$$\mathbf{T}(\mathbf{s}, f_p, f_n)\mathbf{a}(\mathbf{s}, f_n) = \mathbf{a}(\mathbf{s}, f_p). \tag{2.28}$$

The explicit form of $\mathbf{T}(\mathbf{s}, f_p, f_n)$ is

$\mathbf{T}(\mathbf{s}, f_p, f_n) =$

$$
\begin{pmatrix}
\exp\left[-i2\pi \mathbf{sr}_1(f_n - f_p)\right] & 0 & \cdots & 0 \\
0 & \exp\left[-i2\pi \mathbf{sr}_2(f_n - f_p)\right] & \cdots & 0 \\
\vdots & \vdots & \ddots & \vdots \\
0 & \cdots & & \exp\left[-i2\pi \mathbf{sr}_n(f_n - f_p)\right]
\end{pmatrix}.
$$
(2.29)

The matrix is then used to transform the CSDM to account for the frequency mismatch

$$
\mathbf{C}_{CAS}(\mathbf{s}, f_p) = \sum_n \mathbf{T}(\mathbf{s}, f_p, f_n)\mathbf{C}(f_n)\mathbf{T}^H(\mathbf{s}, f_p, f_n).
$$
(2.30)

For $f_p = f_n$ the matrix $\mathbf{T}$ equals the identity matrix and phase information is unchanged, while for $f_p \neq f_n$ the phase information is modified. Contrary to the conventional case, here the CSDM $\mathbf{C}_{CAS}(\mathbf{s}, f_p)$ is slowness dependent, because the knowledge of the slowness is required in order to correct for the frequency mismatch. This means that in order to obtain a bias-free power spectrum, one is required to re-calculate $\mathbf{C}_{CAS}(\mathbf{s}, f_p)$ for every possible slowness $\mathbf{s}$

$$
P(\mathbf{s}, f_p) = \frac{1}{M^2}\mathbf{a}^H(\mathbf{s}, f_p)\mathbf{C}_{CAS}(\mathbf{s}, f_p)\mathbf{a}(\mathbf{s}, f_p).
$$
(2.31)

For every $\mathbf{s}$ the power contribution from a frequency other than f_p is modified by the appropriate phase delay and hence avoids slowness bias due to frequency mixing. This can be viewed as an improved version of the frequency smoothing method described in equation (2.27) for the cost of an increased computational burden. In this derivation of the method, it is assumed that the slowness remains constant over the course of the coherent summation. In cases where this assumption does not hold, bias is introduced. For the case where such slowness variations over the frequency range are known, it can be accounted for. It should be further noted that the concept of the ARF differs for this coherent approach, since signals are focused on one specific slowness at a time. While the power estimation for the projection slowness $\mathbf{s}_p$ is accurate, signals with different slowness present in the CSDM are distorted and contribute in a unique way for each slowness $\mathbf{s}_p$.

The extensions to broadband described above favor different wavefield types. For the analysis of ocean-induced microseisms, the IAS approach is mainly used (e.g., Kværna and Ringdal, 1986; Friedrich et al., 1998; Gerstoft et al., 2006, 2008; Nishida et al., 2008; Zhang et al., 2009; Reading et al., 2014; Euler et al., 2014; Gal et al., 2014, 2015) given that the signal-to-noise ratio in the band of 1–300 s is comparably high. However, for cases with higher noise levels ($\gtrsim$ 1 Hz) CAS approaches can show better performance (e.g., Yoon et al., 2006).

2.3.3 Post-Processing

In this section two methods which aim to improve the resolution of the power spectrum after its initial estimation are reviewed. Both methods fall into the category of deconvolution, where the underlying idea is to reduce the imprint of the ARF in the results. Since the ARF is constructed under the assumption of plane wave propagation, the deconvolution approach requires plane wave propagation to yield meaningful results. If seismic waves do not propagate in accordance with the plane wave model, deconvolution will introduce errors into the final results unless the true ARF is known. Hence the approaches outlined below put additional emphasis on the fulfillment of the underlying path propagation approximations.

Richardson-Lucy Deconvolution

The Richardson-Lucy (RL) deconvolution (Richardson, 1972; Lucy, 1974) was originally derived in astronomy for the purpose of deblurring images. Stars appear blurred on an image due to the telescope optics as they are convolved with a point spread function (PSF) of the instrument (similar to the concept of the ARF). With the knowledge of the instrument's PSF, the idea is to remove the imprint of the PSF from the image.

The data model is defined for an observed image $I(x)$ as

$$I(x) = B(x) * O(x), \tag{2.32}$$

where $B(x)$ is the PSF of the optical system, $O(x)$ is the true noise-free image, $*$ denotes the convolution operator, and x are the image's coordinates. It is desired to obtain $O(x)$ with the knowledge of the PSF and the observed image $I(x)$. This reconstruction can be addressed in a Bayesian framework and leads to an iterative equation for the true image

$$\hat{O}_{i+1}(x) = \hat{O}_i(x) \left[B(x)^T * \frac{I(x)}{B(x) * \hat{O}_i(x)} \right], \tag{2.33}$$

where i is the iteration number and $\hat{O}_0$ denotes the first guess for the solution, e.g., the mean of $I(x)$. It was shown that equation (2.33) deconvolves the PSF from the image $I(x)$ under a maximum likelihood constraint (Shepp and Vardi, 1982).

In cases where an object of interest is located near the edge of the image, artifacts in the form of ringing may be introduced during the iteration procedure (Gibbs oscillations). The artifacts can be reduced by an extension proposed by Bertero and Boccacci (2005), which modifies the original equation to

$$\bar{O}_{i+1}(x) = \bar{w}(x)\,\bar{O}_i(x) \left[\bar{B}(x)^T * \frac{\bar{I}(x)}{\bar{B}(x) * \bar{O}_i(x) + b} \right]. \qquad (2.34)$$

The objects denoted with a bar have been extended by zero padding, b is a regularization parameter that equals the background noise, and $\bar{w}(x)$ is a mask object, which defines the area of influence of the PSF. An in-depth explanation of each object and how it can be merged with f-k beamforming is given by Gal et al. (2016).

The RL deconvolution can be directly transferred and applied for the case of f-k beamforming, where the power spectrum is a convolution between true source location in the wavenumber plane and the ARF. The first successful application of the RL deconvolution in the observation of microseisms was demonstrated by Nishida et al. (2008) with the Hi-net array in Japan, where smearing of the f-k power spectrum induced by Hi-net's elongated station configuration was reduced. Other studies have since then evaluated the performance of the RL deconvolution approach on ambient noise beamforming results in more detail (e.g., Picozzi et al., 2010; Gal et al., 2016).

The maximum likelihood constraint makes the RL deconvolution method very robust, but one feature in particular that differs between seismology and astronomy should be highlighted. PSFs in astronomy generally exhibit small side lobes (small in comparison to seismic ARFs), which decay with an increasing distance from the main lobe. In seismology, the ARF does not decay in this way and side lobes extend into infinity. This can pose an ambiguity in the choice of the mask $\bar{w}(x)$ (Gal et al., 2016), which has a direct influence on the final solution. Hence testing of the $\bar{w}(x)$ for each array configuration is required for an optimal solution. Additionally, the optimal stop criterion is unknown and needs to be selected by the user.

An example demonstrating the capabilities of the RL deconvolution is shown in Figure 2.3. The ARF, f-k, and Capon beam power for a 1-hour-long recording with the Gräfenberg Array on 1 January 2017 00:00:00 UTC is displayed in Figure 2.3a–c. The analysis is carried out at 0.067 Hz (15 s) and the f-k results show one broad main lobe. The Capon beamformer shows improved resolution and can separate two sources which appear as one in the f-k result. The RL deconvolution result (Figure 2.3d), here shown after 50 iterations, shows a move of the main lobe toward one of the maxima found with the Capon beamformer but does not identify the second source to the north.

CLEAN

The CLEAN algorithm, initially developed in radio astronomy (Högbom, 1974), follows the same principle, i.e., the deconvolution of the ARF from an image. The idea is to locate the brightest spot in the image $I(x)$ and remove a fraction of the PSF $P(x)$ associated with the brightness maximum

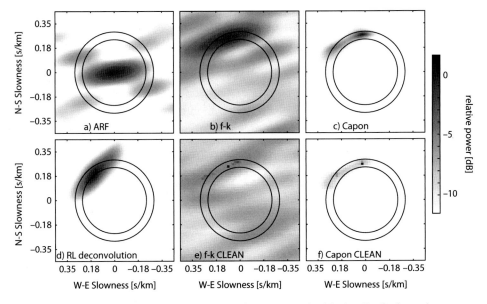

Figure 2.3. Deconvolution approaches demonstrated with the Gräfenberg Array on 1 January 2017 00:00:00 UTC at 0.067 Hz. The top row shows the (a) ARF, (b) f-k, and (c) Capon beam power. The bottom row shows (d) the RL deconvolution calculated from the f-k result, (e) the CLEAN algorithm in combination with f-k beamforming, and (f) the CLEAN in combination with the Capon beamformer. The two black circles of constant velocity are set at 3 and 4 km/s.

$$\hat{I}^{i+1}(x) = I^i(x) - \phi I^i(x_{max}) P(x_{max}) \quad \longrightarrow \quad \hat{O}(x) = \sum_i \phi I^i(x) P(x_{max}),$$

$$(2.35)$$

where the superscript i denotes the iteration number, ϕ is the CLEAN control parameter which regulates the strength of the removed PSF, and x_{max} points toward the maximum in the image. In the case of $P(x_{max})$, the main lobe of the PSF is centered at x_{max}. Hence one iteratively removes a fraction of the brightness maximum and its associated side lobes from $I^i(x)$. Since the power and location of the removed brightness are known, one can combine the removed brightness to form a representation $\hat{O}(x)$ free of side lobes, which is shown in the right-hand side of equation (2.35).

In its original form, CLEAN performs the deconvolution procedure on the observed image and hence the algorithm is directly transferable to images from f-k beamforming. In the case of beamforming, an alternative approach for the implementation of CLEAN is possible, where the cleaning procedure is directly applied to the CSDM (Wang et al., 2004; Sijtsma, 2007). This has the advantage that CLEAN can be applied to data adaptive beamformers (Gal et al., 2017), which have a data-dependent ARF.

The underlying equation is

$$\mathbf{C}^{i+1}(f) = \mathbf{C}^i(f) - \phi P^i(f, \mathbf{s}_{max})\mathbf{a}(f, \mathbf{s}_{max})\mathbf{a}^H(f, \mathbf{s}_{max}), \qquad (2.36)$$

where $\mathbf{s}_{max}$ is the slowness vector which maximizes P^i. Hence this implementation removes a fraction of the power associated with the strongest signal in P^i. The term $\mathbf{a}\mathbf{a}^H$ generates the appropriate phase delays which need to be removed. This can be alternatively seen as an introduction of a synthetic point source of negative power at the position of the dominant source. In analogy to the original CLEAN, since the removed power is known, one can combine all "cleaned" point sources and combine them to yield the side lobe free power spectrum

$$P_{CLEAN}(f, \mathbf{s}) = \sum_i \phi P^i(f, \mathbf{s}_{max}). \qquad (2.37)$$

A range of stopping criteria can be applied to CLEAN, ranging from power thresholds, to total power decomposition, to user-defined criteria. Regardless of the choice, the iteration procedure should be stopped once an eigenvalue of the CSDM becomes negative. A negative eigenvalue would suggest negative energy present in the CSDM which is nonphysical.

An example for the CLEAN approach is shown in Figure 2.3e,f for the f-k and Capon algorithm. In the f-k case, after CLEAN has removed a major part of the main lobe's beam power, the single maximal peak separates into its two sources. Given that a major part of the power was removed with an incorrect slowness vector, the successive iterations try to compensate for it, which can lead to inaccurate source identification (documented in Gal et al. (2016)). CLEAN in combination with the Capon beamformer (Figure 2.3f) shows highest resolution.

2.3.4 Array Design and Use

The analysis of microseisms by means of beamforming benefits most from arrays with narrow beam main lobes and a strong side lobe suppression. The width of the main lobe is narrowed by increasing the aperture of the array, while the side lobes are more strongly suppressed when as many differing wavenumbers as possible are realized by inter-station pairs. The second point is especially important given that multiple sources from differing azimuths are likely to be present in the estimate wavefield. One could naively assume that increasing the amount of stations in an array automatically results in a better side lobe suppression, but this is only the case if the additional stations are positioned in a way which covers inter-station distances that were not realized by the previous configuration. An example is the regular grid array, which if extended periodically will not improve the side suppression.

Optimal Array

An optimal array configuration for the analysis of seismic ambient noise is realized by the spiral shape (e.g., Kennett et al., 2015). A synthetic example of a spiral array configuration with 101 sensors and an aperture of 220 km is shown in Figure 2.4a. The ARF for this configuration at 0.15 Hz is displayed in Figure 2.4b. The advantage of the spiral configuration is the coverage of a range of inter-station distances, i.e., the sampling of a variety of wavenumber vectors.

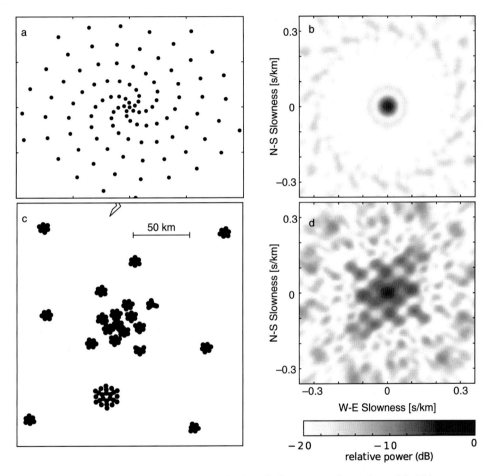

Figure 2.4. First row displays a synthetic spiral array configuration with 101 stations and the corresponding ARF at $f = 0.15$ Hz. The second row displays the Large Aperture Seismic Array configuration with over 300 stations and its corresponding ARF at the same frequency. The aperture of the synthetic array was chosen to match the one of LASA. The spiral array has a better coverage over the wavenumber space, which allows it to strongly suppress the beam side lobes, which is preferable for the analysis of microseisms.

Furthermore, the configuration is an optimal trade-off between amount of stations and side lobe suppression and shows good performance over a range of frequencies with omnidirectional capabilities. This can be visualized when compared with the large aperture seismic array (LASA), which has 300+ stations (Figure 2.4c). The ARF of LASA at 0.15 Hz is shown in Figure 2.4d and displays stronger side lobes than the synthetic array with 1/3 of LASA's stations. For higher frequencies, which LASA was initially designed for, the side lobes still extend 0.1 s/km from the main lobe and are suppressed below -20 dB for higher slownesses. The synthetic array configuration on the other hand shows sparse side lobe maxima with power levels below -15 dB for higher slownesses at 1 Hz, which could be further suppressed if the amount of stations would approach the LASA station count.

Imperfect Arrays

In most cases, the optimal array in the desired location is unavailable and one has to work with available arrays which deviate from the optimal form. This situation occurs with grid-shaped temporary arrays which are commonly deployed for tomography purposes where a constant spatial sampling of the wavefield is desired. For such arrays, the ARF side lobes are large and can strongly influence the beamforming results. To mitigate the strong side lobes, one can adjust station weights to obtain an improved ARF.

The above idea is discussed using a temporary array deployed in Tasmania, Australia (station configuration - Figure 2.5a). The SETA array comprises 42 stations in a relatively grid-like configuration, and its ARF is shown in Figure 2.5b at 0.2 Hz. Six strong side lobes are visible with power levels up to ~ -3 dB, which are the reason for strong artifacts when estimating the ambient noise power spectrum. Instead of using the conventional ARF equation, one can extend it to include station weights (Lacoss et al., 1969), which leads to

$$P_{ARF}(f, \mathbf{s}) = \frac{1}{M^2} \sum_n \sum_m w_n w_m e^{-2\pi i f (\mathbf{s} - \mathbf{s}')(\mathbf{r}_m - \mathbf{r}_n)}. \tag{2.38}$$

The station weights w_n are set to 1 for the conventional case for which the above equation equals equation (2.14). The idea is to search for stations which are positioned at locations not conforming to the regular grid, i.e., stations which sample different wavenumbers, and increase their weight (gain). If this is done for the SETA array, one obtains the modified ARF displayed in Figure 2.5c. The weights of stations MOO and TAU (Figure 2.5a) were empirically increased, which results in a side lobe reduction to ~ -5.5 dB. In this particular example, the weight value of station MOO and TAU was changed from 1 to 4.

The modification of the weights comes at a trade-off, as can be seen in Figure 2.5c where new side lobes appear in contrast to the unweighted case. By

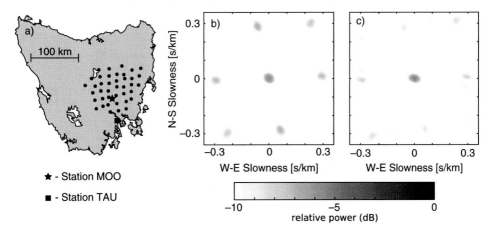

Figure 2.5. Displayed is a temporary seismic array (black dots) deployed in Tasmania, Australia (a), with an approximately mesh-like station configuration. Additionally, two permanent stations are displayed (MOO and TAU) in the area of the temporary array. Mesh-like array configurations suffer from a limited sampling of the wavenumber space and (b) result in strong side lobes visible in the ARF ($f = 0.2$ Hz). By increasing the importance (station weight) of the two permanent stations, one boosts the contribution of underrepresented wavenumbers in the beamformer process, which results in an overall decrease of side lobe power levels. The modification of station weights comes at a trade off and generates new side lobes, but under careful optimization a more suitable ARF can be achieved.

suppressing the large side lobes via the station weight modification, one increases the power of other features in the power spectrum. In this case, the modified ARF is preferred over the original case as side lobe power levels are much smaller. The station weights are applied in the same way for the desired beamformer.

2.4 Three-Component Beamforming

Three-component plane wave beamforming, which is able to estimate plane wave propagation on all three components (i.e., including Love, SV, SH, and L_g waves), is constrained by the same approximations as the one-component case. Hence the estimated signal propagates as a plane wave over all three components in the far-field approximations. Two groups of extensions in which three-component beamforming can be realized are discussed. The first group reduces the two horizontal components to a single dimension by rotating the horizontal components into the desired look direction to separate radial and transverse signals (e.g., Toksöz and Lacoss, 1968; Lacoss et al., 1969; Haubrich and McCamy, 1969; Capon, 1973; Kværna and Doornbos, 1985; Friedrich et al., 1998; Nishida et al., 2008; Brooks et al., 2009; Poggi and Fäh, 2010; Gibbons et al., 2011; Behr et al., 2013) and applies one-component beamforming separately on each component. The second group makes use of all three components simultaneously (e.g., Esmersoy et al., 1985; Wagner, 1996; Wagner and Owens, 1996; Wagner, 1997; Miron et al., 2005; Riahi et al., 2013; Gal et al., 2016; Juretzek and Hadziioannou, 2016) and will be discussed on the basis of two potential implementations in the following section.

2.4.1 Dimensional Reduction

The reduction from two horizontal components to a single dimension may be achieved with

$$
\begin{aligned}
\mathbf{x}_R(t, \theta) &= \mathbf{x}_N(t) \cos \theta + \mathbf{x}_E(t) \sin \theta \\
\mathbf{x}_T(t, \theta) &= -\mathbf{x}_N(t) \sin \theta + \mathbf{x}_E(t) \cos \theta,
\end{aligned}
\tag{2.39}
$$

where the subscripts N, E, R, and T denote the north, east, radial, and transverse components, respectively, and θ is the look direction (back azimuth) of the array. In analogy to the one-component case, the rotated radial and transverse components can be used as an input to any one-component beamformer to generate the respective CSDM

$$
\begin{aligned}
\mathbf{C}_R(f, \theta) &= \frac{1}{L^2} \mathbf{X}_R(f, \theta) \mathbf{X}_R^H(f, \theta) \\
\mathbf{C}_T(f, \theta) &= \frac{1}{L^2} \mathbf{X}_T(f, \theta) \mathbf{X}_T^H(f, \theta)
\end{aligned}
\tag{2.40}
$$

and the radial and transverse power spectra follow as

$$
\begin{aligned}
P_R(f, \mathbf{s}) &= \frac{1}{M^2} \mathbf{a}^H(f, \mathbf{s}) \mathbf{C}_R(f, \theta) \mathbf{a}(f, \mathbf{s}) \\
P_T(f, \mathbf{s}) &= \frac{1}{M^2} \mathbf{a}^H(f, \mathbf{s}) \mathbf{C}_T(f, \theta) \mathbf{a}(f, \mathbf{s}).
\end{aligned}
\tag{2.41}
$$

This approach has the advantage that all one-component beamforming extensions can be directly applied to the horizontal components and the simplicity and robustness remain. A disadvantage is that the intrinsic separation of the sensor data (equation 2.39) is based on the assumption of straight propagation paths and might not hold for large arrays where wavefront bending can occur. Assuming the ray path deviates from its expected direction by 5.7°, 10% signal leakage is introduced from the transverse to the radial component (based on a simple trigonometric calculation) and *vice versa*. Hence, small deviations from the plane wave model in the far field induce a mixing of radial and transverse energy. Additionally, anisotropy of the elastic modulus can also lead to deviations between polarization and propagation direction.

2.4.2 Polarization Beamforming

Polarization beamforming utilizes all three components simultaneously by exploiting the polarization characteristics of the propagating waves. The following two approaches, which employ three-component polarization beamforming, are summarized. The method of Esmersoy et al. (1985), recently applied to ambient noise (e.g., Riahi et al., 2013; Juretzek and Hadziioannou, 2016), extends the Fourier transformed data vector to

$$\mathbf{X}_{3C}(f) = \left[X_{E,1}(f), \ldots, X_{E,M}(f), X_{N,1}(f), \ldots, X_{N,M}(f), X_{Z,1}(f), \ldots, \right.$$
$$\left. X_{Z,M}(f) \right], \tag{2.42}$$

where X_{nm} denotes the Fourier transform of the nth component and mth sensor. The three components are concatenated, where the order of the components is interchangeable, but is required to be consistent for the successive steps. The three-component CSDM leads to

$$\mathbf{C}_{3C}(f) = \frac{1}{L^2} \mathbf{X}_{3C}(f) \mathbf{X}_{3C}^H(f) \tag{2.43}$$

and is a $3M \times 3M$ matrix. A more intuitive form of the 3C CSDM is (dependence on f is omitted)

$$\mathbf{C}_{3C} = \begin{bmatrix} \mathbf{C}_{EE^*} & \mathbf{C}_{EN^*} & \mathbf{C}_{EZ^*} \\ \mathbf{C}_{NE^*} & \mathbf{C}_{NN^*} & \mathbf{C}_{NZ^*} \\ \mathbf{C}_{ZE^*} & \mathbf{C}_{ZN^*} & \mathbf{C}_{ZZ^*} \end{bmatrix}. \tag{2.44}$$

$\mathbf{C}_{3C}$ is a Hermitian matrix as in the one-component case and governs phase delay information between all stations and components.

To estimate the power of the desired polarization it is required to define a propagation model which describes the particle motion of the signal. Esmersoy et al. (1985) obtained the representation via the following equation

$$\hat{\mathbf{a}}(f, \mathbf{s}, \xi) = \mathbf{c}(\xi) \otimes \mathbf{a}(f, \mathbf{s}), \tag{2.45}$$

where $\mathbf{c}(\xi) = [c_E(\xi), c_N(\xi), c_Z(\xi)]^T$ is the normalized complex polarization state (Samson, 1983) with the polarization parameter ξ and $\otimes$ is the Kronecker product. Hence the one-component steering vector is extended to a three-component propagation times the polarization state of the desired signal. The three-component power spectrum for a particular polarization is then

$$P_{3C}(\mathbf{s}, f, \xi) = \frac{1}{M^2} \hat{\mathbf{a}}^H(f, \mathbf{s}, \xi) \mathbf{C}_{3C}(f) \hat{\mathbf{a}}(f, \mathbf{s}, \xi). \tag{2.46}$$

Hence the approach uses all stations and components of an array to evaluate the power contribution of a plane wave with the polarization state $\mathbf{c}(\xi)$. To obtain the complete power representation of all signals present in the wavefield, it is required to sum over all possible polarization states.

Another three-component beamforming framework which was recently applied to microseism analysis (Gal et al., 2016; Liu et al., 2016; Gal et al., 2017) was given by Wagner (1996), where a steering vector matrix is defined as

$$\mathbf{e}(f, \mathbf{s}) = [\mathbf{a}_E(f, \mathbf{s}), \mathbf{a}_N(f, \mathbf{s}), \mathbf{a}_Z(f, \mathbf{s})], \tag{2.47}$$

which in more detail is given as

$$\mathbf{e}(f, \mathbf{s}) = \begin{bmatrix} a_{E1} & \cdots & a_{EM} & 0 & \cdots & 0 & 0 & \cdots & 0 \\ 0 & \cdots & 0 & a_{N1} & \cdots & a_{NM} & 0 & \cdots & 0 \\ 0 & \cdots & 0 & 0 & \cdots & 0 & a_{Z1} & \cdots & a_{ZM} \end{bmatrix}^T. \tag{2.48}$$

In accordance to the previous approach, the components are interchangeable but are required to be consistent for all remaining steps. Here a_{nm} are the conventional steering elements of $\mathbf{a}(f, \mathbf{s})$, where n denotes the component and m the station. Hence all components are populated with the propagation of a plane wave. The steering matrix is then used to obtain the 3×3 polarization covariance matrix

$$\mathbf{Y}_{3C}(f, \mathbf{s}) = \frac{1}{M^2} \mathbf{e}^H(f, \mathbf{s}) \mathbf{C}_{3C}(f) \mathbf{e}(f, \mathbf{s}). \tag{2.49}$$

This projection and the resulting polarization covariance matrix can be viewed as the sum of all signals that propagate over the array as plane waves with a coherent polarization. This is in contrast to the previous approach, where a projection on a single polarization state is used, while here all coherent polarizations between all stations are retained.

Single-station polarization processing (e.g., Vidale, 1986) is applied to obtain the power spectrum by solving the eigenvalue problem for $\mathbf{Y}_{3C}$

$$\mathbf{Y}_{3C}(f, \mathbf{s}) = \mathbf{u}(f, \mathbf{s}) \lambda(f, \mathbf{s}) \mathbb{1} \mathbf{u}^H(f, \mathbf{s}), \tag{2.50}$$

where the unitary matrix $\mathbf{u}(f, \mathbf{s})$ comprises the eigenvectors, $\lambda(f, \mathbf{s})$ are the eigenvalues, and $\mathbb{1}$ is the identity matrix (for readers unfamiliar with the principal component approach, more details can be found in section 2.5.1). The eigen decomposition parametrizes the polarization states into an orthogonal basis (e.g., Samson, 1983; Wagner, 1996), where the eigenvalues denote the contribution of the polarization direction given by its eigenvector (for more detail see section 2.5). The total power spectrum (power from all 3 components) is obtained from the eigenvalues

$$P_{3C}(f, \mathbf{s}) = \sum_{i=0}^{2} \lambda_i(f, \mathbf{s}). \tag{2.51}$$

Although not specifically stated in Wagner (1996), if the average signal power per station is desired the normalization of $\frac{1}{M^2}$ is required for the f-k_{3C} approach. The eigenvector of each polarization direction carries the detailed power contribution from each component, e.g., the eigenvector corresponding to λ_0 (the strongest polarization estimated by the principal component analysis) is

$$\begin{bmatrix} \mathcal{A}_E \\ \mathcal{A}_N \\ \mathcal{A}_Z \end{bmatrix} = |\mathbf{u}_0(\mathbf{s}, f)| = \begin{bmatrix} |\psi_E| \\ |\psi_N| \\ |\psi_Z| \end{bmatrix}. \tag{2.52}$$

The complex numbers ψ_i are the individual contributions to the polarization for each component. The power contribution of each component is obtained as the square of the individual amplitudes $\mathcal{A}$ times the corresponding eigenvalue

$$\begin{bmatrix} P_E \\ P_N \\ P_Z \end{bmatrix} = \begin{bmatrix} \mathcal{A}_E^2 \\ \mathcal{A}_N^2 \\ \mathcal{A}_Z^2 \end{bmatrix} \lambda_0. \tag{2.53}$$

To rotate the components into radial and transverse, one can use the rotation matrix $\tilde{\mathbf{M}}$

$$\tilde{\mathbf{M}} = \begin{bmatrix} \cos\theta & \sin\theta \\ -\sin\theta & \cos\theta \end{bmatrix}, \tag{2.54}$$

to translate the horizontal components of the eigenvector $\psi_{R,T} = \tilde{\mathbf{M}}\psi_{N,E}$, where θ is the angle between north and the radial component, i.e., the back azimuth. After the rotation, inserting into equation (2.52) followed by (2.53) will lead to the beam power for the radial P_R and transverse P_T component respectively. In general only the two largest eigenvalues are considered, as the third is usually dominated by noise (Gal et al., 2016). Adaptive beamformers such as MVDR (Capon, 1969), MUSIC (Schmidt, 1986), and others can be directly implemented into the three-component approaches presented above.

The advantage of polarization three-component beamforming is the increased resolution of the power spectrum given that all three components of a station are simultaneously evaluated for coherent energy propagation. This also results in a higher sensitivity of the approach toward wavefront bending, as the polarization direction at each station is assumed to be the same. If this is not the case, the beam power will start to degrade. While this could be used as a quality control to test if the underlying assumptions are fulfilled, it limits the range at which polarization beamforming is the preferred approach. In the case of wavefront bending, the approach via dimensional reduction might still give meaningful results under the consideration of signal leakage from orthogonal horizontal components.

2.5 Polarization Analysis

Single station polarization analysis is a useful tool for the study of the ambient
noise field where the approximation and/or conditions for beamforming cannot be
fulfilled or where array data are not available. In such a case, a single seismic
station with a three-component sensor can be used to estimate the properties of
the seismic wavefield. The estimated parameters of interest are generally the back
azimuth of the polarization direction, rectilinearity/planarity, angle of incidence,
and phase relationships between the principal axes of particle motion. In the following section, the commonly used methods and the required steps to ensure an
accurate analysis of the ambient noise are introduced.

2.5.1 The Principal Component Analysis Approach

A popular approach is the estimation of the particle motion via the principal component analysis of the data covariance matrix (e.g., Flinn, 1965; Vidale, 1986;
Jurkevics, 1988; Bataille and Chiu, 1991; Schulte-Pelkum et al., 2004) or the cross
spectral density matrix (e.g., Samson, 1983; Park et al., 1987; Koper and Hawley,
2010; Schimmel et al., 2011). The approaches are similar for the time, frequency,
and time-frequency (e.g., S-transform or wavelet transform) domain when following the principal component analysis approach. The underlying principle on the
case of the time domain is demonstrated here. For a three-component sensor

$$\mathbf{x}(t) = [x_E(t), x_N(t), x_Z(t)]^T \tag{2.55}$$

the covariance matrix is a positive semidefinite 3×3 matrix

$$\mathbf{Y} = \frac{1}{L} \sum_{t=1}^{L} \mathbf{x}(t)\mathbf{x}^T(t), \tag{2.56}$$

where the superscript ^{T} is the transpose operator. The diagonal elements are the
variances of each component, while the off-diagonal elements are a measure of
correlation between the two respective components. The matrix is used to estimate
the phase relationship between the three components of a single station. Note, in
the case of complex data, e.g., the analytical signal in time domain or complex
data in frequency domain, the transpose operator ^{T} is replaced by the Hermitian
operator ^{H}. In the time domain, pre-filtering of the seismic data to the desired
frequency range is required followed by an averaging of the covariance matrix
over a time window with length N, where the minimal length of the time window
should allow the presence of a full period of the longest wavelength present in the
frequency band of interest.

To obtain estimates of the particle motion of the wavefield, the covariance matrix is decomposed using principal component analysis (PCA). In this case, either the eigenvalue or singular value decomposition can be used since the matrix is positive semidefinite. Through the PCA, the covariance matrix is transformed into an orthonormal basis which best describes the underlying structure of the recorded data. In the case of a single seismic wave recorded by a three-component sensor, the PCA transforms the data into the principal axes of the particle motion. For multiple and/or non-stationary signals, where the covariance matrix is governed by a mixture of particle motions, the decomposition via the PCA is not able to decompose the signals into its unique particle motions and the resulting orthonormal basis can be an abstract one. For an accurate separation of multiple polarization states, multiple stations are required (e.g., Roueff et al., 2009).

Assuming an accurately decomposed polarization state, it is common practice to sort the eigenvalues and their corresponding eigenvectors so that $\lambda_0 > \lambda_1 > \lambda_2$. The eigenvalues describe the contribution of the principal axes (eigenvectors) of the estimated signal's particle motion (e.g., Samson, 1983; Park et al., 1987), and hence represent power along the principal axes. The computation of various ratios of eigenvalues λ or eigenvectors $\mathbf{u}$ of $\mathbf{Y}$ is required to infer polarization and other properties. The direction of polarization is estimated from the horizontal components of the dominant polarization direction by utilizing the information in the eigenvector

$$\mathbf{u}_0 = [\psi_E, \psi_N, \psi_Z]^T \tag{2.57}$$

$$\Theta = \arctan\left(\frac{\psi_E}{\psi_N}\right), \tag{2.58}$$

which in the case of P, SV, and Rayleigh waves coincides with the back azimuth $\pm 180°$. The $\pm 180°$ ambiguity arises since a single three-component station cannot estimate the propagation direction of a seismic wave from the polarization direction alone, but requires an additional constraint. For body waves one can make use of the dip of the direction of polarization

$$\Theta_V = \arctan\left(\frac{\psi_Z}{\sqrt{\psi_N^2 + \psi_E^2}}\right) \tag{2.59}$$

while Rayleigh waves are usually assumed to be retrograde and their propagation direction is estimated by the phase delay between the vertical and horizontal phase angles. Care should be taken when estimating polarization parameters via the frequency domain, where the polarization ellipsoid is prone to "tip over" and switch between prograde and retrograde (Koper and Hawley, 2010).

Furthermore, the rectilinearity

$$G = 1 - \frac{\lambda_1 + \lambda_2}{2\lambda_0}, \tag{2.60}$$

and planarity

$$P = 1 - \left(\frac{2\lambda_2}{\lambda_0 + \lambda_1} \right) \tag{2.61}$$

of the particle motion state can be defined (Jurkevics, 1988). For $\lambda_0 \gg \lambda_{1,2}$, the polarization is rectilinear (i.e., a pure body wave or Love wave) and is solely parametrized by a single polarization direction. For $\lambda_{0,1} \gg \lambda_2$, the particle motion is parametrized by the two largest eigenvalues and confined in a plane as is the case for a Rayleigh wave.

The approaches in the time-frequency and frequency domain follows the same logic, where the spectral covariance matrix is calculated from $\mathbf{X}(t, f)$ or $\mathbf{X}(f)$, respectively. One difference is that the resulting eigenvectors are complex and to apply the above equations the real part of the eigenvector needs to be used.

2.5.2 Instantaneous Polarization Attributes

The introduction of the analytic signal in seismic signal processing can be used to estimate the instantaneous frequency (Taner et al., 1979) and polarization attributes (Vidale, 1986; Schimmel and Gallart, 2003). The Hilbert transform $\mathcal{H}(x)$, which is used to obtain the phase-shifted function of x by $\frac{\pi}{2}$, is applied to the observed data $x(t)$ and added as the imaginary part to yield the analytic signal representation

$$\tilde{\mathbf{x}}(t) = [x_E(t) + i\mathcal{H}(x_E(t)), x_N(t) + i\mathcal{H}(x_N(t)), x_Z(t) + i\mathcal{H}(x_Z(t))]^T. \tag{2.62}$$

Following the covariance approach, the eigen analysis yields complex eigenvectors which contain the phase information (Vidale, 1986). To obtain the back azimuth of the polarization direction, only the real part of the eigenvector in equation (2.58) is used (as for the time-frequency and the frequency domain). This approach has the advantage that the covariance matrix is not required to be averaged over a portion of the desired wavelength (as is the case in the real valued analysis) and hence allows to estimate the polarization attributes at a given time $\tilde{\mathbf{x}}(t)$. This is possible owing to the Hilbert transform, which is computed on the whole seismic record, i.e., it uses signal information from the whole data window to compute the phase-transformed signal at a time t.

2.5.3 Ambient Noise Polarization Analysis

The nature of the ambient noise field, where multiple incoherent signals are likely to be present at one time, means that the estimation of the polarization is not trivial.

Whether the time, the time-frequency, or the frequency domain approach is more advantageous depends on the properties of the wavefield and the desired task. The frequency domain approach is advantageous if multiple incoherent signals can be separated in the frequency domain, i.e., unique sources occupy different frequency bands. If multiple incoherent signals are separated in time but share the same frequency band, the time or the time-frequency domain approach in combination with the analytic signal is preferable. The time-frequency domain approach combines strengths from both and gives the user more control over the polarization analysis.

All three of the above approaches have been successfully used for the analysis of the seismic ambient noise wavefield. It was shown that in the case of a relatively stationary source (stationary within the length of a data window) the use of long time windows in time and frequency domain can result in stable estimates (e.g., Schulte-Pelkum et al., 2004; Tanimoto et al., 2006; Koper and Hawley, 2010). These studies focused on the secondary microseism peak, which has an increased likelihood of encountering a single strong source compared to even higher frequency bands where polarization analysis with long time windows is unlikely to succeed unless additional constraints such as degree of polarization are employed (e.g., Koper and Burlacu, 2015) or a large number of observations is used to increase confidence.

For non-stationary sources, a better time resolution is required which can be achieved by transforming into the time-frequency domain (e.g., Schimmel et al., 2011). The transformation has the advantage that optimal frequency-dependent windows are used to sample the seismic data, which can take into account the non-stationary evolution of the wavefield.

When analyzing seismic ambient noise data, strategies to increase robustness should be employed regardless of the approach and domain chosen. The first is to use a large sample size of observations to increase robustness of the analysis as statistical outliers are down weighted. Another option is the inclusion of a broader frequency range for the computation of the particle motion, which, as long as the broadband signal has the same spatial origin, will result in a more robust estimation of the particle motion (e.g., Tanimoto et al., 2006). If, on the other hand, multiple sources are likely to enter the polarization analysis due to an extended frequency range, the robustness of the observations is likely to decrease.

Another strategy is to employ a framework which can estimate the purity of an estimated polarization state. This can be estimated for long time windows as a comparison between eigenvalues (Samson, 1983; Koper and Hawley, 2010; Koper and Burlacu, 2015) or for instantaneous polarization attributes with a degree of polarization measure (e.g., Schimmel and Gallart, 2003, 2004; Schimmel et al., 2011; Sergeant et al., 2013; Davy et al., 2015).

2.5.4 Beamforming versus Polarization Analysis on Observed Data

A convenient way to test the polarization analysis is with the help of a three-component array where the results can be verified via beamforming. Additionally, it provides multiple stations which should observe the same sources, and hence show similar particle motions across all stations. Use is made of the Pilbara Array in Australia and the Gräfenberg Array in Germany with 13 three-component stations each.

The first example with the Pilbara Array details observations for two differing frequency bands centered on $f_1 = 0.17 \pm 0.1$ Hz and $f_2 = 0.35 \pm 0.2$ Hz on 31 May 2013 at 00:00:00-01:00:00 UTC. The two frequency bands were chosen to demonstrate the behavior of the polarization analysis for two differing scenarios. At f_1 the Capon beamformer estimates the maximal energy arriving at the array with a back azimuth around $\sim 180°$ and some weaker energy from other southern back azimuths, Figure 2.6a. For the polarization analysis of the same data, a histogram in Figure 2.6b of Rayleigh wave arrivals estimated with the method by Schimmel et al. (2011) is shown. The polarization analysis was used on all 13 stations of the array and only Rayleigh waves which exceed the degree of polarization (DOP) threshold of 0.75 were used for the generation of this figure. There is a good agreement between the back azimuth of the beam power spectrum and the polarization analysis despite the fact that multiple sources are contributing to the observed wavefield. The success can be attributed to the approach via the time-frequency domain, where the DOP constraint is used as a quality control, which allows to pick out short time periods where the polarization can be accurately estimated. For the approaches which average over longer time windows in time and frequency domain (Park et al., 1987; Jurkevics, 1988), the estimated back azimuths of each station show a larger scatter and are predominantly within the bounds of $180° - 250°$ (not shown). The higher scatter can be attributed to the fact that one computes the covariance or cross spectral density matrix from a window with multiple polarization states which can lead to an eigen basis which is not representable of the strongest polarization direction.

At the higher frequency f_2, the Capon beamformer estimates sources from a wide range of back azimuths (Figure 2.6c). The polarization analysis estimates most of the Rayleigh waves from the west, and does not represent the overall state of the wavefield. The DOP constraint, while ensuring robustness in the result, prevents us from observing the complete source distribution in this example, and furthermore leads to relatively few accepted polarization measurements for this

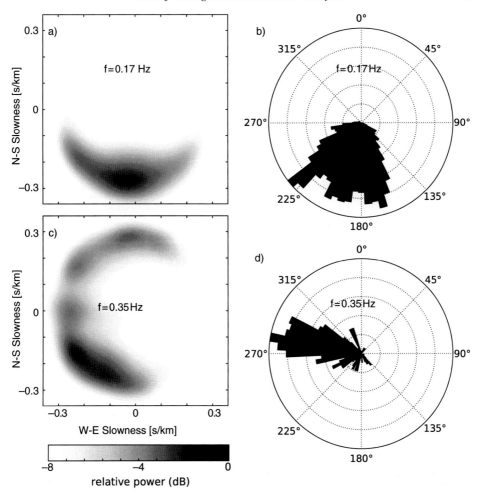

Figure 2.6. Results from (a,c) beamforming and (b,d) polarization analysis in two differing frequency bands for one hour of the ambient noise wavefield at PSAR (Australia). The first column shows the Capon power spectrum estimated from the vertical component and the left column shows a Rayleigh wave occurrence histogram. The radial axis on the histograms was normalized.

particular hour. If the DOP threshold is lowered to 0.5, the strong sources to the south-west are observed, although one can expect higher bias due to mixed polarization states. Hence, one should keep in mind that the DOP filtering procedure does not ensure the observation of the strongest source.

The window averaging techniques (Park et al., 1987; Jurkevics, 1988) estimate a back azimuth between 220° and 250° on approximately half the stations, while the other half shows large scatter and hence would benefit from a larger sample size of observations to increase the confidence in the results (not shown).

The second example with the Gräfenberg Array details observations for three differing frequency bands centered on $f_1 = 0.067 \pm 0.003$ Hz, $f_2 = 0.133 \pm 0.003$ Hz, and $f_3 = 0.25 \pm 0.01$ on 1 January 2017 at 01:00:00-02:00:00 UTC. The results from the f-k analysis are displayed in Figure 2.7a–c. The same data was previously used to discuss deconvolution techniques (Figure 2.3). F-k beamforming gives

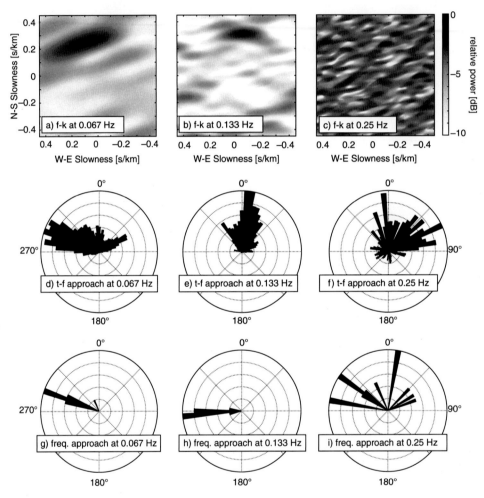

Figure 2.7. Comparison between results from f-k beamforming, time-frequency, and frequency polarization analysis. The top row (a-c) shows f-k beam power results at three differing frequencies. The middle row (d-f) displays the time-frequency approach and the bottom row (g-i) the time averaged frequency approach. The polar histogram plots display all accepted polarization directions above a set DOP for the time-frequency approach, while for the frequency approach the histogram is simply one estimated polarization direction per station.

reasonable results in the first two frequency bands but fails to give a meaningful estimate at 0.25 Hz. This is due to the small number of stations and the relatively large (100 km) aperture of the Gräfenberg Array. The approach is not able to identify plane wave propagation, and hence it is likely that path propagation effects are non-negligible.

The time-frequency polarization approach (Figure 2.7d–f), with a DOP threshold of 0.9, 0.8, and 0.6 for f_1, f_2, and f_3 respectively, shows partial agreement with the f-k beam power results. At f_1, the polarization approach estimates the source location around 300° which is roughly 30° off from the f-k result, but coincides with the source distribution found by the Capon beamformer or the CLEAN algorithm (Figure 2.3). Additionally, arrivals from the north-east are estimated. Looking at the data in more detail, it is found that only 4 stations observe these signals from the north-east and it is likely a local source. Given that it does not seem to propagate as a plane wave, the f-k analysis does not indicate signals at the corresponding back azimuth. At f_2 good agreement with the f-k analysis is found and the main source is well estimated. At f_3 the f-k analysis fails, but the polarization analysis can still yield directional information.

The frequency polarization approach (Park et al., 1987; Jurkevics, 1988; Koper and Hawley, 2010) is displayed in Figure 2.7g–i. Given that the approach averages over time, we obtain only a single back azimuth value per station per hour, and hence the statistics and robustness of the approach are limited. At f_1, an agreement with the time-frequency approach is found, and the frequency approach shows a much smaller spread. At f_2, the results do not agree with the f-k analysis or the time-frequency approach, even though all stations seem to estimate the same back azimuth. A possibility is the mixing of multiple polarizations from differing back azimuths, and hence this approach would require longer time durations to obtain a statistically significant observation. Alternatively, the polarization direction would agree with a Love waves source to the north (confirmed with three-component beamforming). At f_3 the directions of polarization are distributed over a wide range of azimuths and the same logic with longer duration studies applies.

Acknowledgments

Supported by the Australian Research Council under project DP150101005. We would like to thank Lucia Gualtieri, Kiwamu Nishida, and an anonymous reviewer for their comments, which greatly improved this chapter. We thank many colleagues for useful discussions including Keith Koper, Steven Gibbons, Hrvoje Tkalčić, and also Michael Ritzwoller and members of the CIEI at University of Colorado, Boulder, where AMR was hosted as a Fulbright Senior Scholar in 2017.

References

Ardhuin, F., Gualtieri, L., and Stutzmann, E. 2018. Physics of ambient noise generation by ocean waves. Pages – of: Nakata, N., Gualtieri, L., and Fichtner, A. (eds.), *Seismic Ambient Noise*. Cambridge University Press, Cambridge, UK.

Aster, R. C., McNamara, D. E., and Bromirski, P. D. 2010. Global trends in extremal microseism intensity. *Geophysical Research Letters*, **37**(JUL 21).

Baggeroer, A., Kuperman, W., and Mikhalevsky, P. 1993. An overview of matched field methods in ocean acoustics. *IEEE Journal of Oceanic Engineering*, **18**(4), 401–424.

Bataille, K., and Chiu, J. 1991. Polarization analysis of high-frequency, three-component seismic data. *Bulletin of the Seismological Society of America*, **81**(2), 622–642.

Behr, Y., Townend, J., Bowen, M., Carter, L., Gorman, R., Brooks, L., and Bannister, S. 2013. Source directionality of ambient seismic noise inferred from three-component beamforming. *Journal of Geophysical Research*, **118**, 240–248.

Bertero, M., and Boccacci, P. 2005. A simple method for the reduction of boundary effects in the Richardson-Lucy approach to image deconvolution. *Astronomy and Astrophysics*, **437**(1), 369–374.

Brooks, L., Townend, J., Gerstoft, P., Bannister, S., and Carter, L. 2009. Fundamental and higher-mode Rayleigh wave characteristics of ambient seismic noise in New Zealand. *Geophysical Research Letters*, **36**(23), L23303.

Bucker, H. 1976. Use of calculated sound fields and matched-field detection to locate sound sources in shallow-water. *The Journal of the Acoustical Society of America*, **59**(2), 368–373.

Capon, J. 1969. High-resolution frequency-wavenumber spectrum analysis. *Proceedings of the IEEE*, **57**(8), 1408–1418.

———. 1973. Analysis of microseismic noise at LASA, NORSAR and ALPA. *Geophysical Journal of the Royal Astronomical Society*, **35**, 39–54.

Chiou, S.-J., and Bolt, B. 1993. Seismic wave slowness-vector estimation from broad-band array data. *Geophysical Journal International*, **114**(2), 234–248.

Davy, C., Stutzmann, E., Barruol, G., Fontaine, F., and Schimmel, M. 2015. Sources of secondary microseisms in the Indian Ocean. *Geophysical Journal International*, **202**(2), 1180–1189.

Esmersoy, C., Cormier, V. F., and Toksöz, M. N. 1985. Three-Component Array Processing. *The VELA Program: A Twenty-five Year Review of Basic Research*, 565–578.

Euler, G. G., Wiens, D., and Nyblade, A. A. 2014. Evidence for bathymetric control on the distribution of body wave microseism sources from temporary seismic arrays in Africa. *Geophysical Journal International*, **197**(3), 1869–1883.

Farra, V., Stutzmann, E., Gualtieri, L., Schimmel, M., and Ardhuin, F. 2016. Ray-theoretical modeling of secondary microseism *P* waves. *Geophysical Journal International*, **206**(3), 1730–1739.

Featherstone, W., Strangeways, H. J., Zatman, M. A., and Mewes, H. 1997. A novel method to improve the performance of Capon's minimum variance estimator. *Antennas and Propagation, Tenth International Conference on (Conf. Publ. No. 436)*, **1**, 332–335.

Flinn, E. 1965. Signal analysis using rectilinearity and direction of particle motion. *Proceedings of the IEEE*, **53**(12), 1874–1876.

Friedrich, A., Klinge, K., and Krüger, F. 1998. Ocean-generated microseismic noise located with the Graefenberg array. *Journal of Seismology*, **2**(1), 47–64.

Gal, M., Reading, A. M., Ellingsen, S. P., Gualtieri, L., Koper, K. D., Burlacu, R., and Tkalčić, H. 2015. The frequency dependence and locations of short-period microseisms generated in the Southern Ocean and West Pacific. *Journal of Geophysical Research: Solid Earth*, **120**(8), 5764–5781.

Gal, M., Reading, A. M., Ellingsen, S. P., Koper, K. D., and Burlacu, R. 2017. Full wavefield decomposition of high-frequency secondary microseisms reveals distinct arrival azimuths for Rayleigh and Love waves. *Journal of Geophysical Research: Solid Earth*, **122**(6), 4660–4675. 2017JB014141.

Gal, M., Reading, A. M., Ellingsen, S., Koper, K. D., Burlacu, R., and Gibbons, S. J. 2016. Deconvolution enhanced direction of arrival estimation using one- and three-component seismic arrays applied to ocean induced microseisms. *Geophysical Journal International*, **206**(1), 10.1093/gji/ggw150.

Gal, M., Reading, A. M., Ellingsen, S. P., Koper, K. D., Gibbons, S. J., and Näsholm, S. P. 2014. Improved implementation of the fk and Capon methods for array analysis of seismic noise. *Geophysical Journal International*, **198**(2), 1045–1054.

Gerstoft, P., Sabra, K. G., Roux, P., Kuperman, W. A., and Fehler, M. C. 2006. Green's functions extraction and surface-wave tomography from microseisms in southern California. *Geophysics*, **71**(4, S), SI23–SI31.

Gerstoft, P., Shearer, P. M., Harmon, N., and Zhang, J. 2008. Global P, PP, and PKP wave microseisms observed from distant storms. *Geophysical Research Letters*, **35**(23), 1–6.

Gibbons, S. J., Schweitzer, J., Ringdal, F., Kværna, T., Mykkeltveit, S., and Paulsen, B. 2011. Improvements to seismic monitoring of the European Arctic using three-component array processing at SPITS. *Bulletin of the Seismological Society of America*, **101**(6), 2737–2754.

Gualtieri, L., Stutzmann, E., Farra, V., Capdeville, Y., Schimmel, M., Ardhuin, F., and Morelli, A. 2014. Modeling the ocean site effect on seismic noise body waves. *Geophysical Journal International*, **197**(2), 1096–1106.

Haubrich, R. A., and McCamy, K. 1969. Microseisms: Coastal and pelagic sources. *Reviews of Geophysics*, **7**(3), 539–571.

Högbom, J. 1974. Aperture synthesis with a non-regular distribution of interferometer baselines. *Astronomy and Astrophysics Supplement Series*, **15**, 417–426.

Juretzek, C., and Hadziioannou, C. 2016. Where do ocean microseisms come from? A study of Love-to-Rayleigh wave ratios. *Journal of Geophysical Research: Solid Earth*, **121**, 1–16.

Jurkevics, A. 1988. Polarization analysis of three-component array data. *Bulletin of the Seismological Society of America*, **78**(5), 1725–1743.

Kelly, E. J. 1967. Response of seismic signals to wide-band signals. *Lincoln Lab. Tech. Note*, **30**.

Kennett, B. L. N., Stipčević, J., and Gorbatov, A. 2015. Spiral-arm seismic arrays. *Bulletin of the Seismological Society of America*, **105**(4), 2109–2116.

Koper, K. D., and Burlacu, R. 2015. The fine structure of double-frequency microseisms recorded by seismometers in North America. *Journal of Geophysical Research: Solid Earth*, **120**(3), 1677–1691.

Koper, K. D., and Hawley, V. L. 2010. Frequency dependent polarization analysis of ambient seismic noise recorded at a broadband seismometer in the central United States. *Earthquake Science*, **23**, 439–447.

Krim, H., and Viberg, M. 1996. Two decades of array signal processing research. *IEEE Signal Process. Mag*, **13**(4), 67–94.

Kværna, T., and Doornbos, D. 1985. An integrated approach to slowness analysis with array and three-component stations. *NORSAR Semiannual Technical Summary*, **2**(85/86), 60–69.

Kværna, T., and Ringdal, F. 1986. Stability of various f-k estimation techniques. *Norsar Scientific Report*, **1**(86/87), 29–40.

Lacoss, R., Kelly, E., and Toksöz, M. 1969. Estimation of seismic noise structure using arrays. *Geophysics*, **34**(1), 21–38.

Liu, Q., Koper, K. D., Burlacu, R., Ni, S., Wang, F., Zou, C., Wei, Y., Gal, M., and Reading, A. M. 2016. Source locations of teleseismic P, SV, and SH waves observed in microseisms recorded by a large aperture seismic array in China. *Earth and Planetary Science Letters*, **449**, 39–47.

Lucy, L. B. 1974. An iterative technique for the rectification of observed distributions. *The Astronomical Journal*, **79**(6), 745.

McNamara, D., and Boaz, R. 2018. Visualization of the Seismic Ambient Noise Spectrum. Pages — of: Nakata, N., Gualtieri, L., and Fichtner, A. (eds.), *Seismic Ambient Noise*. Cambridge University Press, Cambridge, UK.

Menon, R., Gerstoft, P., and Hodgkiss, W. S. 2014. On the apparent attenuation in the spatial coherence estimated from seismic arrays. *Journal of Geophysical Research: Solid Earth*, **119**, 3115–3132.

Meschede, M., Stutzmann, E., Farra, V., Schimmel, M., and Ardhuin, F. 2017. The Effect of Water Column Resonance on the Spectra of Secondary Microseism P Waves. *Journal of Geophysical Research: Solid Earth*, **122**(10), 8121–8142. 2017JB014014.

Miron, S., Le Bihan, N., and Mars, J. I. 2005. Vector-Sensor MUSIC for Polarized Seismic Sources Localization. *EURASIP Journal on Advances in Signal Processing*, **2005**(1), 74–84.

Nishida, K., Kawakatsu, H., Fukao, Y., and Obara, K. 2008. Background Love and Rayleigh waves simultaneously generated at the Pacific Ocean floors. *Geophysical Research Letters*, **35**(16), L16307.

Park, J., Lindberg, C., and Vernon, F. 1987. Multitaper spectral-analysis of high-frequency seismograms. *Journal of Geophysical Research*, **92**(B12), 12675–12684.

Picozzi, M., Parolai, S., and Bindi, D. 2010. Deblurring of frequency-wavenumber images from small-scale seismic arrays. *Geophysical Journal International*, **181**(1), 357–368.

Poggi, V., and Fäh, D. 2010. Estimating Rayleigh wave particle motion from three-component array analysis of ambient vibrations. *Geophysical Journal International*, **180**(Jan.), 251–267.

Reading, A. M., Koper, K. D., Gal, M., Graham, L. S., Tkalčić, H., and Hemer, M. A. 2014. Dominant seismic noise sources in the Southern Ocean and West Pacific, 20002012, recorded at the Warramunga Seismic Array, Australia. *Geophysical Research Letters*, **41**(10), 3455–3463.

Riahi, N., Bokelmann, G., Sala, P., and Saenger, E. H. 2013. Time-lapse analysis of ambient surface wave anisotropy: A three-component array study above an underground gas storage. *Journal of Geophysical Research: Solid Earth*, **118**(10), 5339–5351.

Richardson, W. 1972. Bayesian-based iterative method of image restoration. *Journal of the Optical Society of America*, **62**(1), 55–59.

Rost, S., and Thomas, C. 2002. Array seismology: Methods and applications. *Reviews of Geophysics*, **40**(3), doi:10.1029/2000RG000100.

Roueff, A., Roux, P., and Réfrégier, P. 2009. Wave separation in ambient seismic noise using intrinsic coherence and polarization filtering. *Signal Processing*, **89**(4), 410–421.

Ruigrok, E., Gibbons, S., and Wapenaar, K. 2017. Cross-correlation beamforming. *Journal of Seismology*, **21**(3), 495–508.

Samson, J. C. 1983. Pure states, polarized waves, and principal components in the spectra of multiple, geophysical time-series. *Geophysical Journal International*, **72**(3), 647–664.

Schimmel, M., and Gallart, J. 2003. The use of instantaneous polarization attributes for seismic signal detection and image enhancement. *Geophysical Journal International*, **155**(2), 653–668.

Schimmel, M., and Gallart, J. 2004. Degree of polarization filter for frequency-dependent signal enhancement through noise suppression. *Bulletin of the Seismological Society of America*, **94**(3), 1016–1035.

Schimmel, M., Stutzmann, E., Ardhuin, F., and Gallart, J. 2011. Polarized Earth's ambient microseismic noise. *Geochemistry, Geophysics, Geosystems*, **12**(7), 1–14.

Schmidt, R. 1986. Multiple emitter location and signal parameter estimation. *Antennas and Propagation, IEEE Transactions on*, **AP-34**(3), 276–280.

Schulte-Pelkum, V., Earle, P. S., and Vernon, F. L. 2004. Strong directivity of ocean-generated seismic noise. *Geochemistry, Geophysics, Geosystems*, **5**(3), 1–13.

Schweitzer, J., Fyen, J., Mykkeltveit, S., Gibbons, S., Pirli, M., Kühn, D., and Kværna, T. 2002. Chapter 9: Seismic Arrays. In Bormann, P. (ed), *IASPEI New Manual of Seismological Observatory Practice*. GeoForschungsZentrum, Potsdam. 52 pp.

Seligson, C. D. 1970. Comments on high-resolution frequency-wavenumber spectrum analysis. *Proceedings of the IEEE*, **58**, 947–949.

Sergeant, A., Stutzmann, E., Maggi, A., Schimmel, M., Ardhuin, F., and Obrebski, M. 2013. Frequency-dependent noise sources in the North Atlantic Ocean. *Geochemistry, Geophysics, Geosystems*, **14**(12), 5341–5353.

Shepp, L. a., and Vardi, Y. 1982. Maximum likelihood reconstruction for emission tomography. *IEEE transactions on medical imaging*, **1**(2), 113–122.

Sijtsma, P. 2007. CLEAN based on spatial source coherence. *International Journal of Aeroacoustics*, 21–23.

Soares, C., and Jesus, S. M. 2003. Broadband matched-field processing: Coherent and incoherent approaches. *The Journal of the Acoustical Society of America*, **113**(5), 2587.

Taner, M. T., Koehler, F., and Sheriff, R. E. 1979. Complex seismic trace analysis. *Geophysics*, **44**(6), 1041–1063.

Tanimoto, T., Ishimaru, S., and Alvizuri, C. 2006. Seasonality in particle motion of microseisms. *Geophysical Journal International*, **166**(1), 253–266.

Thomson, D. 1982. Spectrum estimation and harmonic analysis. *Proceedings of the IEEE*, **70**(9), 1055–1096.

Toksöz, M., and Lacoss, R. 1968. Microseisms: Mode structure and sources. *Science*, **159**(December), 872–873.

Vidale, J. E. 1986. Complex polarization analysis of particle motion. *Bulletin of the Seismological Society of America*, **76**(5), 1393–1405.

Wagner, G. 1996. Resolving diversely polarized, superimposed signals in three-component seismic array data. *Geophysical Research Letters*, **23**(14), 1837–1840.

Wagner, G. 1997. Regional wave propagation in southern California and Nevada: Observations from a three-component seismic array. *Journal of Geophysical Research*, **102**(B4), 8285–8311.

Wagner, G., and Owens, T. 1996. Signal detection using multi-channel seismic data. *Bulletin of the Seismological Society of America*, **86**(1A), 221–231.

Wang, H., and Kaveh, M. 1985. Coherent signal-subspace processing for the detection and estimation of angles of arrival of multiple wideband sources. *Acoustics, Speech and Signal Processing, IEEE Transactions on*, **33**(4), 823–831.

Wang, Y., Li, J., Stoica, P., Sheplak, M., and Nishida, T. 2004. Wideband RELAX and wideband CLEAN for aeroacoustic imaging. *The Journal of the Acoustical Society of America*, **115**(2), 757–767.

Wax, M., Shan, T.-J., and Kailath, T. 1984. Spatio-temporal spectral analysis by eigen-structure methods. *Acoustics, Speech and Signal Processing, IEEE Transactions on*, **32**(4), 817–827.

Westwood, E. 1992. Broadband matched-field source localization. *The Journal of the Acoustical Society of America*, **91**(5), 2777–2789.

Yoon, Y. S., Kaplan, L. M., and McClellan, J. H. 2006. TOPS: New DOA estimator for wideband signals. *IEEE Transactions on Signal Processing*, **54**(6), 1977–1989.

Zhang, J., Gerstoft, P., and Shearer, P. M. 2009. High-frequency P-wave seismic noise driven by ocean winds. *Geophysical Research Letters*, **36**(9), 1–5.

3

Physics of Ambient Noise Generation by Ocean Waves

FABRICE ARDHUIN, LUCIA GUALTIERI, AND
ELÉONORE STUTZMANN

Abstract

At periods between 3 and 300 s, seismic ambient noise is largely generated by ocean waves. Seismic and ocean waves have frequencies in the same range, but seismic waves have wavelengths much longer than those of ocean waves typically by a factor 100 or more. This generation of long waves from short waves is not simple. Two mechanisms can explain some of the observed seismic waves. For seismic waves with periods longer than 10 s, the interference of ocean waves with bottom topography can generally explain the generation of Rayleigh and Love waves. Shorter periods are rather explained by the interference of pairs of ocean wave trains. In this chapter, we focus on these two mechanisms. In both cases, the interference can produce a beating pattern in the pressure field, respectively at the seafloor or at the sea surface. This pattern contains long wavelengths compatible with seismic waves. Surface pressure generates acoustic waves that propagate within the ocean layer and produces a resonance effect within the source region, called "source site effect." We represent the pressure induced by ocean waves with a power spectral density in three dimensions, namely frequency and two horizontal wavenumber components. Because these spectra are broad in the wavenumber domain, the distributed pressure field induced by ocean waves can be replaced by equivalent point forces. Computing these forces from an ocean wave model, the "source site effect," and the seismic propagation within the solid Earth enable one to estimate the power spectral density of seismic ambient noise. These theories of ambient seismic noise generation have been tested quantitatively for body and surface waves.

3.1 Introduction

3.1.1 Context and Motivations

We need to understand the generation of seismic ambient noise in order to locate and model noise sources. This modeling can be applied to the investigation of storms and the ocean wave climate (e.g., Grevemeyer et al., 2000; Gualtieri et al., 2018), or to the seismic tomography and monitoring of the solid Earth. Currently, most seismic ambient noise tomography studies rely on the assumption of a homogeneous distribution of noise sources (e.g., Shapiro et al., 2005; Fichtner and Tsai, 2018; Ritzwoller and Feng, 2018; Shapiro, 2018). This application is based on the hypothesis that the seismic wave field propagates on average isotropically near the receivers (e.g., Snieder, 2004). Knowing the actual distribution of seismic ambient noise sources in space and time might lead to improved tomography (e.g., Tromp et al., 2010) and monitoring of subsurface velocity changes. In this context, one needs a quantitative estimate of the three following elements: (1) the relevant ocean wave properties, (2) the mechanisms to convert ocean waves to seismic sources, and (3) the propagation of seismic waves all the way to a receiver. The first two points define the source timing and location. These depend on subtle ocean wave properties, and are the main topic of this chapter. To be precise, ocean waves are surface gravity waves (sgw for short), and their properties are described in detail in many textbooks (e.g., Dean and Dalrymple, 1991; Boccotti, 2000). Yet, these textbooks do not address the particular properties shown to be sufficient for microseism generation by Hasselmann (1963) and Ardhuin et al. (2015).

In the following paragraphs of this introduction, we review the mechanisms proposed for the three frequency bands that are generally observed in seismic records: the seismic "hum" with periods longer than about 30 s, the primary microseisms with periods between about 30 and 10 s, and the secondary microseisms with a peak generally around 5–7 s. In section 3.2, we present the general properties of ocean waves, and how they can be predicted from the wind field using numerical ocean wave models. In section 3.3, we describe the mechanisms that transform the short wavelengths of ocean waves into the long wavelength of seismic waves. We give key equations of the theory and refer to the specific parts of Longuet-Higgins (1950) and Hasselmann (1963) for more details. For the secondary mechanism, we also give practical expressions of seismic sources of body and surface seismic waves, and describe the composition of the seismic wave field. In section 3.4, we present examples of modeled noise sources across the seismic ambient noise spectrum, with periods ranging from 3 to 300 s.

3.1.2 Ocean Waves and Seismic Waves: Wave-Wave Interaction

In this chapter we will follow the method of wave-wave interactions that is summarized by Hasselmann (1966) and is common in oceanography or plasma physics. This was introduced in seismology by a brilliant postdoctoral fellow at the La Jolla Institute of Geophysics and Planetary Physics, Klaus Hasselmann, who worked there from 1961 to 1963 on many topics related to ocean waves (e.g., Snodgrass et al., 1966), under the guidance of Walter Munk and with Mickael Longuet-Higgins as a guest professor (von Storch and Olbers, 2007). The resulting well-known paper (Hasselmann, 1963) is written with a formalism that is classical in oceanography, but not in seismology. We will highlight here the most important findings of this paper, with a focus on the seismological aspects.

Hasselmann (1963) was the first to formulate a theory for the interference of waves and bottom topography that could explain the primary microseisms, with a few assumptions that we will discuss. As illustrated in Figure 3.1, he replaced waves over a complex bottom topography by a pressure field pushing on a flat seafloor. For the secondary microseisms, he followed Longuet-Higgins (1950) and considered the interference of wave trains propagating in opposing directions. In that case, the complex motion of waves at the sea surface was low-pass filtered to keep large wavelengths. This filtered motion is equivalent to a pressure field given

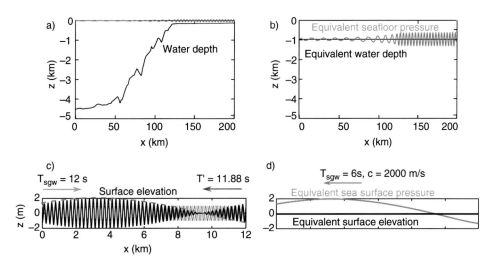

Figure 3.1. For the primary mechanism, the effect of ocean waves propagating over a sloping bottom in (a) is replaced by the action of a pressure on the seafloor in (b). For the secondary mechanism, the motions in wave groups are caused by two opposing wave trains, in (c) at periods 12 and 11.88 s (with horizontal wavenumbers k_{sgw} and k'_{sgw}). These motions are replaced by (d) a surface pressure field acting on a flat sea surface, with a wavenumber $k = k_{sgw} - k'_{sgw}$.

by the wave group envelope acting on a flat sea surface (Hasselmann, 1963, his equation 2.10). That equivalence only holds for ocean waves short compared to the water depth, otherwise there is an extra pressure induced by ocean waves directly at the seafloor (Ardhuin and Herbers, 2013).

For both primary and secondary mechanisms, the resulting pressure fields are distributed over a horizontally homogeneous layered medium, at the ocean bottom and at the ocean surface. Therefore, the seismic source is the pressure fields, defined by ocean waves and their interactions, acting on the free sea surface and on the seafloor. The propagator matrix method (e.g., Gilbert and Backus, 1966; Aki and Richards, 2002) can be used to solve the equation of seismic motion in a layered medium. In this method, the unknowns are the amplitudes of the elastic potentials, and the matrix coefficients are given by the continuity of stresses and displacements at the layer interfaces. A common use of the method is the determination of pairs of frequency f and wavenumber k for which the matrix is singular. These pairs correspond to seismic modes that propagate horizontally and that persist in the absence of a forcing pressure. For non-singular cases, solution amplitudes are constants and are linearly related to the surface amplitude forcing. Therefore the solutions have the same wavenumber and frequency as the forcing pressure field at the sea surface or the seafloor. In the words of Hasselmann (1963): *In all cases appreciable microseisms are generated only by Fourier components of the random exciting fields that have the same phase velocities as [...] modes of the elastic system.*

For example, let us consider ocean-wave interaction that generates an equivalent sinusoidal pressure of amplitude P, $\widehat{p}_{\text{surf}}(x, y, t) = \Re\{P \exp[i(k_x x + k_y y - 2\pi f t)]\}$, where $\Re$ is the real part. In this case, the solution for the pressure at the ocean bottom is $p(z = -D) = \Re\{G_p(|\mathbf{k}|, f)P \exp[i(k_x x + k_y y - 2\pi f t)]\}$, where $G_p(|\mathbf{k}|, f)$ is the propagator matrix. In a simple case of a two-layer medium (ocean and crust), the modulus of $G_p(|\mathbf{k}|, f)$ is shown in Figure 3.2 as a function of seismic frequency f and wavenumber $|\mathbf{k}|$. The dashed lines correspond to P-wave velocity in the water layer (c_w), S-wave velocity in the crust (c_{cs}), and P-wave velocity in the crust (c_{cp}). Rayleigh waves have velocities between c_w and c_{cs}. The dispersion curve of the Rayleigh modes corresponds to infinite values of the propagator matrix G_p (dark bands in Figure 3.2). The spectral width of the pressure derived from ocean waves gives a broad distribution of forcing wavenumber k at any frequency f. Integrating over these k, one finds the seismic response. At velocities above $c = c_{cs}$, seismic waves in the crust can propagate as body waves. At the other extreme, velocities below $c = c_w$ correspond to evanescent waves in both the water layer and the crust. Within the white region on the right, there are acoustic-gravity modes that are particularly relevant for the generation of microbaroms, not covered in this chapter.

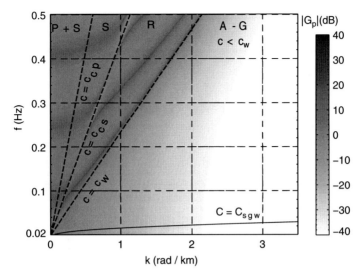

Figure 3.2. Modulus of the propagator matrix $G_p(k, f)$ from the ocean surface to the ocean bottom, considering an ocean layer of $D = 4500$ m depth over a solid half space. The P-wave velocities, S-wave velocities, and densities are: $c_w = 1.47$ km/s, $c_{ws} = 0$ km/s, $\rho_w = 1026$ kg/m^3 in the water layer, and $c_{cp} = 5.2$ km/s, $c_{cs} = 3.0$ km/s, $\rho_c = 3300$ kg/m^3 in the solid half space. The solid line at the bottom shows the much slower phase speed of ocean surface gravity waves c_{swg}. The dashed lines correspond to P-wave velocity in the water layer (c_w), S-wave velocity in the crust (c_{cs}), and P-wave velocity in the crust (c_{cp}).

Only the components of the pressure source that have a slowness that matches the slowness of a seismic phase generate seismic waves that can propagate away from the source region. The other components of the pressure source generate a transient response that only exists when and where ocean waves are present and is like the compliance effect: a quasi-static deformation of the seafloor under long gravity waves (e.g., Crawford et al., 1991).

The question of seismic ambient noise generation is thus a problem of matching the a priori incompatible phase speeds of ocean waves and seismic (or acoustic) waves. The typical velocity of ocean gravity waves goes up as the frequency decreases, from 15 m/s at $f = 0.1$ Hz, to less than 250 m/s for $f < 0.05$ Hz. In contrast, in a two-layer medium (ocean and crust), the typical phase velocity of Rayleigh waves is much faster, between 1.5 and 3.0 km/s in the frequency range 0.1–0.2 Hz.

3.1.3 Secondary and Primary Microseisms

Two known mechanisms can allow a matching of these very different phase velocities. These are the direct coupling between ocean waves and the seafloor (called here primary mechanism), and the nonlinear interaction among ocean waves (called here secondary mechanism). For the primary mechanism, the frequencies of the seismic source and the ocean waves are the same. For the secondary mechanism, seismic sources have a frequency twice that of the ocean waves. For that reason, secondary microseisms (generated by the secondary mechanism) are also called double-frequency microseisms. Primary microseisms (generated by the primary mechanism) are also called single-frequency microseisms.

For a given pressure source at a given frequency, all seismic phases are generated from the same region and propagate into the solid Earth. At that frequency and in the crust, the largest wavelengths correspond to P- and S-waves, while shortest wavelengths correspond to Rayleigh waves. In the period range 3–10 s (secondary microseisms), from the source at the ocean surface, P-waves are multiply reflected in the ocean ("source site effect"). Like organ pipes, the water depth preferentially amplifies modes for which the ratio of the apparent vertical wavelength and the water depth is 1/4, 3/4 ... (Longuet-Higgins, 1950). At different water depths, P, S, and Rayleigh modes are excited differently (e.g., Longuet-Higgins, 1950; Gualtieri et al., 2013, 2014; Meschede et al., 2017).

The quantitative verification of theoretical Rayleigh wave sources is generally satisfactory for secondary microseisms (e.g., Kedar et al., 2008; Ardhuin et al., 2011; Stutzmann et al., 2012; Gualtieri et al., 2013). The theory was extended to body waves by Vinnik (1973), Ardhuin and Herbers (2013), Gualtieri et al. (2014), Farra et al. (2016), and Nishida and Takagi (2016). Few quantitative validations of the theory for body waves have been published so far (e.g., Obrebski et al., 2013; Farra et al., 2016; Nishida and Takagi, 2016; Neale et al., 2017; Meschede et al., 2017).

3.1.4 The Particular Case of the Seismic "Hum"

Several mechanisms have been proposed for the generation of Rayleigh waves or spheroidal modes in the seismic hum (e.g., Fukao et al., 2002; Tanimoto, 2005; Webb, 2007; Uchiyama and McWilliams, 2008; Webb, 2008). Traer and Gerstoft (2014) considered the interaction of ocean waves of frequency f and f' and wavenumber vectors $\mathbf{k}_{sgw}$ and $\mathbf{k}'_{sgw}$ giving a seismic motion of frequency $|f - f'|$, but they did not evaluate the amplitude of the seismic response. We will do it here for a flat seafloor and horizontally homogeneous sediment and crust layers, using the propagator matrix method (Gilbert and Backus, 1966), already mentioned in the previous section.

As noted by Hasselmann (1963), the solution has the same horizontal velocity as the external pressure field. Geometrical analysis shows that the speed of that pressure field is maximum when $\mathbf{k}_{sgw}$ and $\mathbf{k}'_{sgw}$ are aligned. That speed is always less than 2π times the ratio of frequency and wavenumber differences $2\pi|f - f'|/(|k_{sgw} - k'_{sgw}|)$, which is less than the group speed of ocean waves, typically less than 20 m/s. Hence the forcing proposed by Traer and Gerstoft (2014) falls in the white region of Figure 3.2. The difference interaction of two ocean wave trains with frequencies f and f' produces an equivalent external pressure field that has a frequency $|f - f'|$ that may fall in the hum frequency band, but as long as the bottom is flat, it will only produce a response that travels at very slow speeds. This process is similar to the compliance effect (Crawford et al., 1991), and it does not generate seismic waves.

Earlier, Webb (2007) had proposed that interfering ocean waves in shallow water could be an efficient source of hum. However, that analysis ignored the additional wave-induced pressure on the seafloor. That extra term is out of phase with the surface pressure, and the two cancel out when ocean waves have a larger wavelength than the water depth (Ardhuin and Herbers, 2013).

After reviewing these proposed theories, Ardhuin et al. (2015) showed that Rayleigh waves in the seismic hum could be explained quantitatively by the primary mechanism, that is, by the interference of ocean infragravity waves with the seafloor topography. This is thus the theory that we have retained in this chapter.

Love waves or toroidal modes in the seismic hum have been discovered by Kurrle and Widmer-Schnidrig (2008). Fukao et al. (2010) and Saito (2010) proposed that wave propagation over a sloping seafloor creates equivalent tangential stresses with a magnitude compatible with the measured Love wave amplitudes.

3.2 Ocean Waves and Their Spectral Properties

In this section, we present the general properties of ocean waves, and how they can be predicted from the wind field using numerical ocean wave models. When dealing with seismic ambient noise, only the statistical properties of ocean waves are of interest. The combination of all waves in a given region and time is the "sea state." It is usually described by a power spectral density of the sea surface elevation. The investigation of the shape and properties of individual waves will not be covered here, and can be found in ocean engineering textbooks (e.g., Boccotti, 2000). The ocean wave spectrum is dominated by surface gravity waves with significant variations in their power over scales of a few hours. Other surface gravity waves that take the form of short transient wave packets, like tsunamis or ship wakes, will not be discussed here, and we restrict our discussion to wind-generated waves.

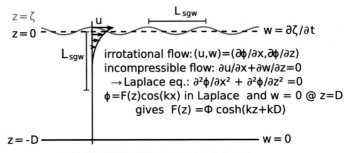

Figure 3.3. Main elements of the Airy theory for linear ocean waves propagating along the x direction, connecting horizontal velocity u and vertical velocity w and their gradients. Horizontal velocities under a crest are shown with arrows. Airy theory assumes irrotational and incompressible flow, and small wave amplitudes compared to the water depth D and wavelength $L = 2\pi/k_{sgw}$. The combination of irrotational and incompressible constraints replaces the mass conservation by Laplace's equation for the velocity potential ϕ. It is Laplace's equation that gives an exponential decay in depth for any cosine pattern on the horizontal axis.

3.2.1 Properties of Linear Surface Gravity Waves

Describing the sea state as a linear superposition of sinusoidal waves is generally very accurate. Hence the linear theory of ocean waves, usually attributed to Airy (1841), plays a central role. Its main characteristics are summarized in Figure 3.3 and can be found in all hydrodynamics textbooks (e.g., Lamb, 1932; Mei, 1989; Dean and Dalrymple, 1991; Boccotti, 2000). We define k_{sgw} as the modulus of the horizontal wavenumber of ocean surface gravity waves (subscript sgw). The conservation of momentum gives the dispersion relation that links the wavelength $2\pi/k_{sgw}$ and the period $T_{sgw} = 2\pi/\omega_{sgw}$, as a function of the water depth D,

$$\omega_{sgw}^2 = gk_{sgw}\tanh(k_{sgw}D), \tag{3.1}$$

where g is the gravitational acceleration. This dispersion relation for linear surface gravity waves has two limits. When $k_{sgw}D \gg 1$, waves are in the deep-water regime, and the phase speed $c_{sgw} = \omega_{sgw}/k_{sgw}$ becomes $c_{sgw} = g/\omega_{sgw}$ and is only a function of period: these ocean waves are dispersive. On the other hand, if $k_{sgw}D \ll 1$, waves are in the shallow-water regime, the phase speed reduces to $c_{sgw} = \sqrt{gD}$, and it is only a function of the water depth: these shallow-water waves are not dispersive. The definition of deep or shallow water is thus given by the ratio of the water depth D and wavelength $2\pi/k_{sgw}$ of the ocean wave.

The phase velocity $c_{sgw} = \omega_{sgw}/k_{sgw}$ and the group velocity

$$U_{sgw} = \partial\omega_{sgw}/\partial k_{sgw} \tag{3.2}$$

are further modified by ocean currents, which introduce a Doppler shift. For the sake of simplicity we shall neglect currents, in spite of their importance for ocean waves (e.g., Ardhuin et al., 2012, 2017).

In the deep ocean, a 10 s ocean wave has a wavelength of 156 m and a phase speed of 15 m/s. Longer ocean waves propagate faster, but due to the limited water depth, the upper limit is around $\sqrt{gD} \simeq 220$ m/s for $D < 5000$ m.

All ocean waves are much slower than seismic waves. For example, a 5 s seismic P-wave travels in the ocean at the speed of sound $\alpha_w \simeq 1500$ m/s and has a wavelength of 7500 m. Because of the mismatch between ocean and seismic wavelength at a given frequency, linear ocean waves in homogeneous media cannot generate seismic waves. Effects of non-homogeneity and non-linearity are described in section 3.3.

3.2.2 Typical Sea States

Wind waves take all their energy from the wind, more or less directly. The phase speed of ocean waves that are directly driven by the wind cannot exceed the wind speed by more than about 20% (Pierson and Moskowitz, 1964). As a result, ocean wave periods T_{sgw} are constrained by the dispersion relation (3.1) to be under 30 s for usual ocean waves. For $T_{\text{sgw}} = 30$ s, the phase speed c_{sgw} is 47 m/s in deep water, corresponding to the maximum wind speed in a category-2 tropical cyclone on the Saffir-Simpson scale. Such large periods are unusual.

The wind speed is not the only factor controlling the wave period. Indeed, in order to reach large periods and heights, waves need time and space to develop. This is because waves grow slowly, on the time scale of days for the highest wind speeds (Hasselmann et al., 1973). As a result, the waves with largest height and periods are not found in tropical cyclones, which move too fast, but in those extra-tropical storms that travel at the group speed of the dominant waves. For example, the largest ever sea state was measured in such a storm in the North Atlantic, with a maximum significant wave height $H_s = 20.1$ m, and a peak period of 25 s (Hanafin et al., 2012). In summary, ocean wave properties are determined by the space and time patterns of the wind speed and direction, and the shoreline geometry. Moreover, sea ice also modifies wave propagation and dissipation by impeding the transfer of energy from wind to waves, and causing an extra dissipation that is particularly strong at higher frequencies (e.g., Squire et al., 1995).

Typical correlation distances of wave spectra range from a few tens of kilometers in coastal regions, to thousands of kilometers in the case of far-propagating swells in the deep ocean (Snodgrass et al., 1966; Delpey et al., 2010). Swells are the waves radiated away from storms, with typically longer periods than wind sea because shorter components dissipate rapidly in the absence of wind forcing. These

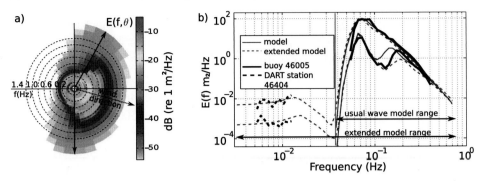

Figure 3.4. (a). Example of directional wave spectrum $E(f_{\mathrm{sgw}}, \theta)$ as a function of azimuth θ and frequency f_{sgw} as recorded by a stereo video system in the Black Sea (Leckler et al., 2015). The high frequency of the peak, $f_p \simeq 0.33$ Hz, is typical of small seas or weak winds, but the directional shape of the spectrum is expected to be similar to that found in the open ocean for similar values of f_{sgw}/f_p. At relative high frequencies ($f_{\mathrm{sgw}}/f_p > 3$) the dominant directions (solid arrows) can be 70° away from the wind direction (dashed arrow). (b) Example of modeled and measured wave spectra $E(f_{\mathrm{sgw}}) = \int E(f_{\mathrm{sgw}}, \theta)d\theta$ in the North-East Pacific, on 5 (top curves) and 24 (bottom curves with 2 peaks) January 2008. Classical wave models provide wave spectra between 0.04 and 0.7 Hz (blue solid line). Ardhuin et al. (2014) have extended that range to include infragravity waves down to $f_{\mathrm{sgw}} = 3 \times 10^{-3}$ Hz (grey dashed line; blue dashed line in the plate section). (A black-and-white version of this figure appears in some formats. For the color version, please refer to the plate section.)

wind-generated waves can interact nonlinearly to produce longer ocean wave periods, up to 300 s. Those long waves are called infragravity waves. Contrary to the shorter wind-sea and ocean swell waves, they are not generated by the wind. Instead their source is the nonlinear interaction of wind-sea or ocean swell waves at the shoreline (Longuet-Higgins and Stewart, 1962; Bertin et al., 2018). Some examples of ocean wave spectra are shown in Figure 3.4 and discussed in the next section.

3.2.3 Numerical Ocean Wave Modeling

Numerical ocean wave models were first developed for navigation safety (Gelci et al., 1957), and are generally based on an evolution equation for the wave spectrum. At each point of the ocean surface, the sea state is represented by a two-dimensional power spectral density $E(f_{\mathrm{sgw}}, \theta)$ that gives the distribution of the surface elevation variance across frequencies f_{sgw} and directions θ. One example is shown in Figure 3.4.a.

Wave energy propagates in all directions θ at the speed given by the group velocity U_{sgw} that is a function of frequency f_{sgw}. Applying equation (3.2) in deep water, $U_{sgw} = g/(4\pi f_{sgw})$ is half of the phase speed. For a flat ocean bottom, the evolution in time t and space (x, y) of the power spectral density, $E(f_{sgw}, \theta)$, is given by Gelci et al. (1957) and WAMDI Group (1988),

$$\frac{\partial}{\partial t} E(f_{sgw}, \theta) + \frac{\partial}{\partial x} \left[\cos\theta \, U_{sgw} \, E(f_{sgw}, \theta) \right] + \frac{\partial}{\partial y} \left[\sin\theta \, U_{sgw} \, E(f_{sgw}, \theta) \right] = S(f_{sgw}, \theta),$$

(3.3)

where the left-hand side represents ocean wave propagation, and the right-hand-side source term $S(f_{sgw}, \theta)$ represents many processes including generation by the wind (e.g., Janssen, 2004), nonlinear wave evolution (Hasselmann, 1962), and dissipation by wave breaking or wave-ice interactions. Both $E(f_{sgw}, \theta)$ and $S(f_{sgw}, \theta)$ vary with the horizontal coordinates x and y, as well as with frequency f_{sgw} and azimuth θ.

For each frequency and direction, the source term $S(f_{sgw}, \theta)$ is also a function of the spectral density $E(f'_{sgw}, \theta')$ evaluated at all the other frequencies f'_{sgw} and directions θ', but at the same location given by x and y. The source term S is also a function of many other parameters, including wind speed and direction, water depth, ice, and ocean bottom properties. Unfortunately, there is no expression for S that is based on first principles. Today's empirical parameterizations are still the topic of active research (e.g., Janssen, 2004; Ardhuin et al., 2010; Romero et al., 2012; Ardhuin et al., 2016).

There are very few measurements of the full 2D ocean wave power spectral density as a function of azimuth θ and frequency f_{sgw}, such as the one shown in Figure 3.4a. As a result, numerical wave models have been calibrated and validated mostly in terms of dominant ocean wave frequency f_p, and significant wave height H_s, defined as

$$H_s = 4\sqrt{\int E(f_{sgw}) \mathrm{d}f}, \quad \text{with} \quad E(f_{sgw}) = \int E(f_{sgw}, \theta) \mathrm{d}\theta.$$

(3.4)

Most in situ wave measurements rely on time series of pressure, velocity, and/or surface elevation. Thus it is usual to work with the 1-D power spectral density of the surface elevation $E(f_{sgw})$. A typical frequency range of measurements and models goes from a minimum around 0.03 Hz to a maximum somewhat below 1 Hz. Numerical wave models generally give good estimates of H_s, but accurate spectral distributions $E(f_{sgw})$ are difficult to achieve. Figure 3.4b shows measured spectra and results of a numerical ocean wave model (The WAVEWATCH III ® Development Group, 2016) off the Oregon coast, where the water depth is 4000 m, on January 5 and 24, 2008. The extended modeled and measured spectra are for the location of the tsunameter DART 46404 system, 200 km to the east of the wave buoy 46005. On 5 January the model reproduces well the single peak at 0.07 Hz.

On 24 January the ocean swell peak at 0.07 Hz is relatively well modeled, while the peak of the wind sea is at 0.15 Hz in the model and 0.18 Hz in the buoy data, due to errors in the wind speed used to drive the wave model. As a result there is a large difference in the energy level at 0.15 Hz. These results depend mostly on the accuracy of winds used to force the wave model.

The application of these ocean wave spectra to seismic noise modeling is described in section 3.4, but we first have to introduce seismic noise generation processes.

3.3 Transforming Short Ocean Waves into Long Seismic Waves

Following Hasselmann (1963), we start from the statistical properties of ocean waves with frequency f_{sgw} and horizontal wavenumber $\mathbf{k}_{sgw}$. From this ocean wave spectrum, we determine the equivalent pressure power spectral density $F_p(k_x, k_y, f)$ at much smaller horizontal wavenumbers $\mathbf{k} = (k_x, k_y)$, which correspond to the horizontal components of seismic wavenumbers. The wavenumber components are $|k_x| << k_{sgw}$ and $|k_y| << k_{sgw}$. Given the typical width of ocean wave spectra, Hasselmann (1963) showed that the pressure power spectral density $F_p(k_x, k_y, f)$ around seismic wavenumbers is independent of $\mathbf{k}$. Therefore, the pressure spectrum is white, i.e., $F_p(k_x, k_y, f) \simeq F_p(f)$. We thus drop the variables k_x and k_y.

At the seismic frequency f, we will call $F_{p,1}(f)$ the pressure power spectral density due to the primary mechanism, which is the interaction between ocean waves and the seafloor topography in shallow water. We will call $F_{p,2}^{surf}(f)$ the power spectral density of surface pressure due to the secondary mechanism, which accounts for the nonlinear interaction of pairs of ocean wave trains at similar frequencies, propagating in nearly opposite directions. Both $F_{p,1}(f)$ and $F_{p,2}^{surf}(f)$ are spectral densities in frequency-wavenumbers space, with units of $Pa^2 \cdot m^2 \cdot s$.

In the following, we will consider the primary mechanism in two cases: (a) a small-scale ocean bottom topography, as a proxy for seamounts in deep ocean or coastal features on the continental shelf (section 3.3.1), and (b) a large-scale slope, as a proxy for a continental slope (section 3.3.2). The secondary mechanism is presented last (section 3.3.3).

3.3.1 Primary Mechanism: (a) Small-Scale Topography

After propositions by Wiechert (1904) and others, the first modern quantitative theory for generation of seismic noise at the same frequency as that of ocean waves was proposed by Hasselmann (1963). First using the Wentzel-Kramers-Brillouin (WKB) method to generalize Airy waves to a varying depth $D(x)$, he obtained the pressure field at the seafloor. Then the pressure power spectral density F_p was determined, focusing on the values at the wavenumbers that match seismic waves. Hasselmann (1963) generally assumed that the seismic response to this bottom pressure is equivalent to the effect of the same pressure acting on a flat seafloor. Hasselmann (1963) also validated his theory using measurements in California. A more recent theory, developed only for Love waves, is due to Fukao et al. (2010) and Saito (2010), who analyzed the ocean wave propagation over sinusoidal and random seafloor topographies and worked out the expression for the tangential stress. Following that analysis, in this section we consider small amplitude variations around a mean depth, namely $D(x) = D_0 + D_1(x)$ with $|D_1| \ll D_0$.

The seismic source due to the primary mechanism can be described as a combination of a tangential stress and a vertical pressure force acting on the local bathymetry. Here, we analyze a wave propagating over a sinusoidal topography that easily generalizes to a random topography of small amplitude, and we give the expression of the source in terms of both bottom pressure and tangential stress.

We employ a 2D coordinate system, composed of a horizontal x and a vertical z axis, with z positive upward. We also consider an ocean wave traveling along the x direction. The ocean surface is at $z = 0$, and the seafloor is at the depth $z = -D$. This configuration is illustrated in Figure 3.5a. The interference of ocean waves (gray in Figure 3.5a) with the undulating seafloor (black in Figure 3.5a) produces a pressure field at the seafloor (gray in Figure 3.5b) that can contain wavelengths long enough (black in Figure 3.5b) to excite seismic waves. The amplitude of this long wave-component pressure is controlled by a modulation factor α.

In this simple case, the water depth $D(x)$ is defined by,

$$D(x) = D_0 + d\cos(k_b x). \tag{3.5}$$

where D_0 is the average ocean depth, d and k_b are the amplitude and the horizontal wavenumber of the ocean bottom topography, respectively. A monochromatic wave of angular frequency $\omega_{\text{sgw}} = 2\pi f_{\text{sgw}}$ and amplitude a propagating in the x direction has a surface elevation

$$\zeta(x, t) = a(x)\cos\left[S(x) - \omega_{\text{sgw}}t\right] \tag{3.6}$$

where the phase is given by

$$S(x) = \int_0^x k_{\text{sgw}}(x')\mathrm{d}x' \tag{3.7}$$

such that $k_{\text{sgw}}(x)$ is adjusted to the local water depth via the dispersion relation (3.1) that we rewrite as

$$\omega_{\text{sgw}}^2 = gk_{\text{sgw}}(x)\tanh\left[k_{\text{sgw}}(x)D(x)\right]. \tag{3.8}$$

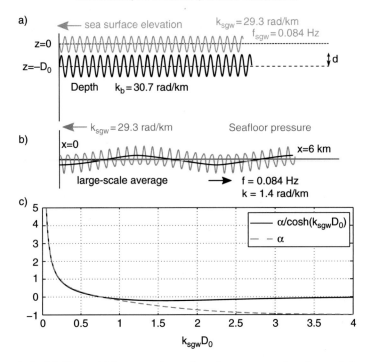

Figure 3.5. (a) Sinusoidal sea surface elevation propagating from right to left (gray) and sinusoidal ocean bottom elevation, having amplitude d (black). Their vertical scale is strongly exaggerated for readability. (b) Seafloor pressure (gray). The pattern of pressure at the seafloor contains long wavelength components (black) that oscillate in time and, in this example, propagate from left to right at a speed that can match those of seismic waves. For animations see http://tinyurl.com/mswanim (last accessed 04/04/2018). (c) Modulation factor α that comes in the bottom pressure spectrum near $k = 0$, as a function of the non-dimensional mean water depth $k_{sgw} D_0$.

The wavenumber $k_{sgw}(x)$ oscillates around the value k_{sgw}^0 given by $(2\pi f_{sgw})^2 = g k_{sgw}^0 \tanh\left[k_{sgw}^0 D_0 \right]$.

The bottom pressure is given by the Airy theory (e.g., Dean and Dalrymple, 1991),

$$p(z = -D) = \frac{\rho g}{\cosh(k_{sgw}(x) D(x))} a(x) \cos\left[S(x) - \omega t \right]. \tag{3.9}$$

The horizontal force per unit horizontal area at the seafloor is equivalent to a tangential stress acting on a flat seafloor and is given by the pressure force projected on the horizontal. Therefore, the tangential stress is given by the product between the bottom pressure (equation (3.9)) and the bottom slope ($-dD/dx$ from equation (3.5)):

$$\tau(z = -D) = -p(z = -D)\frac{dD}{dx} =$$

$$= \frac{-\rho g}{\cosh(k_{\text{sgw}}(x)D(x))}a(x)k_b d \sin(k_b x) \cos[S(x) - \omega t]. \quad (3.10)$$

To get the source of primary microseisms, we must find the components of pressure $p(z = -D)$ and tangential stress $\tau(z = -D)$ at seismic wavelengths, that is, at wavenumbers much smaller than k_{sgw}.

Regarding the tangential stress, the transformation from short to long wavelengths comes from the product between slope and pressure. This product gives an aliasing of the pressure at long wavelengths,

$$\sin(k_b x) \cos[S(x) - \omega t] \simeq \sin(k_b x) \cos(k_{\text{sgw}}^0 x - \omega t)$$

$$= \frac{1}{2}\left\{\sin\left[(k_b - k_{\text{sgw}}^0)x - \omega t\right] + \sin\left[(k_b + k_{\text{sgw}}^0)x - \omega t\right]\right\}.$$
$$(3.11)$$

In equation (3.11), the first term on the right-hand side has a wavenumber $K = k_b - k_{\text{sgw}}^0$ which can be small and can fulfill the condition for seismic wave generation. In other words, when the bottom topography has a wavelength $(2\pi/k_b)$ that is nearly equal to the ocean surface wavelength $(2\pi/k_{\text{sgw}}^0)$, then the tangential stress contains a component with a very small wavenumber K (i.e., with a large wavelength).

Therefore, the tangential stress and pressure on the seafloor contain components with large scale (ls) wavelengths (black in Figure 3.5b) with an amplitude proportional to the amplitude d of the bottom oscillations (Saito, 2010). Retaining only the first term in (3.11) we have,

$$\tau_{ls} \simeq -\frac{\rho g k_b}{\cosh(k_{\text{sgw}}^0 D_0)}\frac{a_0 d}{2}\sin\left[(k_b - k_{\text{sgw}}^0)x - \omega t\right]. \quad (3.12)$$

Equation (3.12) gives the horizontal stress that applies to the equivalent flat bottom and can be used to compute Rayleigh and Love wave responses without having to model the real bottom at the small scale of the depth oscillations.

The long-wavelength pressure at the ocean bottom is more complex to derive because a similar approximation, with a constant wave amplitude a_0, wavelength k_{sgw}^0, and water depth D_0, gives a zero contribution. Thus, we need to take into account the modification of (1) the wave amplitude a, (2) the wavenumber k_{sgw}, and (3) the ocean depth D, which gives a more subtle and generally weaker effect. For a small bottom amplitude d, each of the three modulation effects (due to variations of a, k_{sgw}, and D) gives a pressure that is in the form $p = p_0 + \alpha k_b d \sin(k_b x) \cos(k_0 x)$, where α is a non-dimensional coefficient. When combined, the three effects give a long-wavelength term,

$$p_{ls} = -\frac{\rho g k_b \alpha}{\cosh(k_{sgw}D)} \frac{a_0 d}{2} \cos\left[(k_b - k^0_{sgw})x - \omega t\right] \tag{3.13}$$

The numerical evaluation of α, is shown in Figure 3.5c. In particular, the pressure modulation vanishes ($\alpha = 0$) for $k^0_{sgw}D_0 \simeq 0.76$, as already noted by Ardhuin et al. (2015). We also note that for $k^0_{sgw}D_0 > 0.2$ we have $|\alpha| < 1$, in which case the horizontal force given by equation (3.12) is larger than the vertical force given by equation (3.13). This variable ratio $|\alpha|$ of the vertical and horizontal forces may explain why Love wave amplitudes are sometimes larger than those of Rayleigh waves (Juretzek and Hadziioannou, 2016).

The tangential stress and pressure at seismic wavelengths were derived for sinusoidal seafloor topography and are valid for $d \ll D_0$. In a general case, a random seafloor shape can be decomposed using Fourier analysis. Saito (2010) gives the power spectral density of the horizontal stress ($F_{T,x}$, $F_{T,y}$) as a function of the wave spectrum $E(f_{sgw}, \theta)$, and the bottom elevation spectrum F_B,

$$(F_{T,x}, F_{T,y})(f) = \int_0^{2\pi} \left[\frac{\rho g k_{sgw}}{\cosh(k_{sgw}D)}\right]^2 (\cos^2\theta, \sin^2\theta) E(f_{sgw}, \theta) F_B(\mathbf{k}_b) d\theta,$$

$$\tag{3.14}$$

Given the similarity between (3.12) and (3.13), the power spectral density of the pressure field $F_{p,1}$ is

$$F_{p,1}(f) = \int_0^{2\pi} \left[\frac{\rho g k_{sgw}\alpha}{\cosh(k_{sgw}D)}\right]^2 E(f_{sgw}, \theta) F_B(\mathbf{k}_b) d\theta, \tag{3.15}$$

with $\mathbf{k}_b = k_{sgw}(\cos\theta, \sin\theta)$ in both integrals, corresponding to the bottom spectral components that match the wavenumber of ocean surface waves, and $f = f_{sgw}$.

3.3.2 Primary Mechanism: (b) Large-Scale Slope

In this section, as a proxy for the continental shelf, we consider a water depth $D(x)$ and ocean waves propagating along the x-axis. As we shall see, the same physical mechanism of modulation of short waves and aliasing into long wavelength can explain recorded seismic signals, at least in the hum frequency band.

This seafloor topography can be treated in the Fourier domain as long as the change in water depth is small compared to the water depth itself, corresponding to our hypothesis $|D_1(x)| \ll D_0$ in the previous section. Unfortunately, this is not generally the case, and therefore we use a numerical estimation of the pressure $p(z = -D(x))$ and horizontal force $\tau(z = -(D(x))$ at the seafloor. We first consider a constant slope, s, extending all the way to the shoreline where $D = 0$, with $s = |dD/dx| = 0.01$ (note that Ardhuin et al. (2015) defined s to be in percentage, thus 100 times larger). In this case, the bottom pressure power spectral density was first estimated by Hasselmann (1963). Because the ocean wave spectrum evolves

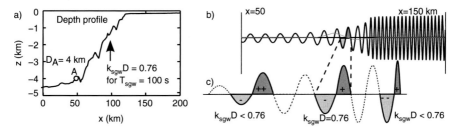

Figure 3.6. (a) Example of depth profile from a West-East transect across the Bay of Biscay at 46.47° N, the point A is chosen where the depth is $D_A = 4$ km. (b) Seafloor pressure along this transect (black curve). The pattern of pressure at the seafloor contains long wavelength components (gray curve) that oscillate in time and that can match those of seismic waves. (c) Schematic of varying bottom pressure amplitude and wavelength that define the large-scale pressure. An increasing amplitude gives a larger integrated pressure for $k_{sgw}D > 0.76$, and a shorter wavelength gives a lower integrated pressure for $k_{sgw}D < 0.76$, and the two effects cancel at $k_{sgw}D = 0.76$.

along the x direction due to shoaling, with wave energy increasing toward shallow water, we choose a reference point A where the wave spectrum is $E_A(f_{sgw}, \theta)$, and the depth at A is D_A (see Figure 3.6a). The pressure spectral density at seismic wavelength is again caused by a change in the wavelength and amplitude of the bottom pressure that are all a function of x, as given by Ardhuin et al. (2015)

$$F_{p,1}(f_{sgw}) = s \frac{\rho_w^2 g^4 \left[E_A(f_{sgw}, \theta_n) + E_A(f_{sgw}, \theta_n + \pi)\right]}{k_A (2\pi f_{sgw})^4 32 \times L_x}. \qquad (3.16)$$

It is interesting to note that the resulting force over an area $dA = L_x \times L_y$ is independent of L_x, as if the source microseism was localized in the x direction (see section 3.3.6). In practice, for any topography $D(x)$, we can compute the Fourier spectrum of the pressure along the topography (Ardhuin et al., 2015), and writing the pressure spectrum in the same form defines an effective slope s that is then a function of the seismic frequency. We have verified that equation (3.16) gives errors under 40% for a constant slope, $f = 0.07$ Hz and $0.005 < s < 0.02$, and an underestimation as large as 70% for $s = 0.04$.

For a given depth profile $D(x)$, the equivalent slope s is a function of the wave frequency. Figure 3.6 shows the example of a depth profile in the Bay of Biscay (Figure 3.6a) and the resulting bottom pressure in black, and (not to scale) the large-scale average pressure in light gray (Figure 3.6b).

This pressure pattern at a scale much larger than the wavelength of ocean waves is due to a change in wavelength and amplitude of the bottom pressure signal. The graphic in Figure 3.6c reproduces the interpretation given by Ardhuin et al. (2015), who found that the local average over neighboring regions of positive (dark gray) and negative (light gray) pressure anomalies gives an oscillating pressure field that is shifted in phase relatively to the ocean wave train (Figure 3.6c). The sign of the phase shift reverses at the depth where $k_{sgw}D = 0.76$ because at that depth the positive and negative contributions cancel out: The increase in amplitude toward shallow water cancels out the reduction in wavelength. This region of phase shift dominates the large-scale spatial-averaged pressure. Hence, the sources are located where $k_{sgw}D \simeq 0.76$, unlike for an oscillating depth, for which there is no source there. The source distribution varies with the details of the bottom topography, but we have found a concentration of sources for depths D where $k_{sgw}D \simeq 0.76$ for realistic depth profiles off Hawaii, Oregon, California, and frequencies below 0.03 Hz.

For all these bottom topographies, $s \simeq 0.06$ is a good approximation for the modulation of infragravity waves at frequency 0.01 Hz, over the slopes of typical continental margins. At higher frequencies, bottom topographies that do not resolve the scales of ocean waves typically give $s < 0.01$ for $f > 0.06$ Hz (Ardhuin et al., 2015, Figure S3).

As expressed by the two forms of the pressure spectrum respectively in equations (3.15) and (3.16), the seismic source comes from the interference of ocean waves and bottom topography. This implies that significant sources must be in water depths not too large compared to the ocean wavelength ($k_{sgw}D < 2$), and in regions where the bottom is not too flat. An example is detailed in section 3.4.

3.3.3 Secondary Mechanism

In this section, we discuss another way to generate seismic waves with very long wavelengths, that is the interaction of ocean waves with other ocean waves. Bernard (1941) found that the dominant microseism peak (between about 1 and 10 s) has a frequency twice that of ocean waves, and hypothesized that standing waves play a role in the generation process. Indeed, Miche (1944) showed that standing waves in an incompressible ocean generate pressure oscillations that are uniform in space. Longuet-Higgins and Ursell (1948) connected these two ideas, leading to the full theory by Longuet-Higgins (1950) who considered standing waves in a compressible ocean, and made a heuristic extension to random waves. Hasselmann (1963) generalized the theory developed by Longuet-Higgins (1950) considering statistical properties of random ocean waves.

From First-Order Airy Waves to Second-Order Motions

Because a detailed derivation would be too long, we only point to the key elements in Longuet-Higgins (1950), Hasselmann (1963), and Ardhuin and Herbers (2013).

The first-order solution of the equation of motion is a superposition of linear Airy waves propagating in all horizontal directions, with wavenumber vectors $\mathbf{k}_{sgw}$. This gives a surface elevation (e.g., Dean and Dalrymple, 1991),

$$\zeta_1 = \sum_{\mathbf{k}_{sgw},s} a^s_{1,\mathbf{k}_{sgw}} e^{i(\mathbf{k}_{sgw}\cdot\mathbf{x}-s\omega_{sgw}t)}, \tag{3.17}$$

where $s = +1$ or $s = -1$ is a sign index, and $a^s_{1,\mathbf{k}_{sgw}}$ is the surface elevation amplitude of waves propagating in the direction of $s\mathbf{k}_{sgw}$. The corresponding velocity potential is given by the Airy theory (e.g., Dean and Dalrymple, 1991)

$$\phi_1 = \sum_{\mathbf{k}_{sgw},s} -i\frac{sg}{\omega_{sgw}}\frac{\cosh[k_{sgw}(z+D)]}{\cosh[k_{sgw}D]} a^s_{1,\mathbf{k}_{sgw}} e^{i(\mathbf{k}_{sgw}\cdot\mathbf{x}-s\omega_{sgw}t)} \tag{3.18}$$

In reality, waves are slightly nonlinear due to the momentum conservation equation, which takes the form of Bernoulli's equation, and surface kinematic boundary conditions. Following Longuet-Higgins (1950), we then consider the second-order solution of the equation of motion, which corresponds to a nonlinear correction of the surface elevation, ζ_2, such that $|\zeta_2| \ll |\zeta_1|$, with an associated velocity potential $\phi_2 \ll \phi_1$:

$$\zeta \simeq \zeta_1 + \zeta_2, \quad \phi \simeq \phi_1 + \phi_2. \tag{3.19}$$

We note that ϕ_2 is a solution to Laplace's equation in three dimensions.

The combination of Bernoulli's equation with the surface kinematic boundary condition $d\zeta/dt = w$ gives, at $z = 0$, the equation (3.22) in Hasselmann (1962). At second order, this is the equation (2.13) in Ardhuin and Herbers (2013), in which the second-order potential is obtained from the linear first-order solutions given by equations (3.17) and (3.18),

$$\frac{\partial^2\phi_2}{\partial t^2} + g\frac{\partial\phi_2}{\partial z} = -g\left[\nabla\phi_1\cdot\nabla\zeta_1 + \zeta_1\frac{\partial^2\phi_1}{\partial z^2}\right] - \frac{\partial\,|\nabla\phi_1|^2}{2\partial t} - \zeta_1\frac{\partial^3\phi_1}{\partial t^2\partial z}. \tag{3.20}$$

As noted by Hasselmann (1963, his equation 2.10), this surface boundary condition is equivalent to a surface pressure field acting on a flat sea surface at $z = 0$,

$$\left(\frac{\partial^2}{\partial t^2} + g\frac{\partial}{\partial z}\right)\phi_2 = -\frac{1}{\rho_w}\frac{\partial\widehat{p}_{2,\text{surf}}}{\partial t}. \tag{3.21}$$

with the equivalent pressure defined in terms of products of the linear wave terms,

$$\widehat{p}_{2,\text{surf}} = \rho_w \sum_{\mathbf{k}_{sgw},s,\mathbf{k}'_{sgw},s'} W\left(\mathbf{k}_{sgw}, s, \mathbf{k}'_{sgw}, s'\right) a^s_{1,\mathbf{k}_{sgw}} a^{s'}_{1,\mathbf{k}'_{sgw}} e^{i\Theta(\mathbf{k}_{sgw},\mathbf{k}'_{sgw},s,s')} \tag{3.22}$$

with the phase of interacting wave Θ, and coupling coefficient W given by Hasselmann (1962, his equation 4.3) or by Ardhuin and Herbers (2013, their equation A1).

The terms that can produce fast components in the pressure field correspond to ocean waves with opposite wavenumbers, $s = s'$ and $\mathbf{k}_{sgw} + \mathbf{k}'_{sgw} \simeq 0$. In this case, W reduces to (Ardhuin and Herbers, 2013, equation 2.25),

$$W\left(\mathbf{k}_{sgw}, 1, -\mathbf{k}_{sgw}, 1\right) = -2\omega_{sgw}^2 \left[1 + \frac{1}{4\sinh(k_{sgw}D)}\right], \qquad (3.23)$$

and the phase of the equivalent pressure has a frequency that is twice that of the ocean waves,

$$\Theta(\mathbf{k}_{sgw}, \mathbf{k}'_{sgw}, 1, 1) = \left(\mathbf{k}_{sgw} + \mathbf{k}'_{sgw}\right) \cdot \mathbf{x} - 2s\omega_{sgw}t. \qquad (3.24)$$

The most simple second-order solution corresponds to a pair of interacting wave trains in deep water. For the sake of simplicity, here we consider ocean waves propagating along the x axis, with positive wavenumbers k_{sgw} and k'_{sgw}, and the same amplitude a, as shown in Figure 3.7. The surface elevation field associated with these two waves is given by

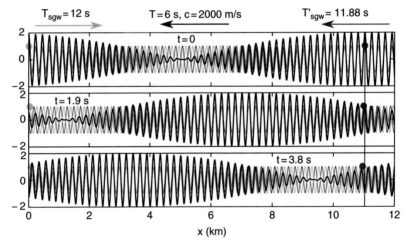

Figure 3.7. Example of ocean wave groups given by the superposition of two opposing deep water monochromatic wave trains of nearly equal periods T and T'. The curves show the surface elevation of the individual wave trains (light and dark gray) and their combination (black). The top, middle, and bottom panels show these waves at three different times. The group with length 12 km propagates in the same direction as the dark gray wave train, which has the shortest wavelength. The gray dots are attached to a wave crest of each wave train, and move 100 times slower than the black group. After Ardhuin and Herbers (2013). For animations see http://tinyurl.com/mswanim (last accessed 04/04/2018).

$$\zeta = a\cos(k_{\text{sgw}}x - \omega_{\text{sgw}}t) + a\cos(k'_{\text{sgw}}x + \omega_{\text{sgw}}t). \tag{3.25}$$

This beating wave pattern is a generalization of the purely standing wave of Longuet-Higgins (1950) when $k_{\text{sgw}} = k'_{\text{sgw}}$. In the more general form used here, the interference of the two wave trains makes groups that travel at the speed

$$U_\Delta = (\omega_{\text{sgw}} + \omega'_{\text{sgw}})/\Delta_k \tag{3.26}$$

which becomes very large when $\Delta_k = k_{\text{sgw}} - k'_{\text{sgw}}$ goes to zero.

Trigonometric identities transform sums in equation (3.25) into products

$$\zeta = 2a\cos\left[\frac{\Delta_k}{2}x - \frac{\omega_{\text{sgw}} + \omega'_{\text{sgw}}}{2}t\right]\cos\left[\frac{k_{\text{sgw}} + k'_{\text{sgw}}}{2}x - \frac{\Delta_\omega}{2}t\right], \tag{3.27}$$

where $\Delta_\omega = \omega_{\text{sgw}} - \omega'_{\text{sgw}}$. The first cosine varies on large spatial scales and gives the envelope of the signal, and the second cosine has the wavelength of the ocean waves. Figure 3.7 illustrates the surface elevation of these wave groups in black, and the two components with k_{sgw} and k'_{sgw} wavenumbers in gray.

For deep water, we have $kD \gg 1$ and $W = -2\omega_{\text{sgw}}^2$. Replacing equation (3.17) with (3.25), and keeping only the long wavelength components, equation (3.22) becomes

$$\widehat{p}_{2,\text{surf}} = -2\rho_w a^2 \omega_{\text{sgw}}^2 \cos[(k_{\text{sgw}} - k'_{\text{sgw}})x - (\omega_{\text{sgw}} + \omega'_{\text{sgw}})t]. \tag{3.28}$$

Equation (3.28) is the generalization for $k_{\text{sgw}} \simeq k'_{\text{sgw}}$ of Longuet-Higgins' (1950) famous example of the standing pattern of pressure fluctuation in the case of $k_{\text{sgw}} = k'_{\text{sgw}}$ (his equation 175).

In the general case, taking wave trains with wavenumber vectors $\mathbf{k}_{\text{sgw}}$ and $\mathbf{k}'_{\text{sgw}}$, we get waves with wavenumber vector $\mathbf{K} = \mathbf{k}_{\text{sgw}} + \mathbf{k'}_{\text{sgw}}$, which can be in any horizontal direction, and phase speed $(\omega_{\text{sgw}} + \omega'_{\text{sgw}})/K$. In fact, ocean wave spectra $E(f_{\text{sgw}}, \theta)$ are generally broad, with a width δ_f in frequency and δ_θ in directions. As a result, all the pairs of interacting ocean waves produce wavenumber vectors $\mathbf{K}$ corresponding to all the seismic phases that can be generated by a surface pressure field. Also, the radiation pattern is isotropic: the same amount of seismic energy is radiated in all horizontal directions. This would not happen for an extremely narrow ocean wave spectrum, and it requires that the width of the ocean spectrum in the two wavenumber directions is $\delta_k = 4\pi f_{\text{sgw}}\delta_f/g$ and that the wavenumbers of ocean waves $k_{\text{sgw}}\delta_\theta$ are much larger than the wavenumbers of seismic waves. For 5 s Rayleigh waves in the ocean $k < 9 \times 10^{-4}$ rad/m, meaning that we should have $\delta_f/f_{\text{sgw}} \gg 0.01$, and $\delta_\theta \gg 1$ deg. These conditions are generally met for ocean waves, and the variation of radiated power with azimuth may only vary by 1 dB for the narrowest ocean swell spectra.

Water Compressibility and Source Site Effect

So far, we did not need to consider the compressibility of the ocean. Solving for the motion inside the water column, compressibility can only be ignored close to the sea surface, as verified with laboratory experiments for standing waves by Cooper and Longuet-Higgins (1951), and in the ocean for random waves by Herbers et al. (1992), including finite depth effects.

The incompressible hypothesis ($\nabla \cdot \mathbf{u} = 0$) is valid as long as the time t necessary for a disturbance to reach the ocean seafloor and come back to the ocean surface is small compared to the period T_{sgw} of the ocean wave. In other words, the ocean is practically incompressible when the depth D is small compared to half of the wavelength of the ocean wave. Thus, in the secondary microseism period band ($T = 1 - 10$ s, that is $T_{sgw} = 2 - 20$ s), the incompressible hypothesis is valid for

$$\text{if } T_{sgw} = 2 \text{ s (i.e., } T = 1 \text{ s)} : t = \frac{2D}{\alpha_w} \ll T_{sgw} \rightarrow D \ll \frac{\alpha_w T_{sgw}}{2} = 1.5 \text{ km}$$

$$\text{if } T_{sgw} = 20 \text{ s (i.e., } T = 10 \text{ s)} : t = \frac{2D}{\alpha_w} \ll T_{sgw} \rightarrow D \ll \frac{\alpha_w T_{sgw}}{2} = 15 \text{ km}$$

where $\alpha_w = 1.5$ km/s is the velocity of sound waves in water. Considering realistic ocean depth of several km, D has same order of magnitude of $\frac{\alpha_w T_{sgw}}{2}$, and therefore the compressibility of the ocean should be taken into account.

Assuming an incompressible ocean would imply to neglect the propagation of compressional (P) seismic waves within the ocean. Assuming a compressible ocean and a free surface condition, P-waves are multiply reflected between the ocean surface and the seafloor, where they are also transmitted and converted to other seismic phases. These seismic phases generated at the seafloor travel in the solid Earth up to a seismic station. The so-called "source site effect" (see section 3.3.6) accounts for the compressibility of the ocean.

Power Spectral Density of Surface Pressure

In the following, we consider a compressible ocean. The pressure field associated with the first-order ocean wave motion decays exponentially with depth over a scale that is $1/k_{sgw}$. In the case of random waves and for deep water waves ($kD \gg 1$), the nonlinear interaction of ocean gravity waves generates a random pressure field, whose power spectral density at frequency f ($f = 2f_{sgw}$) (e.g., Hasselmann, 1963; Ardhuin et al., 2011; Farra et al., 2016) is given by

$$F_{p,2}^{surf}(f) = (2\pi)^2 \rho_w^2 g^2 f E^2(f_{sgw}) I(f_{sgw}). \tag{3.29}$$

where $E(f_{sgw})$ is the power spectral density of the sea surface elevation, which can be measured by surface buoys moored in the ocean, as shown in Figure 3.4. The spectrum $E(f_{sgw})$ is relatively well known, even though numerical ocean wave models have typical errors of 50%. On the contrary, the parameter $I(f_{sgw})$ is poorly known. It is the non-dimensional ocean gravity wave energy distribution as a function of the ocean wave azimuth θ and frequency f_{sgw}

$$I(f_{\text{sgw}}) = \int_0^\pi \frac{E(f_{\text{sgw}}, \theta)E(f_{\text{sgw}}, \theta + \pi)}{E^2(f_{\text{sgw}})} d\theta. \tag{3.30}$$

There are very few direct measurements of $I(f_{\text{sgw}})$ (Leckler et al., 2015; Peureux et al., 2018), and some indirect estimates from acoustic data (e.g., Farrell and Munk, 2010; Duennebier et al., 2012; Peureux and Ardhuin, 2016).

3.3.4 Ocean-Wave Conditions for Secondary Microseism Sources

The double-frequency sources, with an amplitude proportional to $E^2(f_{\text{sgw}})I(f_{\text{sgw}})$ as given by equation (3.29), are associated with particular ocean wave properties. The very large variability of the directional wave integral $I(f_{\text{sgw}})$ means that only specific wave conditions can produce a significant source amplitude. Ardhuin et al. (2011) introduced three broad classes of wave conditions in order to help their understanding. These are illustrated in Figure 3.8.

The first class corresponds to the presence of ocean waves at directions very oblique relative to the main wave direction. This is typical of frequencies above the dominant ocean wave frequency f_p, that is, at frequencies higher than 0.1 Hz in Figure 3.4a. These sources are very common but fairly weak, with $I(f_{\text{sgw}})$ only significant for $f_{\text{sgw}} > 3f_p$ (Duennebier et al., 2012; Peureux and Ardhuin, 2016).

The second class corresponds to ocean waves reflected by the coast that interact with incoming waves. Class II sources are weak because the reflection coefficient of ocean waves at shorelines is generally less than 5% for the wave energy (Ardhuin and Roland, 2012), giving small values of $I(f_{\text{sgw}})$. However, they are numerous because potentially they exist all along the coasts.

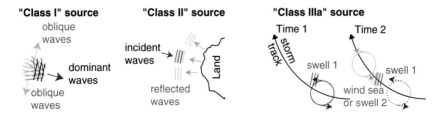

Figure 3.8. Schematic of three classes of wave conditions that can lead to a significant magnitude of double-frequency microseism sources. A class "IIIb" corresponds to waves from a given storm, interfering with swells from a different storm, as in the case presented by Obrebski et al. (2012).

The third class occurs when two wave systems have the same dominant frequency and opposite directions. These conditions give the largest values of $I(f_{sgw})$ and thus the strongest sources. This can happen when a moving storm overruns its previously generated ocean swell (class IIIa), or when ocean swell from a distant storm encounters waves from an unrelated storm (class IIIb). The event discussed by Obrebski et al. (2012) is a class IIIb source. Class IIIb events are rare for seismic frequencies below 0.1 Hz, because there are very few storms powerful enough to produce a significant amount of energy at frequencies below 0.05 Hz, and thus the likelihood that two such storms coexist is small.

These classes are only meant to facilitate the understanding of sources, and are naturally fuzzy. For example, a rapidly turning wind may be somewhere between class I and class IIIa.

3.3.5 Seismic Wavefield of Secondary Microseisms

The seismic displacement recorded at a station can be computed as the convolution between the Green's function and the source. In the case of secondary microseisms, the source is a distributed pressure field on the ocean surface. For a spatially homogeneous wave field, this source is equivalent to a localized force that oscillates in time. Therefore, one can consider a Green's function due to a vertical force at the ocean surface recorded at a distant station. From this source, P-waves propagate down to the ocean bottom, and they are partially transmitted to the crust as P- and S-waves and partially reflected back to the ocean surface as P-waves. At the ocean surface, P-waves are again reflected down to the ocean bottom. Surface waves are generated by the interference of multiply reflected P-waves and evanescent P- and S-waves below the seafloor. In the far field, a seismometer records all these seismic waves.

For each monochromatic forcing with wavenumber k and frequency f, the solution of the equation of motion can be expressed by elastic potentials. This is illustrated for modes associated with Rayleigh waves in Figure 3.9. In the water, the seismic propagation can be described by modes, which correspond to the superposition of oblique acoustic waves, with a take-off angle, relative to the vertical, that varies from about 30 degrees to 90 degrees that is horizontal (Figure 3.9a). The structure of these modes as a function of depth is shown in Figure 3.9b,c.

3.3.6 Distributed Pressure Field and Localized Force Equivalence

When all the ocean wave components are combined, we get a forcing with a broad (practically white) spectrum in the wavenumber space, the same as given by a point force. Hence the action of the pressure over an area dA and a frequency band df

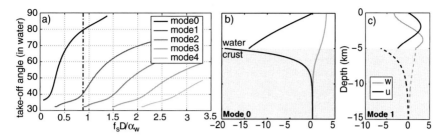

Figure 3.9. (a) Take-off angle of the two acoustic waves that combine to make acoustic modes in the water. The vertical profiles of the velocity amplitude for modes 0 and 1 and $f D/\alpha_w = 0.88$ are shown in b and c. For modes 1 and higher, the vertical velocity w is zero at the nodes of the vertical standing acoustic wave. The horizontal velocity u is also zero at the zero-pressure levels, in particular at the surface.

is equivalent to that of a point force that oscillates in time (e.g., Longuet-Higgins, 1950; Gualtieri et al., 2013). The rms amplitude of the vertical force is obtained by equating the force due to a uniform pressure acting on a square surface of size a and taking the limit of a small a (Gualtieri et al., 2013),

$$F_z(f, dA, df) = 2\pi \sqrt{F_p(f) dA df}. \tag{3.31}$$

where 2π allows for moving from the wavenumber domain to the spatial domain.

To illustrate the corresponding wavefield, we show the result of simulations performed by using the spectral-element method (similarly to the work by Gualtieri, 2014; Gualtieri et al., 2015). We consider a flat seafloor and a pressure source at the surface of the ocean, as shown in Figure 3.10. In order to evaluate the contribution of the different seismic phases separately, we take a Ricker source time function with a cut-off frequency of 1 Hz. At this frequency, interferences between P- and S-waves occur locally, then surface waves are generated only at given points for a given time and the different phases that compose the wavefield can be isolated. At lower frequencies, we are not able to distinguish these seismic phases anymore, since they interfere constructively, generating surface waves. Letters in Figure 3.10 refer to the seismic phases generated by this source. Several seismic phases can be identified: (a) the direct and (b) the reflected P-wave in the fluid, (c) the transmitted P-wave, (d) the P-to-S converted wave, (e) the refracted S-wave in the crust at the critical take-off angle $\theta_{P_c}^{S*} = \arcsin(\beta_c/\alpha_c)$, (f) the refracted P-wave in the fluid at the critical take-off angle $\theta_{P_w}^{P*} = \arcsin(\alpha_w/\alpha_c)$, and (g) the refracted P-wave in the fluid at the critical take-off angle $\theta_{P_w}^{S*} = \arcsin(\alpha_w/\beta_c)$. The refracted waves are also called head-waves. The kinetic energy associated with the direct P-wave (a) propagates in the ocean along an arc of a circle wavefront. In this snapshot, the downgoing P-wave (b) has been reflected only once at the seafloor. Such an arc

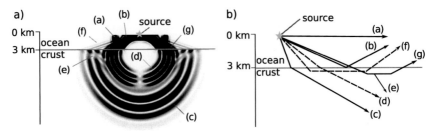

Figure 3.10. Seismic phases generated by a noise-like source (star) at the ocean surface in terms of (a) kinetic energy and (b) seismic rays (zoom in the ocean). Kinetic energy is computed by using a 2D spectral element method. Seismic phases are denoted by letters: (a) direct and (b) reflected P-wave in the fluid, (c) transmitted P-wave, (d) P-to-S converted wave, (e) refracted S-wave in the solid (take-off angle $\theta_{P_c}^{S*}$), (f) refracted P-wave in the fluid (take-off angle $\theta_{P_w}^{P*}$) and (g) refracted P-wave in the fluid (take-off angle $\theta_{P_w}^{S*}$). Take-off angles are defined in the text. Adapted from Gualtieri (2014).

wavefront is confined between the ocean surface and the seafloor and, as waves get further from the source, the wavefront becomes similar to a vertical segment (see Figure 3.11). Furthermore, the kinetic energy related to the P-to-S wavefield (d) is null for vertical transmissions since the S-wave transmission coefficient between a liquid and a solid interface is zero. Because of the free surface condition, a total reflection of P-waves occurs at the ocean surface for the seismic phases denoted as (b), (f), and (g) in Figure 3.10.

Figure 3.11 shows four snapshots of the seismic wavefield for varying time. While the seismic wavefield propagates away from the source, multiply reflected P-waves fill the ocean. The direct P-wave, denoted as (a) in Figure 3.10, can be recognized at $x \simeq 65$ km when $t = 10$ s, at $x \simeq 50$ km when $t = 20$ s, at $x \simeq 35$ km when $t = 30$ s, and at $x \simeq 20$ km when $t = 40$ s. The refracted P-wave in the water layer, denoted as (f) in Figure 3.10, and its reflection at the free surface are clearly visible, for example, between $x \simeq 50$ km and $x \simeq 60$ km when $t = 10$ s. A large part of kinetic energy remains confined within the ocean. The amount of kinetic energy associated with body waves transmitted across the seafloor (seismic phases (c), (d), (e) in Figure 3.10) decreases with increasing time, while quite strong finger-shaped patterns are confined just below the seafloor at some given points (e.g., $x = 60$ km when $t = 10$ s, $x = 20$ km for $t = 30$ s, and many others). These finger-shaped patterns represent the regions where P-waves and S-waves interfere and surface waves are generated by their constructive interference.

In the secondary microseism band ($T = 3 - 10$ s), the seismic phases detailed above interfere constructively and generate a strong energy packet, which

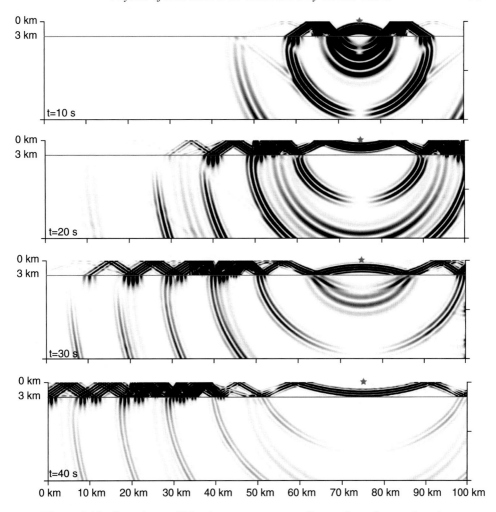

Figure 3.11. Snapshots of kinetic energy across a flat seafloor for varying time. The seismic wavefield is generated by a source at the surface of the ocean.

propagates away from the source with an apparent horizontal velocity of about 1 km/s (Gualtieri et al., 2015). This feature is associated with the Airy phase of Rayleigh wave traveling below the seafloor. The Airy phase corresponds to the minimum group velocity of Rayleigh waves, and it is associated to the maximum amplitude (Ewing et al., 1958). In an oceanic environment, the Airy phase lies in the secondary microseism frequency band (Gualtieri et al., 2015). Several studies have observed this slow phase on ocean bottom records having group velocity around 1 km/s (e.g., Press and Ewing, 1948; Ritzwoller and Levshin, 2002; Harmon et al., 2012; Tian and Ritzwoller, 2015).

Body Waves

The sum of multiply reflected P-waves in the ocean generates a resonance effect on the seismic wavefield below the seafloor in the source region, called "source site effect." In this paragraph, we deal with body waves. Using ray theory, Gualtieri et al. (2014) computed the P-wave source site effect C_P (dimensionless) as the sum of plane waves multiply reflected at the free surface and at the seafloor

$$C_P(\theta_{P_w}, D, \omega) = \frac{T_P(\theta_{P_w})}{1 + R(\theta_{P_w}) \, e^{i\Phi_w(D,\omega,\theta_{P_w})}}. \tag{3.32}$$

where θ_{P_w} is the take-off angle of a P-wave at the source, D is the water depth, ω is the angular seismic frequency, $R(\theta_{P_w})$ and $T_P(\theta_{P_w})$ are the P-wave reflection and transmission coefficients at the seafloor, respectively, and Φ_w is the phase shift due to the propagation within the water layer. A similar expression has been derived by Gualtieri et al. (2014) for S-waves. The source site effect strongly modulates the amplitude of the pressure source as a function of water depth, frequency, and epicentral distance. Farra et al. (2016) gave the theory for computing the seismic displacement power spectral density as the product of the pressure source, the source site effect, the propagation from the ocean bottom to the station, and the receiver site effect. The source site effect was also computed using the propagator matrix method by Ardhuin and Herbers (2013).

Figures 3.12a–c show maps of the source site effect on P-waves integrated over all take-off angles, as a function of bathymetry at seismic periods 4 and 8 s. We observe that at $T = 8$ s period, sources along the ridges are amplified, whereas at $T = 4$ s period, sources on both sides of the ridges are amplified.

Surface Waves

Longuet-Higgins (1950) and Hasselmann (1963) first gave the expressions for computing the spectral amplitude of secondary microseisms as the results of a pressure source over a given area close to the ocean surface. Gualtieri et al. (2013) further used normal modes on a spherical multi-layered Earth model (PREM, Dziewonski and Anderson, 1981), and gave the expressions for computing the spectral amplitude of secondary microseisms as the results of oscillating vertical point forces. Similar expressions were also given by Tanimoto (2013).

Here, we consider a simplified two-layer Earth model. Solutions for the spectrum of the vertical displacement $\delta(k_x, k_y, \omega)$ as a function of wavenumber and frequency at the top of the crust have a spectral amplitude, which is linearly related to the equivalent surface pressure spectrum $\widehat{p}_{2,\mathrm{surf}}(k_x, k_y, \omega)$ by,

$$\delta(k_x, k_y, \omega) = G(k_x, k_y, \omega)\widehat{p}_{2,\mathrm{surf}}(k_x, k_y, \omega). \tag{3.33}$$

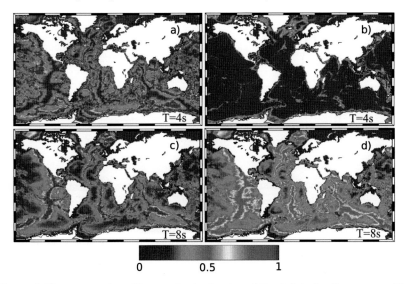

Figure 3.12. source site effect on body (a–c) and Rayleigh (b–d) waves at T = 4 s (top) and T = 8 s (bottom). The source site effect on body waves has been integrated over all take-off angles. The source site effect on Rayleigh waves refers to the fundamental mode. Adapted from Gualtieri (2014). (A black-and-white version of this figure appears in some formats. For the color version, please refer to the plate section.)

Singularities of G correspond to the singularities of G_p in the dark bands in Figure 3.2. These singularities fall in the domain of modes associated with Rayleigh waves for velocities between c_w and c_{cs}. Near the singularity of each mode j, we can write $G(k_x, k_y, \omega) \simeq G'(k_x, k_y, \omega_j)/(\omega^2 - \omega_j^2)$ where $G' = \partial G/\partial \omega$. For a broad spectrum, this form of singularity can be integrated as shown by Hasselmann (1962), giving a linearly growing energy as a function of time. Hasselmann (1963) expressed the source of secondary microseisms as the rate of increase of the variance of the ground displacement δ per unit of propagation distance

$$S_{DF}(\omega) = \frac{4\pi^2 f c_j^2}{\beta_c^5 \rho_c^2} F_{p,2}^{\mathrm{surf}}(k_x, k_y, \omega) \qquad (3.34)$$

where U_j is the group speed of the seismic mode j, and c_j is the source site effect (dimensionless) that depends on frequency, water depth, and seismic mode index j (as shown in Figure 3.13)

$$c_j^2 = \frac{\beta_c^5 \rho_c^2 k_j}{U_j^2 2\pi \omega} \frac{\pi}{2\omega^2} \left| G_j' \right|^2 . \qquad (3.35)$$

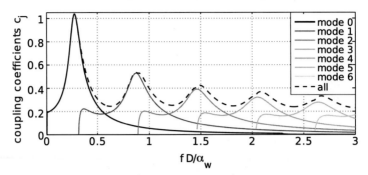

Figure 3.13. Dimensionless source site effect c_j for each Rayleigh wave mode j that amplifies the wave-induced pressure and accounts for compressibility of water. The maxima of c_j correspond approximately to quarter-wavelength resonance typical of organ pipes. The amplitudes of the peaks depend on the impedance ratio of the sea water and crust. Hence the peak amplitude increases with $\rho_c \beta_c / (\rho_w \alpha_c)$. For this figure, we used $\beta_c = 3000$ m s^{-1}.

Figures 3.12b–d show maps of the source site effect on the fundamental mode of Rayleigh waves as a function of bathymetry at seismic periods 4 and 8 s. We observe that, at $T = 4$ s period, very few sources are amplified, whereas at $T = 8$ s, sources on large areas close to the ridges are amplified. Gualtieri et al. (2013) showed that the source site effect on Rayleigh waves can be computed using normal modes as a product between the vertical eigenfunctions at the surface and at the bottom of the ocean divided by the eigenfrequency. Similar expressions in a 2D flat Earth can be also found in Gimbert and Tsai (2015).

A very rough simplification can be obtained by taking U and k independent of j and D (Longuet-Higgins, 1950). In this case, $S_{DF}(f)$ can be taken as a sum of all modes associated with Rayleigh waves in the form

$$S_{DF}(f) = \frac{4\pi^2 f}{\beta_c^5 \rho_c^2} \left(\sum_{i=0}^{\infty} c_j^2 \right) F_{p,2}^{\text{surf}}(\mathbf{k} \simeq 0, f). \quad (3.36)$$

In order to take into account all sources, the ocean can be discretized as a grid of point sources. Each source S_{DF} located at the colatitude-longitude grid point (ϕ_O', λ_O') close to the ocean surface generates Rayleigh waves that propagate along the Earth's surface up to the station. For a given station located at colatitude ϕ_O and longitude λ_O, the power spectral density of the vertical ground displacement is the integral over all grid elementary areas. Taking into account the geometrical spreading and the seismic attenuation Q along each source-receiver path (e.g., Kanamori and Given, 1981), the power spectral density of the vertical displacement is:

$$F_\delta(\lambda_O, \phi_O, f) = \int_{-\pi/2}^{\pi/2} \int_0^{2\pi} \frac{S_{DF}(f)}{R_E \sin \Delta} e^{\frac{-2\pi f \Delta R_E}{UQ}} (R_E^2 \sin \phi_O' d\lambda_O' d\phi_O') \quad (3.37)$$

with R_E the Earth radius, and U the seismic group velocity. The term $(R_E^2 \sin \phi_O' d\lambda_O' d\phi_O')$ is the area of an Earth surface element. The denominator $(R_E \sin \Delta)$ is the geometrical spreading factor on the spherical Earth (e.g., Kanamori and Given, 1981).

3.4 Numerical Modeling of Microseism Sources

After early evaluations by Hasselmann (1963) or Szelwis (1982) of microseism amplitude from simplified wave properties, Kedar et al. (2008) were the first to use a numerical ocean wave model otherwise used for marine meteorology applications. They clearly showed that this type of model could very well predict the space and time variability of class III sources. Ardhuin et al. (2011) included a crude representation of coastal reflection in their ocean wave model, with constant reflection coefficients, and showed that all classes of Rayleigh wave sources could be modeled accurately. Later refinements of the shoreline reflection by Ardhuin and Roland (2012) perform generally better but are limited by a lack of knowledge of shoreline slopes, which could be calibrated using microseismic data.

These features are included in our numerical wave model (The WAVEWATCH III ® Development Group, 2016). Rascle and Ardhuin (2013) have published modeled sources, using global and regional grids, that are updated on a regular basis (ftp://ftp.ifremer.fr/ifremer/ww3/HINDCAST, last accessed 04/04/2018). This database has been used in many studies on surface waves (e.g., Stutzmann et al., 2012; Sergeant et al., 2013; Gualtieri et al., 2013) and in some studies on body waves (e.g., Farra et al., 2016; Nishida and Takagi, 2016; Retailleau et al., 2018). A recent development of this model was the extension into the frequency domain of the infragravity waves, from 0.003 Hz to 0.03 Hz (Ardhuin et al., 2015). This uses an empirical parametrization of the flux of energy from the wind-sea and ocean swell to the infragravity waves that happens at the shoreline (e.g., Ardhuin et al., 2014). So far, the only validation data for these long-period waves is coming from bottom pressure recorders, deployed for tsunami monitoring or as part of marine geoscience experiments (Rawat et al., 2014). The results of wave models in that frequency range should thus be considered with caution.

In the following, we present a few applications using microseisms recorded at single stations. Here we use the method proposed by Hasselmann (1963) in which the energy of seismic waves is computed by summing sources over their propagation path. Alternatively the seismic noise sources can be represented by an equivalent grid of discrete point sources (e.g., Longuet-Higgins, 1950; Gualtieri et al., 2013), and the recorded seismic ambient noise can be seen as the result of the constructive and destructive interference of the wavefields generated by these point sources.

3.4.1 Strong Double-Frequency Sources in the Pacific

Before numerical models were available, correlations and pattern analysis were widely used to evaluate the location of microseism sources (e.g., Friedrich et al., 1998). For example, Bromirski et al. (2005) analyzed in detail mid-ocean sources by combining wave buoy measurements of $E(f)$ and vertical seismometer components on land and at the ocean bottom.

These analyses are often qualitative and sometime may lead to misinterpretation. One example is the discussion of measured microseisms on March 10 and 11, 1993, by Bromirski et al. (2005, their Figure 10). The presence of a primary microseism signal in Figure 3.14a between 0.05 and 0.1 Hz is an evidence of ocean waves interacting with the seafloor. One may conclude that the simultaneous stronger secondary microseism is associated to coastal reflection. On the contrary, our simulations in Figure 3.14b, without any coastal reflection, reproduce very well the double-frequency signal recorded at the Hawaii station of Kipapa (KIP). In that case, coastal reflection only adds a few percent of seismic energy (Figure 3.14c). A detailed analysis of the directional spectrum shows that this event is due to the interaction, around Hawaii, of two remote swells from north and south. These two swells give a source distribution that is maximum to the east and west of the islands (Figure 3.14d).

3.4.2 Ocean Storms, Hum, and Primary Microseisms

In the frequency band of the seismic hum, around 0.01 Hz, we show in Figure 3.14e the modeling of three months of data in 2008 recorded at the Geoscope station SSB. We use the source power spectral density computed by using equation (3.16) for infragravity waves and a seafloor slope $s = 0.06$, as in Ardhuin et al. (2015). That effective slope is justified, for that frequency, by the calculation of bottom pressure modulations over several two-dimensional bottom profiles. The seismic sources are summed along great circles using equations (3.36) and (3.37), where $F_p = F_{p,1}$ as in equation (3.16). In absence of earthquakes, the variability and magnitude of the ground motion is generally well reproduced by the model.

In the primary frequency band, between 0.05 and 0.1 Hz, Hasselmann (1963) did a first evaluation of the seismic response. In order to get amplitudes comparable to measurements, he had to include a factor 3 amplification presumably attributed to wave breaking, together with an *ad hoc* broadening of the directional spectrum

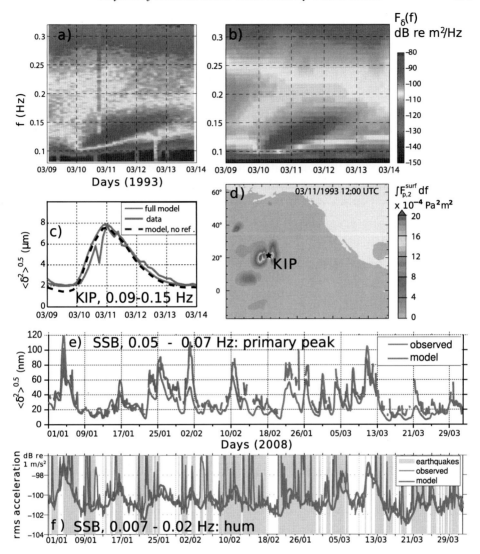

Figure 3.14. (a) Recorded microseismic spectrum (vertical component) at the Geoscope-GSN station KIP, Hawaii, on March 10 and 11, 1993. (b) Modeled spectrum with the secondary mechanism. (c) Time series of measured and modeled root mean square vertical ground displacement (0.09 to 0.15 Hz). (d) Snapshot of equivalent surface pressure due to the secondary mechanism, integrated in the frequency band 0.09 to 0.15 Hz, for the model run without shoreline reflection. The black star shows the location of station KIP. (e) Measured and modeled noise amplitude in the hum band at Geoscope station SSB, with an equivalent seafloor slope of 6% (as in Ardhuin et al., 2015). (f) Same as (e), for the primary microseism frequency band, using an equivalent seafloor slope of 7%. (A black-and-white version of this figure appears in some formats. For the color version, please refer to the plate section.)

to represent three-dimensional seafloor topography effects, and an effective beach slope $s = 0.01$. A more realistic representation of wave breaking is obtained by limiting the wave height to half of the water depth (e.g., Dean and Dalrymple, 1991). This tends to reduce the pressure modulation instead of increasing it. Also, the depth at which sources are significant is not right in the surfzone, but rather corresponds to depths such that $0.3 < kD < 1$, where the large-scale bottom slope is much smaller than 0.01. Our own evaluation at the French station SSB of the Geoscope network shows that the use of equation (3.16) requires an effective slope $s \simeq 0.07$ to reproduce the amplitude of recorded microseisms. This effective slope is much larger than large-scale seafloor topography.

The very large slopes needed in equation (3.16) suggests that the mechanism proposed by Hasselmann (1963) for waves over a constant slope is not sufficient to explain the magnitude of observed primary microseisms. In addition, the topographic features at the scale of ocean waves should probably be included, following Saito (2010). These can be represented by the interactions of the ocean wave spectrum with the bottom topography spectrum, as given by equations (3.15) and (3.14). Further work is needed to test the alternative model given by equations (3.15) and (3.16) for both Love and Rayleigh wave generation in the primary microseism band.

3.5 Summary and Conclusions

A general foundation of the theory of microseism generation by ocean surface gravity waves was given by Hasselmann (1963), relying on previous results by Longuet-Higgins (1950) with further additions in Saito (2010) for Love wave generation and Ardhuin and Herbers (2013) for the effect of shallow water. All these theories have been integrated here in a common framework. Although the first two papers contain lengthy derivations, the general principle is well laid out in the introduction by Hasselmann (1963). The most salient ideas are that

- ocean waves are random and spatially distributed, and best described by the slowly evolving spectrum of the associated ocean surface elevation;
- in a horizontally homogeneous medium that is linear, the response to a pressure oscillating at a given wavenumber and frequency is at the same wavenumber and frequency;
- the ocean gravity wave motion is equivalent to distributed fields of pressure and horizontal stress, acting on a flat seafloor (primary mechanism) and on a flat sea surface (secondary mechanism);
- the primary generation mechanism that explains Rayleigh waves in the hum frequency band consists in the interference of ocean infragravity waves having wavenumber $\mathbf{k}_{sgw}$ with the seafloor, giving seismic waves having a long

wavelength component with wavenumber $\mathbf{k}$ and the same frequency as the ocean waves;

- this same mechanism may explain Rayleigh waves in the primary microseism frequency band;
- the primary generation mechanism with bottom topography of wavenumber $\mathbf{k}_s$ may give both Rayleigh and Love waves with wavenumber $\mathbf{k} = \mathbf{k}_{sgw} + \mathbf{k}_s$; a detailed analysis using high-resolution maps of seafloor topography will be needed to prove this;
- a secondary mechanism that explains Rayleigh and body waves in the secondary microseism peak (usually around periods of 5 s) is the interference of pairs of ocean wave trains with wavenumber $\mathbf{k}_{sgw}$ and $\mathbf{k}'_{sgw} \simeq -\mathbf{k}_{sgw}$ and frequency f_{sgw} giving seismic waves at frequency $2 f_{sgw}$ and $\mathbf{k} = \mathbf{k}_{sgw} + \mathbf{k}'_{sgw}$.

For all these seismic phases, we now have a theoretical framework and a quantitative model based on numerical ocean wave models. These ocean wave models are driven by fields of surface winds evolving in space and time, and provide estimates of the ocean wave spectrum at each node of a spatial grid with resolutions that can range from a few hundred meters at regional scale to 50 km at global scale, evolving over typical time scales of a few hours. These ocean wave models have been developed originally for safety at sea, predicting wave heights. They can be very inaccurate for other parameters such as the subtle properties needed to produce strong seismic noise sources. The recent addition of shoreline reflection, extension to frequencies below 0.03 Hz, and validity above 0.4 Hz are all the topic of ongoing research, driven by the specific need to better understand microseism sources. Other areas of improvement in the forward modeling of microseism, is the use of 3D Earth models for seismic propagation. The comparison of model outcomes and observations will certainly bring further refinements in our understanding of both ocean wave properties and seismic propagation effects.

Acknowledgments

The authors would like to thank Yann Capdeville, Göran Ekström, Véronique Farra, Anne Mangeney, and Frederik J. Simons for fruitful discussions and long-lasting collaborations on ambient seismic noise. The authors are also grateful to the editor Andreas Fichtner and the two anonymous reviewers for providing comments that helped improve this book chapter. F. A. and E. S. acknowledge support from Agence Nationale de la Recherche, via grant ANR-14-CE01-0012 "MIMOSA." L. G. would like to acknowledge support during the preparation of this book chapter from Lamont-Doherty Earth Observatory of Columbia University, King Abdullah University of Science and Technology, and Princeton University.

References

Airy, G. B. 1841. Tides and waves. In *Encyclopedia Metropolitana London*, pp. 241–394.

Aki, K., and Richards, P. G. 2002. *Quantitative seismology*, 2nd edn. Sausalito, California: University Science Books

Ardhuin, F., and Herbers, T. H. C. 2013. Noise generation in the solid Earth, oceans and atmosphere, from nonlinear interacting surface gravity waves in finite depth. *J. Fluid Mech.*, **716**, 316–348.

Ardhuin, F., Balanche, A., Stutzmann, E., and Obrebski, M. 2012. From seismic noise to ocean wave parameters: general methods and validation. *J. Geophys. Res.*, **117**, C05002.

Ardhuin, F., Gualtieri, L., and Stutzmann, E. 2015. How ocean waves rock the Earth: two mechanisms explain seismic noise with periods 3 to 300 s. *Geophys. Res. Lett.*, **42**, 765–772.

Ardhuin, F., Rascle, N., Chapron, B., Gula, J., Molemaker, J., Gille, S. T., Menemenlis, D., and Rocha, C. 2017. Small scale currents have large effects on wind wave heights. *J. Geophys. Res.*, **122**(C6), 4500–4517.

Ardhuin, F., Rawat, A., and Aucan, J. 2014. A numerical model for free infragravity waves: Definition and validation at regional and global scales. *Ocean Modeling*, **77**, 20–32.

Ardhuin, F., and Roland, A. 2012. Coastal wave reflection, directional spreading, and seismo-acoustic noise sources. *J. Geophys. Res.*, **117**, C00J20.

Ardhuin, F., Rogers, E., Babanin, A., Filipot, J.-F., Magne, R., Roland, A., van der Westhuysen, A., Queffeulou, P., Lefevre, J.-M., Aouf, L., and Collard, F. 2010. Semi-empirical dissipation source functions for wind-wave models: part I, definition, calibration and validation. *J. Phys. Oceanogr.*, **40**(9), 1917–1941.

Ardhuin, F., Stutzmann, E., Schimmel, M., and Mangeney, A. 2011. Ocean wave sources of seismic noise. *J. Geophys. Res.*, **116**, C09004.

Ardhuin, F., Sutherland, P., Doble, M., and Wadhams, P. 2016. Ocean waves across the Arctic: attenuation due to dissipation dominates over scattering for periods longer than 19 s. *Geophys. Res. Lett.*, **43**, 5775–5783.

Bernard, P. 1941. Sur certaines proprietes de la houle etudiées a l'aide des enregistrements seismographiques. *Bull. Inst. Oceanogr. Monaco*, **800**, 1–19.

Bertin, X., de Bakker, A., van Dongeren, A., Coco, G., André, G., Ardhuin, F., Bonneton, P., Bouchette, F., Castelle, B., Crawford, W., Deen, M., Dodet, G., Guerin, T., Leckler, F., McCall, R., Muller, H., Olabarrieta, M., Ruessink, G., Sous, D., Stutzmann, E., and Tissier, M. 2018. Infragravity waves: from driving mechanisms to impacts. *Earth-Science Rev.*, **177**, 774–799.

Boccotti, P. 2000. *Wave mechanics for ocean engineering*. Amsterdam: Elsevier.

Bromirski, P. D., Duennebier, F. K., and Stephen, R. A. 2005. Mid-ocean microseisms. *Geochemistry Geophysics Geosystems*, **6**(4).

Cooper, R. I. B., and Longuet-Higgins, M. S. 1951. An experimental study of the pressure variations in standing water waves. *Proc. Roy. Soc. Lond. A*, **206**, 426–435.

Crawford, W. C., Webb, S. C., and Hildebrand, J. A. 1991. Seafloor Compliance Observed by Long-Period Pressure and Displacement Measurements. *J. Geophys. Res.*, **103**(B5), 9895–9916.

Dean, R. G., and Dalrymple, R. A. 1991. *Water wave mechanics for engineers and scientists*. 2nd edn. Singapore: World Scientific. 353 pp.

Delpey, M., Ardhuin, F., Collard, F., and Chapron, B. 2010. Space-time structure of long swell systems. *J. Geophys. Res.*, **115**, C12037.

Duennebier, F. K., Lukas, R., Nosal, E.-M., Aucan, J., and Weller, R. A. 2012. Wind, Waves, and Acoustic Background Levels at Station ALOHA. *J. Geophys. Res.*, **117**, C03017.

Dziewonski, A. M., and Anderson, D. L. 1981. Preliminary reference Earth model. *Phys. Earth and Planet. Int.*, **25**(4), 297–356.

Ewing, W. M., Jardetzky, W. S., and Press, F. 1958. Elastic waves in layered media. *GFF*, **80**(1), 128–129.

Farra, V., Stutzmann, E., Gualtieri, L., Schimmel, M., and Ardhuin, F. 2016. Ray-theoretical modeling of secondary microseism P-waves. *Geophys. J. Int.*, **206**(3), 1730–1739.

Farrell, W. E., and Munk, W. 2010. Booms and busts in the deep. *J. Phys. Oceanogr.*, **40**(9), 2159–2169.

Fichtner, A., and Tsai, V. 2018. Theoretical foundations of noise interferometry. Pages – of: Nakata, N., Gualtieri, L., and Fichtner, A. (eds.), *Seismic Ambient Noise*. Cambridge University Press, Cambridge, UK.

Friedrich, A., Krager, F., and Klinge, K. 1998. Ocean-generated microseismic noise located with the Grafenberg array. *Journal of Seismology*, **2**, 47–64.

Fukao, Y., Nishida, K., and Kobayashi, N. 2010. Seafloor topography, ocean infragravity waves, and background Love and Rayleigh waves. *J. Geophys. Res.*, **115**(10), B04302.

Fukao, Y., Nishida, K., Suda, N., Nawa, K., and Kobayashi, N. 2002. A theory of the Earth's background free oscillations. *Journal of Geophysical Research: Solid Earth*, **107**(B9).

Gelci, R., Cazalé, H., and Vassal, J. 1957. Prévision de la houle. La méthode des densités spectroangulaires. *Bulletin d'information du Comité d'Océanographie et d'Etude des Côtes*, **9**, 416–435.

Gilbert, F., and Backus, G. E. 1966. Propagator matrices in elastic wave and vibration problems. *Geophysics*, **31**(2), 326–332.

Gimbert, F., and Tsai, V. C. 2015. Predicting short-period, wind-wave-generated seismic noise in coastal regions. *Earth and Planetary Science Letters*, **426**, 280–292.

Grevemeyer, I., Herber, R., and Essen, H.-H. 2000. Microseismological evidence for a changing wave climate in the northeast Atlantic Ocean. *Nature*, **408**, 349–351.

Gualtieri, L. 2014. *Modeling the secondary microseismic noise generation and propagation*. Ph.D. thesis, Institut de Physique du Globe de Paris and Università di Bologna.

Gualtieri, L., Camargo, J. S., Pascale, S., Pons, F. M. E., and Ekström, G. 2018. The persistent signature of tropical cyclones in seismic ambient noise. *Earth and Planetary Science Letters*, **484**, 287–294.

Gualtieri, L., Stutzmann, E., Capdeville, Y., Ardhuin, F., Schimmel, M., Mangeney, A., and Morelli, A. 2013. Modeling secondary microseismic noise by normal mode summation. *Geophys. J. Int.*, **193**, 1732–1745.

Gualtieri, L., Stutzmann, E., Capdeville, Y., Farra, V., Mangeney, A., and Morelli, A. 2015. On the shaping factors of the secondary microseismic wavefield. *Journal of Geophysical Research: Solid Earth*, **120**(9), 6241–6262.

Gualtieri, L., Stutzmann, E., Farra, V., Capdeville, Y., Schimmel, M., Ardhuin, F., and Morelli, A. 2014. Modeling the ocean site effect on seismic noise body waves. *Geophys. J. Int.*, **193**, 1096–1106.

Hanafin, J., Quilfen, Y., Ardhuin, F., Sienkiewicz, J., Queffeulou, P., Obrebski, M., Chapron, B., Reul, N., Collard, F., Corman, D., de Azevedo, E. B., Vandemark, D., and Stutzmann, E. 2012. Phenomenal sea states and swell radiation: a comprehensive analysis of the 12-16 February 2011 North Atlantic storms. *Bull. Am. Meteorol. Soc.*, **93**, 1825–1832.

Harmon, N., Henstock, T., Tilmann, F., Rietbrock, A., and Barton, P. 2012. Shear velocity structure across the Sumatran Forearc-Arc. *Geophysical Journal International*, **189**(3), 1306–1314.

Hasselmann, K. 1962. On the non-linear energy transfer in a gravity wave spectrum, part 1: general theory. *J. Fluid Mech.*, **12**, 481–501.

———. 1963. A statistical analysis of the generation of microseisms. *Rev. of Geophys.*, **1**(2), 177–210.

———. 1966. Feynman diagrams and interaction rules of wave-wave scattering processes. *Rev. of Geophys.*, **4**(1), 1–32.

Hasselmann, K., Barnett, T. P., Bouws, E., Carlson, H., Cartwright, D. E., Enke, K., Ewing, J. A., Gienapp, H., Hasselmann, D. E., Kruseman, P., Meerburg, A., Müller, P., Olbers, D. J., Richter, K., Sell, W., and Walden, H. 1973. Measurements of wind-wave growth and swell decay during the Joint North Sea Wave Project. *Deut. Hydrogr. Z.*, **8**(12), 1–95. Suppl. A.

Herbers, T. H. C., Lowe, R. L., and Guza, R. T. 1992. Field observations of orbital velocities and pressure in weakly nonlinear surface gravity waves. *J. Fluid Mech.*, **245**, 413–435.

Janssen, P. 2004. *The interaction of ocean waves and wind.* Cambridge: Cambridge University Press.

Juretzek, C., and Hadziioannou, C. 2016. Where do ocean microseisms come from? A study of Love-to-Rayleigh wave ratios. *J. Geophys. Res.*, **121**, 6741–6756.

Kanamori, H., and Given, J. W. 1981. Use of long-period surface waves for rapid determination of earthquake-source parameters. *Phys. Earth and Planet. Int.*, **27**(1), 8–31.

Kedar, S., Longuet-Higgins, M., Graham, F. W. N., Clayton, R., and Jones, C. 2008. The origin of deep ocean microseisms in the North Atlantic Ocean. *Proc. Roy. Soc. Lond. A*, 1–35.

Kurrle, D., and Widmer-Schnidrig, R. 2008. The horizontal hum of the Earth: A global background of spheroidal and toroidal modes. *Geophys. Res. Lett.*, **35**(2), L06304.

Lamb, H. 1932. *Hydrodynamics.* 6th edn. Cambridge, England: Cambridge University Press.

Leckler, F., Ardhuin, F., Peureux, C., Benetazzo, A., Bergamasco, F., and Dulov, V. 2015. Analysis and interpretation of frequency-wavenumber spectra of young wind waves. *J. Phys. Oceanogr.*, **45**, 2484–2496.

Longuet-Higgins, M. S. 1950. A theory of the origin of microseisms. *Phil. Trans. Roy. Soc. London A*, **243**, 1–35.

Longuet-Higgins, M. S., and Stewart, R. W. 1962. Radiation stresses and mass transport in surface gravity waves with application to "surf beats." *J. Fluid Mech.*, **13**, 481–504.

Longuet-Higgins, M. S., and Ursell, F. 1948. Sea waves and microseisms. *Nature*, **162**, 700.

Mei, C. C. 1989. *Applied dynamics of ocean surface waves.* 2nd edn. Singapore: World Scientific. 740 p.

Meschede, M., Stutzmann, E., Farra, V., Schimmel, M., and Ardhuin, F. 2017. The effect of water-column resonance on the spectra of secondary microseism P-waves. *J. Geophys. Res.*, **122**(10), 8121–8142.

Miche, A. 1944. Mouvements ondulatoires de la mer en profondeur croissante ou décroissante. Première partie. Mouvements ondulatoires périodiques et cylindriques en profondeur constante. *Annales des Ponts et Chaussées*, **114**, 42–78.

Neale, J., Harmon, N., and Srokosz, M. 2017. Monitoring remote ocean waves using P-wave microseisms. *J. Geophys. Res.*, **122**, 470–483.

Nishida, K., and Takagi, R. 2016. Teleseismic S wave microseisms. *Science*, **353**, 919–921.

Obrebski, M., Ardhuin, F., Stutzmann, E., and Schimmel, M. 2012. How moderate sea states can generate loud seismic noise in the deep ocean. *Geophys. Res. Lett.*, **39**, L11601.

————. 2013. Detection of microseismic compressional (P) body waves aided by numerical modeling of oceanic noise sources. *J. Geophys. Res.*, **118**, 4312–4324.

Peureux, C., and Ardhuin, F. 2016. Ocean bottom pressure records from the Cascadia array and short surface gravity waves. *J. Geophys. Res.*, **121**, 2862–2873.

Peureux, C., Benetazzo, A., and Ardhuin, F. 2018. Note on the directional properties of meter-scale gravity waves. *Ocean Science*, **14**, 41–52.

Pierson, Jr, W. J., and Moskowitz, L. 1964. A proposed spectral form for fully developed wind seas based on the similarity theory of S. A. Kitaigorodskii. *J. Geophys. Res.*, **69**(24), 5,181–5,190.

Press, F., and Ewing, M. 1948. A theory of microseisms with geologic applications. *Eos, Transactions American Geophysical Union*, **29**(2), 163–174.

Rascle, N., and Ardhuin, F. 2013. A global wave parameter database for geophysical applications. Part 2: model validation with improved source term parameterization. *Ocean Modeling*, **70**, 174–188. Data is available at ftp://ftp.ifremer.fr/ifremer/ww3/HINDCAST.

Rawat, A., Ardhuin, F., Ballu, V., Crawford, W., Corela, C., and Aucan, J. 2014. Infra-gravity waves across the oceans. *Geophys. Res. Lett.*, **41**, 7957–7963.

Retailleau, L., Landès, M., Gualtieri, L., Shapiro, N. M., Campillo, M., Roux, P., and Guilbert, J. 2018. Detection and analysis of a transient energy burst with beamforming of multiple teleseismic phases. *Geophysical Journal International*, **212**(1), 14–24.

Ritzwoller, M. H., and Feng, L. 2018. Overview of pre- and post-processing of ambient noise correlations. Pages – of: Nakata, N., Gualtieri, L., and Fichtner, A. (eds.), *Seismic Ambient Noise*. Cambridge University Press, Cambridge, UK.

Ritzwoller, M. H., and Levshin, A. L. 2002. Estimating shallow shear velocities with marine multicomponent seismic data. *Geophysics*, **67**(6), 1991–2004.

Romero, L., Melville, W. K., and Kleiss, J. M. 2012. Spectral Energy Dissipation due to Surface Wave Breaking. *J. Phys. Oceanogr.*, **42**(9), 1421–1444.

Saito, T. 2010. Love-wave excitation due to the interaction between a propagating ocean wave and the sea-bottom topography. *Geophys. J. Int.*, **182**, 1515–1523.

Sergeant, A., Stutzmann, E., Maggi, A., Schimmel, M., Ardhuin, F., and Obrebski, M. 2013. Frequency-dependent noise sources in the North Atlantic Ocean. *Geochemistry Geophysics Geosystems*, **14**(12), 5341–5353.

Shapiro, N. 2018. Applications with surface waves extracted from ambient seismic noise. Pages — of: Nakata, N., Gualtieri, L., and Fichtner, A. (eds.), *Seismic Ambient Noise*. Cambridge University Press, Cambridge, UK.

Shapiro, N. M., Campillo, M., Stehly, L., and Ritzwoller, M. H. 2005. High-resolution surface-wave tomography from ambient seismic noise. *Science*, **307**, 1615–1617.

Snieder, R. 2004. Extracting the Green's function from the correlation of coda waves: A derivation based on stationary phase. *Physical Review E*, **69**(4), 046610.

Snodgrass, F. E., Groves, G. W., Hasselmann, K., Miller, G. R., Munk, W. H., and Powers, W. H. 1966. Propagation of ocean swell across the Pacific. *Phil. Trans. Roy. Soc. London*, **A249**, 431–497.

Squire, V., Dugan, J., Wadhams, P., Rottier, P., and Liu, A. 1995. Of ocean waves and sea ice. *Annu. Rev. Fluid Mech.*, **27**(3), 115–168.

Stutzmann, E., Schimmel, M., and Ardhuin, F. 2012. Modeling long-term seismic noise in various environments. *Geophys. J. Int.*, **191**, 707–722.

Szelwis, R. 1982. Modeling of microseismic surface wave source. *J. Geophys. Res.*, **87**, 6906–6918.

Tanimoto, T. 2005. The oceanic excitation hypothesis for the continuous oscillations of the Earth. *Geophys. J. Int.*, **160**, 276–288.

———. 2013. Excitation of microseisms: views from the normal-mode approach. *Geophysical Journal International*, ggt185.

The WAVEWATCH III ® Development Group. 2016. *User manual and system documentation of WAVEWATCH III ® version 5.16*. Tech. Note 329. NOAA/NWS/NCEP/MMAB, College Park, MD, USA. 326 pp. + Appendices.

Tian, Y., and Ritzwoller, M. H. 2015. Directionality of ambient noise on the Juan de Fuca plate: Implications for source locations of the primary and secondary microseisms. *Geophysical Journal International*, **201**(1), 429–443.

Traer, J., and Gerstoft, P. 2014. A unified theory of microseisms and hum. *J. Geophys. Res.*, **119**(4), 3317–3339.

Tromp, J., Luo, Y., Hanasoge, S., and Peter, D. 2010. Noise cross-correlation sensitivity kernels. *Geophysical Journal International*, **183**(2), 791–819.

Uchiyama, Y., and McWilliams, J. C. 2008. Infragravity waves in the deep ocean: Generation, propagation, and seismic hum excitation. *J. Geophys. Res.*, **113**, C07029.

Vinnik, L. P. 1973. Sources of microseismic P waves. *Pure Appl. Geophys.*, **103**(1), 282–289.

von Storch, H., and Olbers, D. 2007. *Interview with Klaus Hasselmann*. Tech. rept. 2007/5. GKSS, Geesthacht, Germany.

WAMDI Group. 1988. The WAM Model - A third generation ocean wave prediction model. *J. Phys. Oceanogr.*, **18**, 1775–1810.

Webb, S. C. 2007. The Earth's "hum" is driven by ocean waves over the continental shelves. *Nature*, **445**, 754–756.

———. 2008. The Earth's "hum": the excitation of Earth normal modes by ocean waves. *Geophys. J. Int.*, **174**, 542–566.

Wiechert, E. 1904. Discussion, Verhandlung der zweiten Internationalen Seismologischen Konferenz. *Beitrage zur Geophysik*, **2**, 41–43.

4

Theoretical Foundations of Noise Interferometry

ANDREAS FICHTNER AND VICTOR C. TSAI

Abstract

The retrieval of a deterministic signal from recordings of a quasi-random ambient seismic field is the central goal of noise interferometry. It is the foundation of numerous applications ranging from noise source imaging to seismic tomography and time-lapse monitoring. In this chapter, we offer a presentation of theoretical approaches to noise interferometry, complemented by a critical discussion of their respective advantages and drawbacks.

The focus of this chapter is on interstation noise correlations that approximate the Green's function between two receivers. We explain in detail the most common mathematical models for Green's function retrieval by correlation, including normal-mode summation, plane-wave decomposition, and representation theorems. While the simplicity of this concept is largely responsible for its remarkable success, each of these approaches rests on different but related assumptions such as wavefield equipartitioning or a homogeneous distribution of noise sources. Failure to meet these conditions on Earth may lead to biases in traveltimes, amplitudes, or waveforms in general, thereby limiting the accuracy of the method.

In contrast to this well-established method, interferometry without Green's functional retrieval does not suffer from restrictive conditions on wavefield equipartitioning. The basic concept is to model the interstation correlation directly for a given power-spectral density distribution of noise sources and for a suitable model of the Earth that may be attenuating, heterogeneous, and anisotropic. This approach leads to a coupled problem where both structure and sources affect data, much like in earthquake tomography. Observable variations of the correlation function are linked to variations in Earth structure and noise sources via finite-frequency sensitivity kernels that can be used to solve inverse problems. While being mathematically and computationally more complex, interferometry without Green's function retrieval has produced promising initial results that make successful future applications likely.

We conclude this chapter with a summary of alternative approaches to noise interferometry, including interferometry by deconvolution, multi-dimensional deconvolution, and iterated correlation of coda waves.

4.1 Introduction

Noise is unwanted, unpleasant, or loud sound, according to the *Cambridge English Dictionary*. This unflattering description carries a historical connotation of useless-ness that hardly reflects the wide range of present-day applications that exploit the omnipresent noise field of the Earth. Seismic noise on Earth is not random *sensu stricto*. Its propagation is governed by the laws of physics that imprint structure on the wavefield. This structure can be exploited using interferometry that "turns noise into signal" (Curtis et al., 2006).

While the pioneering work of Aki (1957) suggested early on that statistical prop-erties of seismic noise may be used to infer Earth structure, practical efforts to extract interpretable information were often found to be more challenging than sug-gested by seemingly simple theories (e.g., Claerbout, 1968; Cole, 1995). Although seismic interferometry requires only two seismic stations, or a single station for autocorrelations, the ambient wavefield only became widely used in seismic tomography when large, dense arrays started to offer unprecedented illumination of the Earth's crust (e.g., Sabra et al., 2005; Shapiro et al., 2005).

Today, the large majority of ambient noise tomography results are ostensibly based on Green's function retrieval, that is, the assumption that the Green's func-tion between two stations is approximated by the time-averaged cross-correlation of noise. By treating correlation-based Green's function approximations as conven-tional earthquake or active-source signals, already established tomographic meth-ods can be used to estimate subsurface properties. Green's function retrieval can be observed in laboratory experiments (e.g., Malcolm et al., 2004; van Wijk, 2006; Nooghabi et al., 2017), and it can be justified theoretically using normal-mode, plane-wave, representation-theorem, or purely numerical approaches (e.g., Lobkis and Weaver, 2001; Snieder, 2004a; Wapenaar and Fokkema, 2006; Cupillard, 2008; Tsai, 2009; Cupillard and Capdeville, 2010; Boschi et al., 2013).

During the past decade, interferometry based on interstation correlations of ambient noise has become a standard tool. In seismic exploration, interferome-try is an inexpensive alternative to active-source imaging and monitoring (e.g., Bussat and Kugler, 2011; Mordret et al., 2013, 2014; de Ridder et al., 2014; de Ridder and Biondi, 2015; Nakata et al., 2015; Delaney et al., 2017). Improved coverage in regions that are less well illuminated by earthquakes has advanced regional crustal studies (e.g., Sabra et al., 2005; Shapiro et al., 2005; Stehly et al., 2009; Shapiro, 2018), continental- and global-scale tomography (e.g., Lin et al.,

2008; Zheng et al., 2011; Verbeke et al., 2012; Saygin and Kennett, 2012; Nishida and Montagner, 2009; Kao et al., 2013; Haned et al., 2016), and the imaging of deep internal discontinuities (e.g., Poli et al., 2012; Boué et al., 2013; Poli et al., 2015) and the inner core (e.g., Lin et al., 2013; Huang et al., 2016). Furthermore, the omnipresence of ambient noise has enabled the long-term investigation of time-dependent subsurface structures along active fault zones, beneath volcanoes, and within geothermal reservoirs (e.g., Brenguier et al., 2008a,b; Obermann et al., 2013, 2014, 2015; Hillers et al., 2015; Sens-Schönfelder and Brenguier, 2018; Snieder et al., 2018).

In general, theories for Green's function retrieval rely on the assumption that the ambient wavefield is equipartitioned, meaning that all propagation modes are equally strong and statistically uncorrelated. Equipartitioning may arise either directly through the action of uncorrelated and homogeneously distributed noise sources, or indirectly through sufficiently strong multiple scattering.

In the Earth, wavefields are not generally equipartitioned for various reasons. Scattering may be too weak and attenuation too strong to produce significant multiple scattering; and the distribution of noise sources is strongly heterogeneous, and time-variable (see Chapters 1–3 [McNamara et al., 2018; Gal and Reading, 2018; Ardhuin et al., 2018]). While the frequency-dependent arrival times of fundamental-mode surface waves are empirically found to be rather robust, other components of the wavefield suffer from not meeting the requirements for Green's function retrieval. Well-documented problems include travel time and amplitude errors, incorrect higher-mode surface waves, the presence of spurious arrivals, and the weakness or complete absence of body waves (e.g., Halliday and Curtis, 2008; Tsai, 2009; Kimman and Trampert, 2010; Froment et al., 2010; Forghani and Snieder, 2010; Fichtner, 2014; Kästle et al., 2016). This has, so far, prevented the application of finite-frequency and full-waveform inversion techniques that exploit complete waveforms for the benefit of improved resolution (e.g., Igel et al., 1996; Pratt, 1999; Friederich, 2003; Yoshizawa and Kennett, 2004; Zhou et al., 2006; Fichtner et al., 2009; Tape et al., 2010).

Concepts to circumvent these drawbacks and to establish a variant of interferometry without Green's function retrieval originated in helioseismology, though wavefields in the Sun are far more equipartitioned than in the Earth (Woodard, 1997; Gizon and Birch, 2002; Hanasoge et al., 2011). Interferometry without Green's function retrieval makes no assumptions on the distribution of noise sources or the properties of the wavefield. Instead, interstation correlations can be computed for any distribution of noise sources and any type of medium. Information on both the noise sources and structure of the Earth may then be extracted through the comparison with observed correlations and the computation of sensitivity kernels (e.g., Tromp et al., 2010; Hanasoge, 2013; Hanasoge and Branicki, 2013; Fichtner et al., 2017a; Sager et al., 2018).

In the following sections, we summarize the most prevalent theories for Green's function retrieval by correlation, which are based on normal-mode summation, plane-wave decomposition, and representation theorems, respectively. Subsequently, we outline the basic concepts of interferometry without Green's function retrieval. Finally, we discuss alternative approaches to seismic interferometry, as well as possible future directions of research.

4.2 Normal-Mode Summation

Perhaps the simplest context in which a theoretical result exists linking correlations of noise recorded at two points to the Green's function between those points is for acoustic normal modes in a finite body. It is this case that we shall consider first, and for which we will give a fairly complete derivation. For this case, the equation governing motion is the acoustic wave equation,

$$\frac{1}{\kappa(\mathbf{x})}\frac{\partial^2}{\partial t^2}u(\mathbf{x},t) - \frac{\partial}{\partial x_i}\left(\frac{1}{\rho(\mathbf{x})}\frac{\partial}{\partial x_i}u(\mathbf{x},t)\right) = f(\mathbf{x},t), \qquad (4.1)$$

where $\mathbf{x}$ is the position vector, κ is bulk modulus, ρ density, and f the external forcing that excites the pressure wavefield u. Throughout this chapter we employ the summation convention, meaning that summation of repeated indices is implicit. The various normal modes that solve equation (4.1) with vanishing right-hand side can be expressed generally as

$$u_p(\mathbf{x},t) = s_p(\mathbf{x})\cos(\omega_p t + \phi_p), \qquad (4.2)$$

where $s_p(\mathbf{x})$ is the spatial mode shape or eigenfunction, ω_p is the frequency of the mode, and ϕ_p is the phase of the mode. For the derivation provided below, it is not necessary to know the precise shape of each mode. The only property used will be that the modes are orthogonal in both space and time. For spatial orthogonality, this implies that $\int s_m(\mathbf{x})s_n(\mathbf{x})dx = 0$ if $m \neq n$; for temporal orthogonality, this implies that $\int \cos(\omega_m t + \phi_m)\cos(\omega_n t + \phi_n)dt = 0$ if $\omega_m \neq \omega_n$. It should be noted that in realistic media, many modes can share the same or close to the same frequency, leading to a more complex result as will be discussed below (Tsai, 2010). As a simple example that may help with intuition, one may consider the homogeneous one-dimensional string with fixed end points at $x = 0$ and $x = L$, for which the spatial modes can be explicitly written as $\sin(n\pi x/L)$ with corresponding frequency $\omega_n = n\pi c/L$, with c being the homogeneous wave speed. Spatial orthogonality reduces to the result that $\int \sin(m\pi x/L)\sin(n\pi x/L)dx = 0$ for $m \neq n$ (e.g., Haberman, 2013).

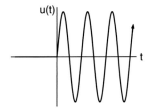

u(t)

t

Figure 4.1. Schematic illustration of the Green's function from equation (4.3).

Part of the reason that a normal-mode framework is particularly straightforward is that (1) Green's functions are particularly easy to describe, and (2) the noise correlation simplifies due to the orthogonality property. For each mode, the spatial orthogonality relation can be used to show that the Green's function is described by (e.g., Snieder, 2004b)

$$
G_p(\mathbf{x}, \mathbf{x}_s, t) = \begin{cases} \frac{s_p(\mathbf{x})s_p(\mathbf{x}_s)}{\omega_p} \sin(\omega_p t), & \text{for } t > 0, \\ 0, & \text{for } t < 0, \end{cases} \tag{4.3}
$$

which is schematically shown in Figure 4.1. In other words, each mode oscillates as a sinusoid starting at $t = 0$, and the full Green's function is a sum over these components, $G = \sum_{p=0}^{\infty} G_p$ (e.g., Gilbert, 1970; Haberman, 2013).

Defining the normalized cross-correlation of two arbitrary functions f and g as

$$
C[f(t), g(t)] = \lim_{T \to \infty} \frac{1}{2T} \int_{-T}^{T} f(\tau)g(\tau + t)d\tau, \tag{4.4}
$$

then for an arbitrary general wavefield $u(\mathbf{x}, t) = \sum_p A_p u_p(\mathbf{x}, t)$, the temporal orthogonality property immediately results in many cross terms disappearing in the cross-correlation. Specifically, if all ω_p are different (see below for the more general case) then direct calculation shows that the cross-correlation of recordings at positions $\mathbf{x}_A$ and $\mathbf{x}_B$ reduces to

$$
C(\mathbf{x}_A, \mathbf{x}_B) = C[u(\mathbf{x}_A, t), u(\mathbf{x}_B, t)] = \frac{1}{2} \sum_{p=0}^{\infty} A_p^2 s_p(\mathbf{x}_A) s_p(\mathbf{x}_B) \cos(\omega_p t). \tag{4.5}
$$

Comparing equations (4.3) and (4.5), it is thus clear that there is a potential relationship between the Green's function and the cross-correlation. The most commonly quoted version of this is that if $A_p = \alpha/\omega_p$, with α an arbitrary constant, then

$$
\frac{d}{dt} C(\mathbf{x}_A, \mathbf{x}_B) = -\frac{\alpha^2}{2}[G(\mathbf{x}_A, \mathbf{x}_B, t) - G(\mathbf{x}_A, \mathbf{x}_B, -t)]. \tag{4.6}
$$

This statement says that if energy happens to be equipartitioned between all of the modes and hence amplitudes are weighted inversely with frequency, then the time derivative of the cross-correlation between two stations is equal to the sum of a Green's function and a time-reversed Green's function between those same two points, up to a normalization factor. Alternatively, if modal amplitudes are all equal such that $A_p = \alpha$, then

$$C(\mathbf{x}_A, \mathbf{x}_B) = \frac{\alpha^2}{2} \frac{d}{dt} [G(\mathbf{x}_A, \mathbf{x}_B, t) - G(\mathbf{x}_A, \mathbf{x}_B, -t)]. \tag{4.7}$$

In other words, if modal amplitudes are all equal, then the cross-correlation between two stations is equal to the time derivative of the sum of a Green's function and its time-reversed version between those same two points, again up to a normalization factor.

The two relationships (4.6) and (4.7) appear different, but in fact express the same identity. Since a time derivative results in a phase advance of 90°, it is clear that both identities express the fact that each modal component of the cross-correlation is expected to be phase advanced by 90° with respect to each modal component of the Green's function. This is also clear simply by comparing the sine term of equation (4.3) with the cosine term of equation (4.5).

It is important to note two related points. Firstly, so far, the identities hold for a deterministic process, and do not require any "noise" property to hold. Secondly, however, as mentioned above, the identities require all modes to have different frequencies. This is clearly not expected in general. For example, on the Earth, two waves arriving from two different azimuths can easily have frequency content that is similar. Resolution to this issue occurs if cross-terms from modes with the same frequency still somehow cancel out, despite not satisfying the temporal orthogonality property. For "noise" sources, this may occur if the sources are uncorrelated over long timescales. For example, take the case above of two waves created by two different wave sources, which could be two different ocean storms, at two different azimuths. If the two sources randomly change their relative phase over time, rather than staying correlated over all time as a deterministic signal would, and if there is at least a small amount of damping in the system such that waves excited at one instance in time do not keep contributing forever, then eventually the cross-correlation between these two different sources would be expected to cancel out and sum to zero. This cancellation of different uncorrelated noise sources has been discussed by many authors (e.g., Lobkis and Weaver, 2001; Tsai, 2010), most of whom treat each independent time period as a separate "realization" of an ensemble, and these authors have shown that the cancellation of noise occurs proportionally to the square root of time. While the above argument is a heuristic one, a mathematically rigorous derivation is possible that demonstrates that spatially uncorrelated sources can result in uncorrelated modes (e.g., Lobkis and Weaver,

2001), but this is beyond the scope of this work. The final result is an identity similar to equation (4.6) that states

$$\frac{d}{dt} C[u^N(\mathbf{x}_A, t), u^N(\mathbf{x}_B, t)] = -\frac{\alpha^2}{2}[G(\mathbf{x}_A, \mathbf{x}_B, t) - G(\mathbf{x}_A, \mathbf{x}_B, -t)], \qquad (4.8)$$

if $A_p = \alpha/\omega_p$ and u^N is a "noise" field for which each mode still vibrates sinusoidally in time, but is temporally uncorrelated with all other modes over long enough times, with the properties discussed in the previous paragraph. This is perhaps the most commonly quoted Green's function-noise correlation identity.

So far, only the simplest case of acoustic waves has been considered. However, all of the arguments made above also work for modes of the elastic wave equation with only minor modifications. For example, for vertical-component Rayleigh modes, one can still write a mode sum in the same manner, and equation (4.8) still holds in the same way.

The primary issue with the normal mode derivation is that the assumptions are extremely limiting. For example, modes on Earth are certainly never close to equipartitioned for at least two reasons. For one, the majority of noise sources on Earth are expected to be at or near the Earth's surface (e.g., ocean waves, wind, industrial activity, traffic) and such surface sources excite fundamental-mode waves much more strongly than they excite higher-order modes. Thus, these different types of modes will have vastly different amounts of energy (and will have very different modal amplitudes), implying that the conditions for the identities to hold are not satisfied. Moreover, even within the subset of fundamental-mode waves, the fact that large noise sources, like ocean storms, are concentrated spatially in certain areas of the Earth implies that waves from certain directions will be much stronger and will therefore not be close to equipartitioned. Since the assumptions resulting in equation (4.8) are not valid on the Earth, the result is not expected to be generally exact. Moreover, within the normal-mode framework, it is not clear how to evaluate how closely to achieving (4.8) noise correlations are likely to be for a realistic distribution of seismic noise on Earth.

4.3 Plane Waves

Given that the assumptions of the normal-mode derivation are not expected to hold on the Earth, it is worthwhile to consider other reasons why the identity of equation (4.8) still might hold or at least approximately hold. Our path forward is to consider the case of plane surface waves incident on a pair of stations, $\mathbf{x}_A$ and $\mathbf{x}_B$, as before. One may note that for waves with wavelengths that are a small fraction of the radius of a weakly heterogeneous Earth, there is an equivalence between plane waves and normal modes. Plane waves are simply the propagating mode produced by a far-field source that corresponds to the standing wave of the normal-mode framework. For a homogeneous halfspace, it may be useful to note that the equipartition assumption would imply an equal amplitude of energy incoming from all azimuths (e.g., Weaver and Lobkis, 2004). Due to this equivalence, it would be natural to assume that a similar Green's function cross-correlation identity should exist specifically for this case in which "noise" energy is equally incident from all directions. While using only far-field sources means that near-field sources are excluded from this description, one may argue that as long as the body (like the Earth) is large enough and attenuation is relatively small, the far-field contribution might be expected to dominate the energy from near-field sources, thus leading to the expectation that an identity might exist even when using far-field sources only.

To derive this identity, one can again simply write down the Green's function for surface waves in a homogeneous, acoustic, non-attenuating medium and compare it to the correlation of far-field plane-wave sources. The Green's function for waves in a homogeneous 2D medium can be expressed in the frequency domain as

$$G(\mathbf{x}, \mathbf{x}_s, \omega) = \frac{-i}{4} \mathcal{H}_0^{(2)}(kr), \tag{4.9}$$

where $\mathcal{H}_0^{(2)}$ is a Hankel function of order zero of the second kind, $k = \omega/c$ is wavenumber, $\mathbf{x}_s$ is the source position, and $r = |\mathbf{x} - \mathbf{x}_s|$. (See, e.g., Morse and Feshbach (1953) or Snieder (2004b) for derivations of equation (4.9) though these use slightly different Fourier conventions than assumed here.) Again, we write only the result for the 2D acoustic case, but anticipate the elastic case having the same form, for instance, for Rayleigh waves in a laterally homogeneous 3D medium. The far-field, time-domain version of this expression (e.g., Watson, 2008) is perhaps more easily recognized:

$$G(\mathbf{x}, \mathbf{x}_s, t) = \sqrt{\frac{1}{8\pi kr}} \cos[\omega t - (kr + \pi/4)]. \tag{4.10}$$

On the other hand, a general sum of far-field plane waves can be expressed as

$$u(\mathbf{x}, t) = \frac{1}{2\pi} \int_0^{2\pi} A(\theta) \cos(\omega t - kr) \, d\theta \tag{4.11}$$

where the plane-wave density of sources at each azimuth is given by $A(\theta)$, and r is the distance from the far-field source to $\mathbf{x}$. Just as in the normal-mode case, if we again assume that the plane wave sources from different azimuths are all uncorrelated, then eventually all cross-terms in the cross-correlation will cancel out and sum to zero, leaving only

$$C[u(\mathbf{x}_A, t), u(\mathbf{x}_B, t)] = \frac{1}{8\pi^2} \int_0^{2\pi} A(\theta)^2 \cos(\omega t - k\Delta x \cos\theta) \, d\theta, \tag{4.12}$$

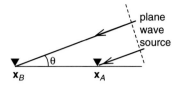

Figure 4.2. Schematic illustration of a plane wave reaching receiver positions $\mathbf{x}_A$ and $\mathbf{x}_B$.

where Δx is the distance between $\mathbf{x}_A$ and $\mathbf{x}_B$, and θ is the azimuth of the source relative to the line between $\mathbf{x}_A$ and $\mathbf{x}_B$ (see Figure 4.2). Note that unlike in the normal-mode case, there is a phase shift for each plane wave that accounts for the time delay between when the wave arrives at $\mathbf{x}_B$ and $\mathbf{x}_A$.

At this point, it is useful to recognize that the integral can be computed exactly if $A(\theta) = \alpha$ is constant, and can be approximated using the stationary-phase approximation even if $A(\theta)$ is not constant with azimuth. Qualitatively, the stationary-phase approximation can be understood as saying that the primary contributions to an oscillatory integral, like equation (4.12), occur from points where the oscillation phase is stationary, and all other contributions approximately cancel out due to there being equal positive and negative contributions from nearby points (e.g., Båth, 1968; Bender and Orszag, 1999). The end result is that the final integral can be approximated by substituting constant values of the non-oscillatory part of the integrand at the stationary points and integrating only over these stationary points. In the case of equation (4.12), there are two stationary-phase points at $\theta = 0°$ and $\theta = 180°$, so that the stationary-phase argument can be used to approximate the original integral as

$$C(\mathbf{x}_A, \mathbf{x}_B, t) \propto \int_{-\pi/2}^{\pi/2} A_+^2 \cos(\omega t - k\Delta x \cos\theta)\, d\theta$$
$$+ \int_{\pi/2}^{3\pi/2} A_-^2 \cos(\omega t - k\Delta x \cos\theta)\, d\theta, \qquad (4.13)$$

where $A_+ \equiv A(0°)$, $A_- \equiv A(180°)$, and the two primary contributions come from the two stationary-phase points at azimuths in line with the two stations. Equation (4.13) can then be evaluated by direct integration (without a further stationary-phase approximation to the integral) as

$$C(\mathbf{x}_A, \mathbf{x}_B, t) \propto A_+^2 [J_0(k\Delta x)\cos(\omega t) + H_0(k\Delta x)\sin(\omega t)]$$
$$+ A_-^2 [J_0(k\Delta x)\cos(\omega t) - H_0(k\Delta x)\sin(\omega t)], \qquad (4.14)$$

where J_0 and H_0 are Bessel functions and Struve functions of order zero, respectively (Watson, 2008). Although Struve functions are not a common special function, they are closely related to Bessel functions, and can be thought of as pairing with Bessel functions much like sine and cosine pair together. Since the stationary-phase approximation is a high-frequency approximation, as long as ω is high enough or $A(\theta)$ is close enough to constant and nonzero at the stationary-phase points, equation (4.14) will be an accurate approximation.

Considering each of the two terms of equation (4.14) separately, and taking a far-field approximation for convenience, then

$$C^{\pm}(\mathbf{x}_A, \mathbf{x}_B) \propto A_{\pm}^2 \sqrt{\frac{2}{\pi k r}} \cos[\omega t \mp (kr - \pi/4)], \qquad (4.15)$$

where $C^{\pm}$ refer to the positive ($\theta = 0°$) and negative ($\theta = 180°$) contributions to the integral, respectively. As in the normal-mode case, comparing equation (4.10) with equation (4.15) shows that there is a strong similarity between these expressions and an identity can be made by taking the time derivative of one term, that is

$$\frac{d}{dt} C^+(\mathbf{x}_A, \mathbf{x}_B, t) \propto -\omega A_+^2 G(\mathbf{x}_A, \mathbf{x}_B, t), \qquad (4.16)$$

and a similar identity holds for the second term. Thus, we see that direct evaluation of the cross-correlation and Green's function for surface waves yields a similar identity as in the normal-mode case, as expected. It may also be noted that the identity could have also been written as $C^+ \propto A_+^2 dG/dt$, analogously to equation (4.7).

Unlike the normal-mode framework, this explicit 2D plane-wave framework is also useful in assessing the degree to which a non-uniform noise source distribution causes departures from the expected result. Specifically, one could input arbitrary azimuthal distributions $A(\theta)$ of surface-wave noise sources into equation (4.12) and simply calculate how both the resulting traveltimes and waveforms from noise correlation depart from the expected Green's function traveltimes and waveforms. The integral in equation (4.12) can be evaluated numerically without using the stationary-phase approximation, and requires no assumptions about smoothness as long as sources from different azimuths are uncorrelated. Tests of this nature have previously been done (e.g., Tsai, 2009, 2011; Weaver et al., 2011) and suggest that while traveltimes are typically only affected to second order, due to the stationary phase regions still dominating, amplitudes and hence waveforms can easily be affected to first order.

4.4 Representation Theorems

An alternative justification of Green's function retrieval by interstation correlation can be derived using representation theorems that relate a wavefield to its sources via the Green's function (e.g., Aki and Richards, 2002). In the following paragraphs we will outline the basics of this theory, borrowing essential concepts from the

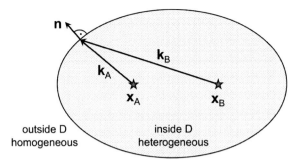

Figure 4.3. Illustration of the two receivers $\mathbf{x}_A$ and $\mathbf{x}_B$ located inside the potentially heterogeneous domain D. The medium at and outside the boundary ∂D is assumed to be homogeneous so that waves only propagate outwards and not inwards. Furthermore, the wavenumber vectors $\mathbf{k}_A$ and $\mathbf{k}_B$ are assumed to be approximately parallel to the boundary normal vector $\mathbf{n}$, which requires a nearly spherical shape of the boundary. To justify the assumption of purely outward-propagating plane waves, the medium also has to be "sufficiently" homogeneous within a region along the inside of the boundary.

work of Wapenaar (2004) and Wapenaar and Fokkema (2006). For educational reasons we will start with the simpler, though for most of the Earth unrealistic, acoustic case, before transitioning to fully elastic wave propagation.

4.4.1 Acoustic Waves

As illustrated in Figure 4.3, we consider a domain D with boundary ∂D within which the two receivers $\mathbf{x}_A$ and $\mathbf{x}_B$ are located. For convenience, we work in the frequency domain, where the acoustic wave equation (4.1) takes the form

$$- \omega^2 \frac{1}{\kappa(\mathbf{x})} u_A(\mathbf{x}, \omega) - \frac{\partial}{\partial x_i} \left(\frac{1}{\rho(\mathbf{x})} \frac{\partial}{\partial x_i} u_A(\mathbf{x}, \omega) \right) = f_A(\mathbf{x}, \omega), \qquad (4.17)$$

with the source f_A exciting the wave state u_A. A second source f_B is assumed to excite the wave state u_B that satisfies the wave equation

$$- \omega^2 \frac{1}{\kappa(\mathbf{x})} u_B(\mathbf{x}, \omega) - \frac{\partial}{\partial x_i} \left(\frac{1}{\rho(\mathbf{x})} \frac{\partial}{\partial x_i} u_B(\mathbf{x}, \omega) \right) = f_B(\mathbf{x}, \omega). \qquad (4.18)$$

Multiplying the complex conjugate of equation (4.17) with u_B, and equation (4.18) with the complex conjugate of u_A, yields the following pair of equations,

$$- \omega^2 \frac{1}{\kappa(\mathbf{x})} u_A^*(\mathbf{x}) u_B(\mathbf{x}) - u_B(\mathbf{x}) \frac{\partial}{\partial x_i} \left(\frac{1}{\rho(\mathbf{x})} \frac{\partial}{\partial x_i} u_A^*(\mathbf{x}) \right) = u_B(\mathbf{x}) f_A^*(\mathbf{x}), \quad (4.19)$$

$$- \omega^2 \frac{1}{\kappa(\mathbf{x})} u_A^*(\mathbf{x}) u_B(\mathbf{x}) - u_A^*(\mathbf{x}) \frac{\partial}{\partial x_i} \left(\frac{1}{\rho(\mathbf{x})} \frac{\partial}{\partial x_i} u_B(\mathbf{x}) \right) = u_A^*(\mathbf{x}) f_B(\mathbf{x}), \quad (4.20)$$

where we omitted dependencies on ω in the interest of a more succinct notation. Equations (4.19) and (4.20) are valid under the assumption that the acoustic medium is not attenuating, meaning that the frequency-domain bulk modulus κ is a real-valued quantity.

Subtracting equation (4.20) from equation (4.19), and integrating over an arbitrary domain D, gives

$$
\int_D \left[u_A^*(\mathbf{x}) \frac{\partial}{\partial x_i} \left(\frac{1}{\rho(\mathbf{x})} \frac{\partial}{\partial x_i} u_B(\mathbf{x}) \right) - u_B(\mathbf{x}) \frac{\partial}{\partial x_i} \left(\frac{1}{\rho(\mathbf{x})} \frac{\partial}{\partial x_i} u_A^*(\mathbf{x}) \right) \right] d^3\mathbf{x}
$$
$$
= \int_D \left[u_B(\mathbf{x}) f_A^*(\mathbf{x}) - u_A^*(\mathbf{x}) f_B(\mathbf{x}) \right] d^3\mathbf{x}. \tag{4.21}
$$

Applying the product rule of differentiation to the left-hand side of equation (4.21) can be written as

$$
\int_D \left[\frac{\partial}{\partial x_i} \left(u_A^*(\mathbf{x}) \frac{1}{\rho(\mathbf{x})} \frac{\partial}{\partial x_i} u_B(\mathbf{x}) \right) - \frac{\partial}{\partial x_i} \left(u_B(\mathbf{x}) \frac{1}{\rho(\mathbf{x})} \frac{\partial}{\partial x_i} u_A^*(\mathbf{x}) \right) \right] d^3\mathbf{x}
$$
$$
- \int_D \left[\frac{\partial}{\partial x_i} u_A^*(\mathbf{x}) \frac{1}{\rho(\mathbf{x})} \frac{\partial}{\partial x_i} u_B(\mathbf{x}) - \frac{\partial}{\partial x_i} u_B(\mathbf{x}) \frac{1}{\rho(\mathbf{x})} \frac{\partial}{\partial x_i} u_A^*(\mathbf{x}) \right] d^3\mathbf{x}
$$
$$
= \int_D \left[u_B(\mathbf{x}) f_A^*(\mathbf{x}) - u_A^*(\mathbf{x}) f_B(\mathbf{x}) \right] d^3\mathbf{x}. \tag{4.22}
$$

Realizing that the terms under the second integral in equation (4.22) cancel, we invoke Gauss's theorem to transform the left-hand side into a surface integral,

$$
\int_{\partial D} \frac{1}{\rho(\mathbf{x})} \left[u_A^*(\mathbf{x}) \frac{\partial}{\partial x_i} u_B(\mathbf{x}) - u_B(\mathbf{x}) \frac{\partial}{\partial x_i} u_A^*(\mathbf{x}) \right] n_i(\mathbf{x}) \, d^2\mathbf{x}
$$
$$
= \int_D \left[u_B(\mathbf{x}) f_A^*(\mathbf{x}) - u_A^*(\mathbf{x}) f_B(\mathbf{x}) \right] d^3\mathbf{x}, \tag{4.23}
$$

where n_i is the i-component of the surface normal vector. So far, the sources f_A and f_B are generic, and therefore equation (4.23) represents a general relation between two acoustic wave states in a lossless medium. Specifying the sources as $f_A(\mathbf{x}) = \delta(\mathbf{x} - \mathbf{x}_A)$ and $f_B(\mathbf{x}) = \delta(\mathbf{x} - \mathbf{x}_B)$ determines the solutions; $u_A(\mathbf{x})$ and $u_B(\mathbf{x})$ are the Green's functions $G(\mathbf{x}, \mathbf{x}_A)$ and $G(\mathbf{x}, \mathbf{x}_B)$, respectively. Inserting this into equation (4.23) yields

$$
2i \, \mathrm{Im} \, G(\mathbf{x}_A, \mathbf{x}_B) = G(\mathbf{x}_A, \mathbf{x}_B) - G^*(\mathbf{x}_B, \mathbf{x}_A)
$$
$$
= \int_{\partial D} \frac{1}{\rho(\mathbf{x})} \left[G^*(\mathbf{x}, \mathbf{x}_A) \frac{\partial}{\partial x_i} G(\mathbf{x}, \mathbf{x}_B) - G(\mathbf{x}, \mathbf{x}_B) \frac{\partial}{\partial x_i} G^*(\mathbf{x}, \mathbf{x}_A) \right] n_i(\mathbf{x}) \, d^2\mathbf{x}, \tag{4.24}
$$

where we also invoked reciprocity, $G(\mathbf{x}_A, \mathbf{x}_B) = G(\mathbf{x}_B, \mathbf{x}_A)$ (e.g., Aki and Richards, 2002). Equation (4.24) is not yet particularly useful for interferometry.

It requires further simplifications based on assumptions and approximations that need to be assessed on a case-by-case basis.

The first set of assumptions is that the domain is homogeneous around and outside the boundary ∂D that we require to be nearly spherical in shape and sufficiently far from both $\mathbf{x}_A$ and $\mathbf{x}_B$. This allows us to approximate the Green's function at the boundary ∂D as a plane wave propagating exclusively *out* of the domain and not into the domain, that is

$$G(\mathbf{x}, \mathbf{x}_X) \approx A_X \, e^{-ik\mathbf{n}\cdot\mathbf{x}}, \quad \text{with either } X = A \text{ or } X = B, \tag{4.25}$$

and the amplitude of the wave state is A_X. The wavenumber k satisfies the dispersion relation $k = \omega/c$, with the acoustic velocity $c = \sqrt{\kappa/\rho}$. This setup is illustrated in Figure 4.3. Under the plane-wave assumption, the spatial derivative $\partial/\partial x_i$ results in multiplication by $-ikn_i$ for $G(\mathbf{x}, \mathbf{x}_B)$ and ikn_i for $G^*(\mathbf{x}, \mathbf{x}_A)$. This condenses equation (4.24) to

$$\text{Im}\, G(\mathbf{x}_A, \mathbf{x}_B) \approx -\frac{\omega}{c'\rho'} \int_{\partial D} G(\mathbf{x}, \mathbf{x}_B) G^*(\mathbf{x}, \mathbf{x}_A) \, d^2\mathbf{x}, \tag{4.26}$$

where ρ' and c' are density and velocity evaluated at the domain boundary ∂D. To relate the Green's functions in equation (4.26) to the propagation of ambient seismic noise, we consider noise sources, $N(\mathbf{x})$, distributed along the boundary ∂D. When averaged, we assume sources at position $\mathbf{x}$ to be temporally uncorrelated with sources at position $\mathbf{y}$, that is

$$S(\mathbf{x})\, \delta(\mathbf{x} - \mathbf{y}) = \langle N(\mathbf{x})N^*(\mathbf{y})\rangle, \tag{4.27}$$

where $\langle\, .\, \rangle$ denotes averaging. The frequency-domain quantity $S(\mathbf{x})$ is the power-spectral density distribution of the noise sources, which we assume to be spatially homogeneous, and denoted simply by S. Multiplying the left-hand sides and the right-hand sides of equations (4.26) and (4.27), and integrating over the boundary ∂D, gives

$$S \,\text{Im}\, G(\mathbf{x}_A, \mathbf{x}_B) \approx -\frac{\omega}{c'\rho'} \langle \iint_{\partial D} G(\mathbf{x}, \mathbf{x}_B) N(\mathbf{x})\, G^*(\mathbf{y}, \mathbf{x}_A) N^*(\mathbf{y}) \, d^2\mathbf{x}\, d^2\mathbf{y}\rangle. \tag{4.28}$$

The surface δ-function has been eliminated by the integration. In equation (4.28) we recognize two representation theorems for a wavefield u excited by the noise sources N and recorded at $\mathbf{x}_A$ and $\mathbf{x}_B$, respectively:

$$u(\mathbf{x}_{A,B}) = \int_{\partial D} G(\mathbf{x}, \mathbf{x}_{A,B}) N(\mathbf{x}) \, d\mathbf{x}. \tag{4.29}$$

Combining equations (4.28) and (4.29), we obtain our final result,

$$G(\mathbf{x}_A, \mathbf{x}_B) - G^*(\mathbf{x}_A, \mathbf{x}_B) \approx -\frac{2i\omega}{Sc'\rho'} C(\mathbf{x}_B, \mathbf{x}_A). \tag{4.30}$$

where $C(\mathbf{x}_B, \mathbf{x}_A) = \langle u(\mathbf{x}_B)u^*(\mathbf{x}_A)\rangle$ is the ensemble averaged correlation function in the frequency domain. Equation (4.30) states that the correlation $C(\mathbf{x}_B, \mathbf{x}_A)$ approximates the Green's function between the receivers, $G(\mathbf{x}_A, \mathbf{x}_B)$ minus its time-reversed version $G^*(\mathbf{x}_A, \mathbf{x}_B)$ up to a scaling factor. The required assumptions are: (1) absence of attenuation, (2) a medium that is homogeneous at and outside the boundary ∂D, as well as within a sufficiently wide region along the inside of the boundary, (3) a sufficiently large distance between the receivers $\mathbf{x}_A$ and $\mathbf{x}_B$ to the domain boundary ∂D, and (4) homogeneously distributed and decorrelated noise sources in the sense of equation (4.27). Items (2) and (3) are needed to justify the approximation of the Green's function in terms of purely outward-propagating plane waves.

4.4.2 Elastic Waves

To derive an analogue of equation (4.30) for elastic waves propagating through the solid Earth's interior, we adapt the concepts used in the previous paragraphs on acoustic waves. As we will see, however, additional complications arise due to the presence of more than one wave type, that is, at least P and S-waves for the simplest case of a homogeneous isotropic medium. These complications will require us to make more severe assumptions and simplifications. The propagation of elastic waves is governed by the elastic wave equation, written in the frequency domain as

$$-\omega^2 \rho(\mathbf{x})u_i(\mathbf{x}, \omega) - \frac{\partial}{\partial x_j}\left[c_{ijkl}(\mathbf{x})\frac{\partial}{\partial x_k}u_l(\mathbf{x}, \omega)\right] = f_i(\mathbf{x}, \omega). \qquad (4.31)$$

Again omitting dependencies on ω, the i-component of the Green's function due to a force in p-direction, $G_{ip}(\mathbf{x}, \mathbf{x}_A)$, is defined as solution of equation (4.31) when the right-hand side is point-localized in time at $t = 0$ and space at $\mathbf{x} = \mathbf{x}_A$, that is

$$f_i(\mathbf{x}) = \delta_{ip}\,\delta(\mathbf{x} - \mathbf{x}_A) \qquad \rightarrow \qquad u_i(\mathbf{x}) = G_{ip}(\mathbf{x}, \mathbf{x}_A), \qquad (4.32)$$

where δ_{ip} denotes the Kronecker delta. Using equations (4.31) and (4.32), we can follow exactly the same steps as in the acoustic case: (1) Define two states, A and B, (2) multiply the equations for these two states by the other state, respectively, (3) subtract the resulting equations, (4) integrate over the volume D, and (5) apply Gauss's theorem. Again under the assumption that c_{ijkl} is real-valued, meaning that the medium is not attenuating, we find the elastic version of equation (4.24):

$$2i\,\mathrm{Im}\,G_{pq}(\mathbf{x}_A, \mathbf{x}_B) = \int_{\partial D}\left[G_{ip}^*(\mathbf{x}, \mathbf{x}_A)c_{ijkl}(\mathbf{x})\frac{\partial}{\partial x_k}G_{lq}(\mathbf{x}, \mathbf{x}_B)\right.$$

$$\left. - G_{iq}(\mathbf{x}, \mathbf{x}_B)c_{ijkl}(\mathbf{x})\frac{\partial}{\partial x_k}G_{lp}^*(\mathbf{x}, \mathbf{x}_A)\right]n_j(\mathbf{x})\,d^2\mathbf{x}.$$

$$(4.33)$$

As in the acoustic case, equation (4.33) requires simplifications to eliminate spatial derivatives and space-dependent medium properties inside the integral. Again, assuming that ∂D is far from the stations, and that the medium is homogeneous and isotropic along and outside the boundary, we may approximate the Green's functions $G_{lp}(\mathbf{x}, \mathbf{x}_A)$ by a plane wave propagating exclusively outwards and parallel to the boundary normal $\mathbf{n}$,

$$G_{lp}(\mathbf{x}, \mathbf{x}_A) \approx P_{lp,A}e^{-ik_P\mathbf{n}\cdot\mathbf{x}} + S_{lp,A}e^{-ik_S\mathbf{n}\cdot\mathbf{x}}. \tag{4.34}$$

Since the medium is assumed isotropic, equation (4.34) contains a P-wave with polarization vector $P_{lp,A}$ and wavenumber $k_P = \omega/c_P$, and an S-wave with polarization vector $S_{lp,A}$ and wavenumber $k_S = \omega/c_S$. Taking the spatial derivative $\partial/\partial x_k$ of equation (4.34),

$$\partial_k G_{lp}(\mathbf{x}, \mathbf{x}_A) \approx -in_k \left(P_{lp,A}k_Pe^{-ik_P\mathbf{n}\cdot\mathbf{x}} + S_{lp,A}k_Se^{-ik_S\mathbf{n}\cdot\mathbf{x}} \right). \tag{4.35}$$

The appearance of two wave propagation modes generally leads to cross-terms of P- and S-waves when (4.35) is substituted back into equation (4.33). These cross-terms can only be eliminated with additional assumptions and approximations, several of which have been proposed in the literature (e.g., Wapenaar, 2004; Wapenaar and Fokkema, 2006). Here, we will follow probably the simplest line of arguments, based on the relative size of the P- and S-wave contributions in equation (4.35). For a typical crust we have $c_P/c_S \approx \sqrt{3}$ (e.g., Dziewoński and Anderson, 1981; Kennett et al., 1995). Using the expression for P- and S-wave amplitudes in a homogeneous medium (e.g., Aki and Richards, 2002) yields $S_{lp,A}k_S/P_{lp,A}k_P \approx 5.4$, meaning that the P-wave contribution can be ignored relatively safely. Therefore,

$$\partial_k G_{lp}(\mathbf{x}, \mathbf{x}_A) \approx -in_k S_{lp,A}k_Se^{-ik_S\mathbf{n}\cdot\mathbf{x}}. \tag{4.36}$$

With the help of equation (4.36), we can simplify (4.33) to

$$\text{Im}\, G_{pq}(\mathbf{x}_A, \mathbf{x}_B) \approx -k_S \int_{\partial D} G_{iq}(\mathbf{x}, \mathbf{x}_B) \left[n_k(\mathbf{x})c_{ijkl}(\mathbf{x})n_j(\mathbf{x}) \right] G_{lp}^*(\mathbf{x}, \mathbf{x}_A)\, d^2\mathbf{x}. \tag{4.37}$$

Since the medium is assumed isotropic along ∂D, we can expand the term in square brackets in terms of the Lamé parameters λ and μ,

$$n_k c_{ijkl}n_j = n_k(\lambda\delta_{ij}\delta_{kl} + \mu\delta_{ik}\delta_{jl} + \mu\delta_{il}\delta_{jk})n_j = (\lambda + \mu)n_i n_l + \mu\delta_{il}. \tag{4.38}$$

Noticing that the product of the normal vector n_l with the transversely polarized S-wave Green's function G_{lp}^* vanishes (because the wave is assumed to propagate parallel to the surface normal), equation (4.37) can now be modified to

$$\text{Im}\, G_{pq}(\mathbf{x}_A, \mathbf{x}_B) \approx -k_S\mu' \int_{\partial D} G_{iq}(\mathbf{x}, \mathbf{x}_B)\delta_{il}G_{lp}^*(\mathbf{x}, \mathbf{x}_A)\, d^2\mathbf{x}, \tag{4.39}$$

with μ' being the constant shear modulus at the boundary ∂D. It remains to introduce vector-valued noise sources $N_l(\mathbf{x})$ that excite the wavefield

$$u_i(\mathbf{y}) = \int_{\partial D} G_{il}(\mathbf{y}, \mathbf{x}) N_l(\mathbf{x}) \, d^2\mathbf{x}. \tag{4.40}$$

Assuming that the sources are on average uncorrelated with vanishing cross-terms in the sense of

$$\langle N_i(\mathbf{x}) N_l^*(\mathbf{y}) \rangle = S \delta_{il} \delta(\mathbf{x} - \mathbf{y}), \tag{4.41}$$

equation (4.39) collapses into an equation for the ensemble-averaged interstation correlation $C_{qp}(\mathbf{x}_B, \mathbf{x}_A) = \langle u_q(\mathbf{x}_B) u_p^*(\mathbf{x}_A) \rangle$:

$$G_{pq}(\mathbf{x}_A, \mathbf{x}_B) - G_{pq}^*(\mathbf{x}_A, \mathbf{x}_B) \approx -\frac{2i\omega\mu'}{Sc_S} C_{qp}(\mathbf{x}_B, \mathbf{x}_A). \tag{4.42}$$

In analogy to equation (4.30), equation (4.42) also relates the correlation of the wavefield at positions $\mathbf{x}_A$ and $\mathbf{x}_B$ to the interstation Green's function $G_{pq}(\mathbf{x}_A, \mathbf{x}_B)$. To arrive at this result, we had to make new assumptions in addition to those made in the acoustic case already. These include (1) isotropy along and outside the domain boundary ∂D, (2) the absence of off-diagonal elements in the noise source power-spectral density, in the sense of equation (4.41), and (3) the dominance of S-waves, which allowed us to neglect P-wave propagation. The latter assumption may, for instance, be replaced by assumptions on the relative strength of S- and P-wave sources, without changing the final result (Wapenaar and Fokkema, 2006).

A difficulty of the representation theorem approach outlined above is a quantification of the extent to which the various approximations are actually met. This applies, in particular, to the plane wave approximations that require a hardly quantifiable degree of homogeneity around the domain boundary. As a consequence of the approximations and assumptions, the retrieval of the Green's functions on the left-hand side of equation (4.42) is in practice never exact.

Finally, we note that the previous derivations can also be performed in 2D where the waves may be interpreted as analogs of single-mode surface waves propagating on the Earth's surface. This would merely require the use of 2D Green's functions in the derivations.

4.5 Interferometry Without Green's Function Retrieval

While noise interferometry based on Green's function retrieval is one of the great successes of seismological research in the past 15 years, the fact that few of the assumptions needed for its theoretical justification are met in the Earth remains a concern. Failure to meet these assumptions leads to well-documented errors in traveltimes, amplitudes, and waveforms that may become serious problems in application where high precision is needed, for instance, in time-lapse monitoring of fault zones, volcanoes, and reservoirs.

An alternative to interferometry by Green's function retrieval, with potential to circumvent these issues, was proposed in helioseismology, before interferometry emerged as a major research field in geophysics (Woodard, 1997; Gizon and Birch, 2002). The fundamental idea is to take a correlation function as what it is, not trying to approximate a Green's function. Being a deterministic time series, the correlation function is related via sensitivity kernels to the variations in noise sources and Earth structure that we are eventually interested in. This leads to a coupled inverse problem where both sources and structure need to be constrained, similar to earthquake tomography. Alternatively, information on noise sources from ocean-wave models (e.g., Ardhuin et al., 2011; Gualtieri et al., 2013, 2014; Ardhuin et al., 2015; Gualtieri et al., 2015; Farra et al., 2016) may be incorporated.

Starting again with the acoustic case, we will outline key elements of this theory in the following paragraphs. Subsequently, we briefly summarize the generalization to elastic wave propagation, and provide a range of examples. A more detailed treatment may be found in Tromp et al. (2010), Hanasoge (2013), or Fichtner et al. (2017a).

4.5.1 Modeling Correlation Functions

We start with the acoustic representation theorem (4.29) that expresses the frequency-domain interstation correlation in terms of the sources N and the Green's function $G(\mathbf{m})$ for a suitable Earth model $\mathbf{m}$. Multiplying $u(\mathbf{x}_A)$ with $u^*(\mathbf{x}_B)$ and using the representation theorem yields an expression for the time-domain correlation function,

$$u(\mathbf{x}_A)u^*(\mathbf{x}_B) = \int\int_{\partial D} G(\mathbf{x}, \mathbf{x}_A)G^*(\mathbf{y}, \mathbf{x}_B)N(\mathbf{x})N^*(\mathbf{y})\, d\mathbf{x}\, d\mathbf{y}. \qquad (4.43)$$

The domain boundary ∂D is taken to be the surface of the Earth where most noise sources are located. However, the integral can be extended to a volume, if needed. The dependence of G on $\mathbf{m}$ is omitted to avoid clutter. When the noise sources are uncorrelated in the sense of equation (4.27), the ensemble average of equation (4.43) is given by

$$C(\mathbf{x}_A, \mathbf{x}_B) = \langle u(\mathbf{x}_A)u^*(\mathbf{x}_B)\rangle = \int_{\partial D} G(\mathbf{x}, \mathbf{x}_A)G^*(\mathbf{x}, \mathbf{x}_B)S(\mathbf{x})\, d\mathbf{x}. \qquad (4.44)$$

Equation (4.44) constitutes a deterministic relation between the correlation $C(\mathbf{x}_A, \mathbf{x}_B)$, the Earth model $\mathbf{m}$, and the noise-source power spectral density $S(\mathbf{x})$, which is allowed to be spatially variable. It implies that $C(\mathbf{x}_A, \mathbf{x}_B)$ can be

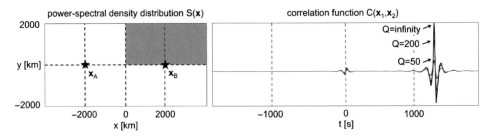

Figure 4.4. Simulation of correlation functions. The source power-spectral density is nonzero only in the gray-shaded region in the left panel. The resulting correlation functions for various Q values are shown to the right for a frequency band from 10 and 30 mHz. (Figure modified from Fichtner (2015).)

modeled without the need to simulate long-duration noise wavefields, and without constraints on the medium properties and the distribution of noise sources.

Figure 4.4 illustrates the simulation of noise correlations in a two-dimensional domain with homogeneous bulk modulus $\kappa = 2.7 \cdot 10^{10}$ N/m^2, density $\rho = 3000$ kg/m^3, and Q taking the values 50, 200, and ∞. Noise sources in the form of nonzero power-spectral density S are located in the gray-shaded region. Due to the heterogeneous source distribution, the correlation function is mostly one-sided with an additional low-amplitude phase around $t = 0$.

To estimate S and $\mathbf{m}$, the simulated correlation $C(\mathbf{x}_A, \mathbf{x}_B)$ must be compared to an observed correlation $C^o(\mathbf{x}_A, \mathbf{x}_B)$. In the interest of simplicity, we assume for the moment that this comparison is done through the computation of the L_2 waveform misfit,

$$\chi = \frac{1}{2} \int [C(\mathbf{x}_A, \mathbf{x}_B, \omega) - C^o(\mathbf{x}_A, \mathbf{x}_B, \omega)]^2 \, d\omega. \tag{4.45}$$

Using Parseval's theorem, equation (4.45) can also be written in the time-domain form

$$\chi = \frac{1}{2} \int [C(\mathbf{x}_A, \mathbf{x}_B, t) - C^o(\mathbf{x}_A, \mathbf{x}_B, t)]^2 \, dt, \tag{4.46}$$

which is, however, less convenient for our purpose. In response to infinitesimal perturbations of the noise sources δS and the Earth model $\delta \mathbf{m}$, the simulated correlation is perturbed from C to $C + \delta C$, which in turn induces a perturbation of the misfit from χ to $\chi + \delta \chi$. Knowing the relation between the perturbations δS and $\delta \mathbf{m}$ would allow us to construct models of noise sources S and Earth structure $\mathbf{m}$ such that the misfit χ is minimized. This relation between model and misfit perturbations can be written in terms of Fréchet or sensitivity kernels that we will derive in the following paragraphs.

4.5.2 Sensitivity Kernels for Noise Sources

Using equation (4.45), the misfit perturbation $\delta\chi = \chi(C+\delta C) - \chi(C)$ is given by

$$\delta\chi = \int [C(\mathbf{x}_A, \mathbf{x}_B) - C^o(\mathbf{x}_A, \mathbf{x}_B)]\,\delta C(\mathbf{x}_A, \mathbf{x}_B)\,d\omega. \qquad (4.47)$$

Assuming that perturbations of the correlation function, δC, arise from perturbations in the power-spectral density, δS, we find a relation between $\delta\chi$ and δS with the help of the forward modeling equation (4.44):

$$\delta\chi = \int_{\partial D} \int [C(\mathbf{x}_A, \mathbf{x}_B) - C^o(\mathbf{x}_A, \mathbf{x}_B)]\,G(\mathbf{x}, \mathbf{x}_A)G^*(\mathbf{x}, \mathbf{x}_B)\delta S(\mathbf{x})\,d\omega\,d\mathbf{x}. \qquad (4.48)$$

To simplify equation (4.48), we define the noise source kernel K_s as

$$K_s(\mathbf{x}) = \int [C(\mathbf{x}_A, \mathbf{x}_B) - C^o(\mathbf{x}_A, \mathbf{x}_B)]\,G(\mathbf{x}, \mathbf{x}_A)G^*(\mathbf{x}, \mathbf{x}_B)\,d\omega. \qquad (4.49)$$

The relation between a perturbation of the noise sources, δS, and the resulting perturbation of the misfit, $\delta\chi$, is now given by

$$\delta\chi = \int_{\partial D} K_s(\mathbf{x})\,\delta S(\mathbf{x})\,d\mathbf{x}. \qquad (4.50)$$

The kernel K_s captures the spatial sensitivity of the measurement to the noise sources. In regions where K_s attains large positive values, a positive perturbation of the sources leads to an increase of the misfit, and vice versa. Solving an inverse problem for the noise sources involves finding perturbations δS of an initial noise source model S such that the misfit is minimized.

The waveform misfit introduced in equation (4.45) is only one of many possible ways to quantify the difference between observed and simulated correlation functions. Other misfits may be better suited for specific applications. One example is the travel time misfit used in transmission tomography. Regardless of the specific choice, the misfit variation, $\delta\chi$, can generally be written in the form

$$\delta\chi = \int f\,\delta C(\mathbf{x}_A, \mathbf{x}_B)\,d\omega, \qquad (4.51)$$

with a frequency-dependent function f, called the adjoint source (e.g., Fichtner et al., 2017a). In the specific case of the L_2 waveform difference, given in equation (4.47), we have $f = C(\mathbf{x}_A, \mathbf{x}_B) - C^o(\mathbf{x}_A, \mathbf{x}_B)$.

Examples of source kernels for travel time measurements (Luo and Schuster, 1991) on the large-amplitude arrival in Figure 4.4 are shown in Figure 4.5 as a function of attenuation and bandwidth. The kernels have the shape of hyperbolic jets, also known as end-fire lobes. As intuitively expected, the kernel decays more quickly with distance from station $\mathbf{x}_B$ as attenuation increases. More oscillatory

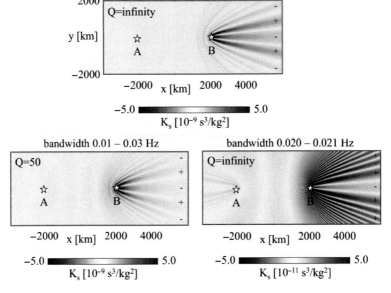

Figure 4.5. Noise source kernels for travel time measurements on the large-amplitude wave in Figure 4.4. The first-order feature is a hyperbolic jet behind station $\mathbf{x}_B$ known as end-fire lobe. Details depend on attenuation, bandwidth and Earth structure. (Figure modified from Fichtner (2015).)

features appear for measurements in a narrower frequency band. It follows that smooth perturbations of S will only contribute when located roughly within the first Fresnel zone along $y = 0$.

4.5.3 Sensitivity Kernels for Earth Structure

The relation between the misfit χ and perturbations in Earth structure, $\delta\mathbf{m}$, are slightly more difficult to derive because $\mathbf{m}$ is not explicit in the forward modeling equation (4.44), but implicit inside the Green's functions. Applying the product rule to equation (4.44), we find the variation of the correlation function in terms of variations of the Green's function,

$$\delta C(\mathbf{x}_A, \mathbf{x}_B) = \int_{\partial D} \left[\delta G(\mathbf{x}, \mathbf{x}_A) \, G^*(\mathbf{x}, \mathbf{x}_B) + G(\mathbf{x}, \mathbf{x}_A) \, \delta G^*(\mathbf{x}, \mathbf{x}_B) \right] S(\mathbf{x}) \, d\mathbf{x}.$$

$$(4.52)$$

With the help of the acoustic wave equation (4.17) we can eliminate δG from equation (4.52). Choosing the force term in the acoustic wave equation (4.1) to be point-localized, that is, $f_A(\mathbf{x}) = \delta(\mathbf{x} - \mathbf{x}_A)$, we obtain the governing equation of the Green's function,

$$-\omega^2 \frac{1}{\kappa(\mathbf{x})} G(\mathbf{x}, \mathbf{x}_A) - \frac{\partial}{\partial x_i} \left(\frac{1}{\rho(\mathbf{x})} \frac{\partial}{\partial x_i} G(\mathbf{x}, \mathbf{x}_A) \right) = \delta(\mathbf{x} - \mathbf{x}_A). \qquad (4.53)$$

Perturbing the bulk modulus from κ to $\kappa + \delta\kappa$, and keeping in mind that G depends on κ as well, we obtain the perturbation equation

$$\omega^2 \frac{\delta\kappa(\mathbf{x})}{\kappa^2(\mathbf{x})} G(\mathbf{x}, \mathbf{x}_A) - \omega^2 \frac{1}{\kappa(\mathbf{x})} \delta G(\mathbf{x}, \mathbf{x}_A) = 0. \tag{4.54}$$

Solving equation (4.54) for $\delta G(\mathbf{x}, \mathbf{x}_A)$, and using the corresponding expression for $\delta G(\mathbf{x}, \mathbf{x}_B)$, equation (4.52) transforms to

$$\delta C(\mathbf{x}_A, \mathbf{x}_B) = \int_{\partial D} \frac{\delta\kappa(\mathbf{x})}{\kappa(\mathbf{x})} \left[G(\mathbf{x}, \mathbf{x}_A) G^*(\mathbf{x}, \mathbf{x}_B) + G(\mathbf{x}, \mathbf{x}_A) G^*(\mathbf{x}, \mathbf{x}_B) \right] S(\mathbf{x})\, d\mathbf{x}. \tag{4.55}$$

Substituting (4.55) into the general expression for the misfit variation (4.51), we finally obtain

$$\delta\chi = \int_{\partial D} K_\kappa(\mathbf{x})\, \delta\kappa(\mathbf{x})\, d\mathbf{x}, \tag{4.56}$$

with the structure kernel

$$K_\kappa(\mathbf{x}) = \int \frac{f\, S(\mathbf{x})}{\kappa(\mathbf{x})} \left[G(\mathbf{x}, \mathbf{x}_A) G^*(\mathbf{x}, \mathbf{x}_B) + G(\mathbf{x}, \mathbf{x}_A) G^*(\mathbf{x}, \mathbf{x}_B) \right] d\omega. \tag{4.57}$$

The kernel K_κ captures the first-order relation between variations in the bulk modulus κ and the misfit χ. It depends on the noise source power-spectral density S and the adjoint source f, which is determined by the specific choice of a misfit functional. Using the same line of arguments as above, a sensitivity kernel for variations in density, $\delta\rho$, can be derived. Sensitivity kernels for derived medium properties, for instance the acoustic velocity $c = \sqrt{\kappa/\rho}$, follow from the Jacobian rule.

Continuing the examples from the previous figures, Figure 4.6 shows sensitivity kernels for travel time measurements on the large-amplitude arrival at positive times with respect to acoustic velocity. Similar to finite-frequency kernels for surface waves from earthquakes or active sources (e.g., Friederich, 2003; Zhou et al., 2004; Yoshizawa and Kennett, 2005), sensitivity is mostly located between the receiver pair. An additional contribution right of station $\mathbf{x}_B$ results from the heterogeneous distribution of the noise sources.

4.5.4 The Elastic Case

While we have limited ourselves to the acoustic case for pedagogical reasons so far, the previous developments can be translated almost one-to-one to elastic wave propagation that is more relevant for the Earth. Indeed, using the elastic version of the representation theorem (4.40), the cross-correlation matrix can be written in analogy to equation (4.44) as

$$C_{pq}(\mathbf{x}_A, \mathbf{x}_B) = \int_{\partial D} G_{pi}(\mathbf{x}_A, \mathbf{x})\, G^*_{qj}(\mathbf{x}_B, \mathbf{x})\, S_{ij}(\mathbf{x})\, d\mathbf{x}, \tag{4.58}$$

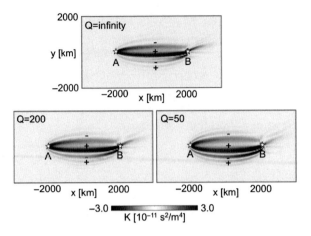

Figure 4.6. Sensitivity kernels for measurements of travel time on the large-amplitude waveform in Figure 4.4 with respect to acoustic velocity c. In contrast to the source kernels, sensitivity is primarily located between the receivers, with an additional contribution right of station $\mathbf{x}_B$ that results from the heterogeneous noise source distribution. The kernels broaden with increasing attenuation, that is, increasing dominance of lower frequencies. (Figure modified from Fichtner (2015).)

where S_{ij} is the power spectral density matrix of the noise sources as a function of space and frequency. Following exactly the same steps taken in sections 4.5.2 and 4.5.3, we can derive sensitivity kernels for noise sources and Earth structure (e.g., Tromp et al., 2010; Fichtner, 2014; Fichtner et al., 2017a),

$$\delta\chi = \int_{\partial D} K_{s,ij}(\mathbf{x})\, S_{ij}(\mathbf{x})\, d\mathbf{x} + \int_{D} K_i(\mathbf{x})\delta m_i(\mathbf{x})\, d\mathbf{x}, \qquad (4.59)$$

where m_i represents all parameters of an elastic Earth model, for instance P and S velocities, density, and attenuation.

Examples for noise source and Earth structure kernels on the global scale are shown in Figure 4.7. The noise source kernel is computed for the measurement correlation asymmetry of the causal and acausal fundamental-mode surface wave at long periods from 150 to 300 s (Ermert et al., 2016, 2017). Note that the kernel focuses at the antipoles of the two stations. This phenomenon reflects the fact that noise sources at the station antipoles radiate waves that focus at the stations, thereby affecting the correlation amplitudes particularly strongly. The Earth structure kernel is for the measurement of surface wave energy in the causal branch of the correlation function, and with respect to the Lamé parameter λ. Noise sources for this example are homogeneously distributed in the oceans but zero on land.

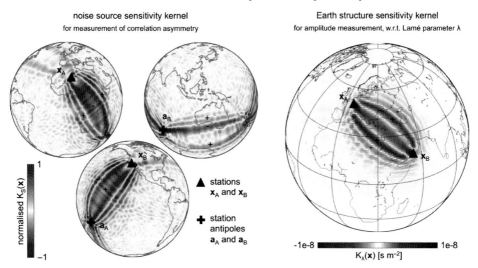

Figure 4.7. Global-scale sensitivity kernels for noise sources and Earth structure. The source distribution used to compute correlation functions is homogeneous and nonzero only in the oceans, and S20RTS (Ritsema et al., 1999) has been used as Earth model. Left: Noise source sensitivity kernel for the measurement of correlation asymmetry between the causal and acausal surface waves at long periods from 150 to 300 s. Note the characteristic focusing of the kernel at the station antipoles. Right: Earth structure sensitivity kernel for the measurement of surface wave energy on the causal branch with respect to the Lamé parameter λ.

These sensitivity kernels provide a quantitative link between measurements on noise correlations and variations in noise sources and Earth structure. With slightly higher algebraic effort, the kernel expressions can be adapted to misfit functionals that may be more suitable than the L_2 waveform misfit of equation (4.45) that we have largely chosen for convenience. Misfit functionals that enable a more differentiated extraction of time- and frequency-dependent differences in phase and amplitude may be found in the recent waveform inversion literature (e.g., Luo and Schuster, 1991; Gee and Jordan, 1992; Fichtner et al., 2008; van Leeuwen and Mulder, 2010; Brossier et al., 2010; Bozdağ et al., 2011; Rickers et al., 2012). The design of suitable measurements that allow us to solve the coupled source/structure inverse problem is an area of active research.

4.6 Discussion

In the following, we provide a summary of the previous mathematical developments in the light of practical applications. Furthermore, we give a brief overview of alternative approaches to interferometry that we do not cover in detail in this chapter due to limited space.

4.6.1 Green's Function Retrieval

The overwhelming majority of noise interferometry studies is based on Green's function retrieval by interstation correlation. The undeniable success of the method is due to its simplicity on multiple levels: The computation of correlation functions is an easy mathematical operation that lends itself very well to implementation on modern supercomputers that allow us to process large quantities of data in reasonable amounts of time (Fichtner et al., 2017b). While the theoretical requirements of Green's function retrieval are generally not met on Earth, at least the frequency-dependent traveltimes (dispersion) of fundamental-mode surface waves are empirically reliable and for most applications sufficiently relatable to the traveltimes of the Green's function. Furthermore, with such an approach, existing techniques for the inversion of earthquake or active-source data can be used to invert noise correlations without the need to develop genuinely new inversion technologies.

The drawback of Green's function retrieval lies in the absence of a theory that applies to the Earth with all its unavoidable complexities, including 3D heterogeneity, attenuation, anisotropy, non-equipartitioned noise, and noise sources that are heterogeneously distributed and highly variable in time. The consequences are, as mentioned in the introduction to this chapter, that traveltimes, amplitudes, and waveforms may be incorrect, and that modern inversion techniques that exploit complete waveforms for improved resolution are therefore not applicable.

In addition to practical limitations of Green's function retrieval, there is a philosophical dilemma. A physical theory is generally valid as long as there are no observations that it cannot explain (e.g., Popper, 1935; Tarantola, 2006). In this regard, the theories for Green's function recovery presented earlier in this chapter are *sensu stricto* invalid because real-data noise correlations typically carry clear signs of not being Green's functions (e.g., spurious arrivals, missing phases). From a pragmatic, though not from a science philosophy, point of view, this problem may be overcome to some extent by ignoring those parts of a noise correlation dataset that are not plausible Green's function approximations.

4.6.2 Interferometry Without Green's Function Retrieval

Interferometry without Green's function retrieval, presented in section 4.5, is designed to function without assumptions on the nature of the ambient noise wavefield and the properties of the Earth and noise sources. A correlation function is modeled on the basis of the noise-source power-spectral density distribution that may vary in time and space. Observable variations in the correlation function can then be related to variations in Earth structure and noise sources, either through model space sampling or via finite-frequency sensitivity kernels.

In addition to being conceptually clean, interferometry without Green's function retrieval offers the opportunity to exploit waveforms in the correlation function that one would not expect in the Green's function. This includes, for instance, early arrivals prior to the P-wave that result from noise sources outside the stationary phase regions.

The most significant drawback of interferometry without Green's function retrieval, which has so far limited its widespread application in geophysics, is increased mathematical and computational complexity. The forward problem cannot be solved simply by computing a Green's function between two receivers, and traditional inversion methods that ignore finite-frequency effects may not be used. Another price to pay for being able to exploit any waveform without restrictive assumptions is the need to account for sources and structure at the same time, which unavoidably increases the model space dimension and the nullspace. As in earthquake tomography, good results can only be achieved when sources and structure are constrained simultaneously (e.g., Valentine and Woodhouse, 2010). This also applies to interferometry based on Green's function retrieval, where a distribution of sources that is incorrectly assumed to be homogeneous may degrade the tomographic images. As an alternative to the joint inversion for source and structure, information from physical noise source models that relate ocean wave height to the noise-source power spectral density may be incorporated in the future (e.g., Ardhuin et al., 2011; Gualtieri et al., 2013, 2014; Ardhuin et al., 2015; Gualtieri et al., 2015; Farra et al., 2016).

While being more involved and still in its infancy, interferometry without Green's function retrieval is producing promising initial results that make future successful applications seem possible (e.g., Hanasoge, 2013; Basini et al., 2013; Delaney et al., 2017; Stehly and Boué, 2017; Ermert et al., 2017; Sager et al., 2018).

4.6.3 The Importance of Processing

In our previous mathematical developments we entirely ignored an essential element of practical ambient noise interferometry: data processing (see Chapter 5 [Ritzwoller and Feng, 2018]). Since the ambient noise wavefield is "polluted" by earthquake signals and excited by sources that are imperfect from the perspective of Green's function recovery, numerous processing and stacking schemes have been proposed in order to obtain more plausible Green's function approximations. These include the averaging of causal and acausal correlation branches, spectral whitening, time-domain running averages, and frequency-domain normalization (e.g., Bensen et al., 2007; Groos et al., 2012), as well as one-bit normalization (e.g., Larose et al., 2004; Shapiro and Campillo, 2004; Cupillard et al., 2011; Hanasoge and Branicki, 2013), phase-weighted stacks based on the Hilbert transform

(Schimmel and Paulssen, 1997; Schimmel et al., 2011) or the S transform (Baig et al., 2009), directional balancing (Curtis and Halliday, 2010), Welch's method of overlapping time windows (Welch, 1967; Seats et al., 2012), the application of curvelet de-noising filters (Stehly et al., 2011), or a sequence of selection and noise suppression filters (e.g., Boué et al., 2014; Nakata et al., 2015).

Despite the large number of noise interferometry studies, a clear consensus on the best processing scheme, however defined, has not emerged. This indicates that the optimal processing is dependent on the specifics of a particular dataset and on the type of information that one wishes to extract. Often, processing is to some extent subjective. Differences in processing can lead to significant differences in the correlation functions (Bensen et al., 2007), which leaves an unavoidable imprint on the sensitivity to Earth structure and noise sources (Fichtner, 2014; Stehly and Boué, 2017). An example of how processing can modify the correlation function is presented in Figure 4.8.

That processing modifies correlation waveforms and their sensitivity is not a problem in itself. However, it may become a problem when not properly taken into account in the forward modeling. From earthquake tomography it is well known that observations and synthetics must be processed in exactly the same way in order to ensure that the sometimes subtle differences between them are indeed

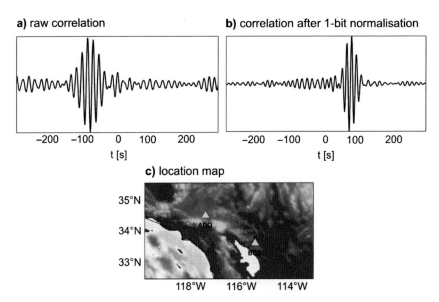

Figure 4.8. Modification of the correlation function between the Californian stations ADO and BC3. (a) Raw correlation function computed without pre-processing of the individual recordings. (b) Correlation function computed after one-bit normalization of the recordings. (c) Location map of stations ADO and BC3. (Figure modified from Fichtner et al. (2017a).)

due to sources or structure. Noise interferometry currently does not operate at the same level of precision, meaning that operations such as various normalizations or elaborate stacking are not taken into account in the forward problem solutions. This may lead to incorrect inferences of structure and sources. While a theory that incorporates noise correlation processing does exist (Fichtner et al., 2017a), it still awaits application to real data.

4.6.4 Alternative Approaches

The early recognition that the theoretical prerequisites of Green's function retrieval are not met on Earth led to the development of a large number of alternative approaches to ambient noise interferometry, only a small fraction of which can be summarized here.

Interferometry by deconvolution (e.g., Snieder and Şafak, 2006; Vasconcelos and Snieder, 2008a,b) replaces the frequency-domain multiplication in correlation by a division. As a result, the interferogram is less dependent on the wavefield source and still an approximation of a Green's function.

Similarly, interferometry by multi-dimensional deconvolution (e.g., Wapenaar et al., 2008; Wapenaar and van der Neut, 2010; Wapenaar et al., 2011) tries to correct for the lack of wavefield equipartitioning. Based on the realization that the virtual source of a noise correlation is not point localized but smeared into a point-spread function, the point-spread function is deconvolved in order to obtain a better Green's function approximation.

Instead of correcting for the imperfections of the noise sources, iterated correlation tries to exploit coda waves (e.g., Stehly et al., 2008). The main rationale is that the coda of a noise correlation may represent a wavefield that is closer to being equipartitioned than the original noise wavefield. Re-correlating the coda should then provide an improved Green's function approximation. The extent to which this approach relies on a homogeneous distribution of scatterers in the Earth remains to be fully explored.

4.7 Conclusions

During the past decade, ambient noise interferometry has profoundly changed the field of seismology. It has substantially increased tomographic resolution in regions without conventional wavefield sources, and it has enabled continuous monitoring of the subsurface.

The large majority of interferometry applications is based on Green's function retrieval by interstation correlation of noise. The success of this approach rests on the easy implementation of the noise correlation operation and on the possibility to reuse inversion methods for earthquake and active-source data almost without modification. Green's function retrieval can be justified theoretically using a variety of techniques, including normal-mode summation, plane-wave superposition, and representation theorems. The results are generally similar and based on related assumptions of wavefield equipartitioning, the homogeneous distribution of decorrelated noise sources, and the absence of attenuation. Since these assumptions are never satisfied on Earth, noise correlations only approximate Green's functions. The extent to which differences between the two are practically relevant is somewhat application-specific and may be assessed on a case-by-case basis.

Interferometry without Green's function retrieval aims to overcome these limitations by dropping any assumption on equipartitioning, noise source distribution, and wave propagation physics. Noise correlations are taken as what they are, without any attempt to approximate a Green's function. Measurements on noise correlations are related to noise sources and Earth structure via sensitivity kernels that allow us to solve inverse problems. This gain of generality comes at the price of increased computational complexity and the need to solve an inherently coupled inverse problem for both noise sources and Earth structure. First, this will require that source and structure inversion use noise correlations averaged over exactly the same time interval, because noise sources may change over time. Second, it is understood that the independent resolution of noise sources and Earth structure will require adequate coverage, the availability of which is a current topic of research.

The future of ambient noise interferometry is naturally hard to predict. While the basics of interferometry by Green's function retrieval have become standard, continued innovation will likely depend on our ability to exploit more information in a more reliable way. This will involve a combination of both improvements of data processing schemes and improvements of the forward and inverse modeling physics.

Acknowledgments

The authors are grateful to Laura Ermert, Patrick Paitz, Korbinian Sager, Roel Snieder, and an anonymous reviewer for fruitful discussions that helped us to improve this chapter.

References

Aki, K. 1957. Space and time spectra of stationary stochastic waves, with special reference to microtremors. *Bull. Earthq. Res. Inst., Univ. Tokyo*, **35**, 415–457.

Aki, K., and Richards, P. 2002. *Quantitative Seismology*. University Science Books.

Ardhuin, F., Gualtieri, L., and Stutzmann, E. 2018. Physics of ambient noise generation by ocean waves. Pages – of: Nakata, N., Gualtieri, L., and Fichtner, A. (eds.), *Seismic Ambient Noise*. Cambridge University Press, Cambridge, UK.

———. 2015. How ocean waves rock the Earth: Two mechanisms explain microseisms with periods 3 to 300 s. *Geophys. Res. Lett.*, **42**(3), 765–772.

Ardhuin, F., Stutzmann, E., Schimmel, M., and Mangeney, A. 2011. Ocean wave sources of seismic noise. *J. Geophys. Res.*, **116**, doi:10.1029/2011JC006952.

Båth, M. 1968. *Mathematical aspects of seismology*. Elsevier Publishing Company, Amsterdam, London, New York.

Baig, A. M., Campillo, M., and Brenguier, F. 2009. Denoising seismic noise cross correlations. *J. Geophys. Res.*, **114**, doi:10.1029/2008JB006085.

Basini, P., Nissen-Meyer, T., Boschi, L., Casarotti, E., Verbeke, J., Schenk, O., and Giardini, D. 2013. The influence of nonuniform ambient noise on crustal tomography in Europe. *Geochem. Geophys. Geosys.*, **14**, 1471–1492.

Bender, C. M., and Orszag, S. A. 1999. *Advanced Mathematical Methods for Scientists and Engineers: Asymptotic Methods and Perturbation Theory*. Springer, New York.

Bensen, G. D., Ritzwoller, M. H., Barmin, M. P., Levshin, A. L., Lin, F., Moschetti, M. P., Shapiro, N. M., and Yang, Y. 2007. Processing seismic ambient noise data to obtain reliable broad-band surface wave dispersion measurements. *Geophys. J. Int.*, **169**, 1239–1260.

Boschi, L., Weemstra, C., Verbeke, J., Ekström, G., Zunino, A., and Giardini, D. 2013. On measuring surface wave phase velocity from station-station cross-correlation of ambient signal. *Geophys. J. Int.*, **192**, 346–358.

Boué, P., Poli, P., Campillo, M., Pedersen, H., Briand, X., and Roux, P. 2013. Teleseismic correlations of ambient noise for deep global imaging of the Earth. *Geophys. J. Int.*, **194**, 844–848.

Boué, P., Roux, P., Campillo, M., and Briand, X. 2014. Phase velocity tomography of surface waves using ambient noise cross correlation and array processing. *J. Geophys. Res.*, **119**, 519–529.

Bozdağ, E., Trampert, J., and Tromp, J. 2011. Misfit functions for full waveform inversion based on instantaneous phase and envelope measurements. *Geophys. J. Int.*, **185**, 845–870.

Brenguier, F., Campillo, M., Haziioannou, C., Shapiro, N. M., Nadeau, R. M., and Larose, E. 2008a. Postseismic relaxation along the San Andreas fault at Parkfield from continuous seismological observations. *Science*, **321**, 1478–1481.

Brenguier, F., Shapiro, N. M., Campillo, M., Ferrazzini, V., Duputel, Z., Coutant, O., and Nercessian, A. 2008b. Towards forecasting volcanic eruptions using seismic noise. *Nat. Geosci.*, **1**, 126–130.

Brossier, R., Operto, S., and Virieux, J. 2010. Which data residual norm for robust elastic frequency-domain full waveform inversion? *Geophysics*, **75**, R37–R46.

Bussat, S., and Kugler, S. 2011. Offshore ambient-noise surface-wave tomography above 0.1 Hz and its applications. *The Leading Edge*, **May 2011**, 514–524.

Claerbout, J. F. 1968. Synthesis of a layered medium from its acoustic transmission response. *Geophysics*, **33**, 264–269.

Cole, S. P. 1995. *Passive seismic and drill-bit experiments using 2-D arrays*. Ph.D. thesis, The Stanford Exploration Project, Stanford University.

Cupillard, P. 2008. *Simulation par la méthode des éléments spectraux des formes donde obtenues par corrélation de bruit sismique*. PhD thesis, Institut de Physique du Globe de Paris.

Cupillard, P., and Capdeville, Y. 2010. On the amplitude of surface waves obtained by noise correlation and the capability to recover the attenuation: a numerical approach. *Geophys. J. Int.*, **181**, 1687–1700.

Cupillard, P., Stehly, L., and Romanowicz, B. 2011. The one-bit noise correlation: a theory based on the concepts of coherent and incoherent noise. *Geophys. J. Int.*, **184**, 1397–1414.

Curtis, A., and Halliday, D. 2010. Directional balancing for seismic and general wavefield interferometry. *Geophysics*, **75**, doi: 10.1190/1.3298736.

Curtis, A., Gerstoft, P., Sato, H., Snieder, R., and Wapenaar, K. 2006. Seismic interferometry - Turning noise into signal. *The Leading Edge*, **25**, 1082–1092.

de Ridder, S. A. L., and Biondi, B. L. 2015. Ambient seismic noise tomography at Ekofisk. *Geophysics*, **80**, B167–B176.

de Ridder, S. A. L., Biondi, B. L., and Clapp, R. G. 2014. Time-lapse seismic noise correlation tomography at Valhall. *Geophys. Res. Lett.*, **41**, 6116–6122.

Delaney, E., Ermert, L., Sager, K., Kritski, A., Bussat, S., and Fichtner, A. 2017. Passive seismic monitoring with non-stationary noise sources. *Geophysics*, **82**, 10.1190/geo2016–0330.1.

Dziewoński, A. M., and Anderson, D. L. 1981. Preliminary reference Earth model. *Phys. Earth Planet. Inter.*, **25**, 297–356.

Ermert, L., Sager, K., Afanasiev, M., Boehm, C., and Fichtner, A. 2017. Ambient noise source inversion in a heterogeneous Earth - Theory and application to the Earth's hum. *J. Geophys. Res.*, **122**, doi:10.1002/2017JB014738.

Ermert, L., Villasenor, A., and Fichtner, A. 2016. Cross-correlation imaging of ambient noise sources. *Geophys. J. Int.*, **204**, 347–364.

Farra, V., Stutzmann, E., Gualtieri, L., and Ardhuin, F. 2016. Ray-theoretical modeling of secondary microseism P waves. *Geophys. J. Int.*, **206**, 1730–1739.

Fichtner, A. 2014. Source and processing effects on noise correlations. *Geophys. J. Int.*, **197**, 1527–1531.

———. 2015. Source-structure trade-offs in ambient noise correlations. *Geophys. J. Int.*, **202**, 678–694.

Fichtner, A., Kennett, B. L. N., Igel, H., and Bunge, H.-P. 2008. Theoretical background for continental- and global-scale full-waveform inversion in the time-frequency domain. *Geophys. J. Int.*, **175**, 665–685.

———. 2009. Full seismic waveform tomography for upper-mantle structure in the Australasian region using adjoint methods. *Geophys. J. Int.*, **179**, 1703–1725.

Fichtner, A., Ermert, L., and Gokhberg, A. 2017b. Seismic noise correlation on heterogeneous supercomputers. *Seis. Res. Lett.*, 88, 1141–1145.

Fichtner, A., Stehly, L., Ermert, L., and Boehm, C. 2017a. Generalised interferometry - I. Theory for inter-station correlations. *Geophys. J. Int.*, **208**, 603–638.

Forghani, F., and Snieder, R. 2010. Underestimation of body waves and feasibility of surface wave reconstruction by seismic interferometry. *The Leading Edge*, **July 2010**, 790–794.

Friederich, W. 2003. The S-velocity structure of the East Asian mantle from inversion of shear and surface waveforms. *Geophys. J. Int.*, **153**, 88–102.

Froment, B., Campillo, M., Roux, P., Gouédard, P., Verdel, A., and Weaver, R. L. 2010. Estimation of the effect of nonisotropically distributed energy on the apparent arrival time in correlations. *Geophysics*, **75**, SA85–SA93.

Gal, M., and Reading, A. M. 2018. Beamforming and polarization analysis. Pages – of: Nakata, N., Gualtieri, L., and Fichtner, A. (eds.), *Seismic Ambient Noise*. Cambridge University Press, Cambridge, UK.

Gee, L. S., and Jordan, T. H. 1992. Generalized seismological data functionals. *Geophys. J. Int.*, **111**, 363–390.

Gilbert, F. 1970. Excitation of the normal modes of the Earth by earthquake sources. *Geophys. J. R. astr. Soc.*, **22**, 223–226.

Gizon, L., and Birch, A. C. 2002. Time-distance helioseismology: the forward problem for random distributed sources. *Astrophys. J.*, **571**, 966–986.

Groos, J. C., Bussat, S., and Ritter, J. R. R. 2012. Performance of different processing schemes in seismic noise cross-correlations. *Geophys. J. Int.*, **188**, 498–512.

Gualtieri, L., Stutzmann, E., Capdeville, Y., Ardhuin, F., Schimmel, A. M., and Morelli, A. 2013. Modeling secondary microseismic noise by normal mode summation. *Geophys. J. Int.*, **193**, 1732–1745.

Gualtieri, L., Stutzmann, E., Capdeville, Y., Farra, V., Mangeney, A., and Morelli, A. 2015. On the shaping factors of the secondary microseismic wavefield. *J. Geophys. Res.*, **120**, 6241–6262.

Gualtieri, L., Stutzmann, E., Farra, V., Capdeville, Y., Schimmel, M., Ardhuin, F., and Morelli, A. 2014. Modeling the ocean site effect on seismic noise body waves. *Geophys. J. Int.*, **197**, 1096–1106.

Haberman, R. 2013. *Applied Partial Differential Equations*. Pearson.

Halliday, D., and Curtis, A. 2008. Seismic interferometry, surface waves and source distribution. *Geophys. J. Int.*, **175**, 1067–1087.

Hanasoge, S. M. 2013. Measurements and kernels for source-structure inversions in noise tomography. *Geophys. J. Int.*, **192**, 971–985.

Hanasoge, S. M., and Branicki, M. 2013. Interpreting cross-correlations of one-bit filtered noise. *Geophys. J. Int.*, **195**, 1811–1830.

Hanasoge, S. M., Birch, A., Gizon, L., and Tromp, J. 2011. The adjoint method applied to time-distance helioseismology. *Astrophys. J.*, **738**, doi:10.1088/0004–637X/738/1/100.

Haned, A., Stutzmann, E., Schimmel, M., Kiselev, S., Davaille, A., and Yelles-Chaouche, A. 2016. Global tomography using seismic hum. *Geophys. J. Int.*, **204**, 1222–1236.

Hillers, G., Husen, S., Obermann, A., Planes, T., Larose, E., and Campillo, M. 2015. Noise-based monitoring and imaging of aseismic transient deformation induced by the 2006 Basel reservoir stimulation. *Geophysics*, **80**, KS51–KS68.

Huang, H.-H., Lin, F.-C., Tsai, V. C., and Koper, K. D. 2016. High-resolution probing of inner core structure with seismic interferometry. *Geophys. Res. Lett.*, in press.

Igel, H., Djikpesse, H., and Tarantola, A. 1996. Waveform inversion of marine reflection seismograms for P impedance and Poisson's ratio. *Geophys. J. Int.*, **124**, 363–371.

Kao, H., Behr, Y., Currie, C. A., Hyndman, R., Townend, J., Lin, F.-C., Ritzwoller, M. H., Shan, S.-J., and He, J. 2013. Ambient seismic noise tomography of Canada and adjacent regions: Part I. Crustal structures. *J. Geophys. Res.*, **118**, 5865–5887.

Kästle, E. D., Soomro, R., Weemstra, C., Boschi, L., and Meier, T. 2016. Two-receiver measurements of phase velocity of ambient-noise and earthquake-based observations. *Geophys. J. Int.*, **207**, 1493–1512.

Kennett, B. L. N., Engdahl, E. R., and Buland, R. 1995. Constraints on seismic velocities in the Earth from traveltimes. *Geophys. J. Int.*, **122**, 108–124.

Kimman, W., and Trampert, J. 2010. Approximations in seismic interferometry and their effects on surface waves. *Geophys. J. Int.*, **182**, 461–476.

Larose, E., Derode, A., Campillo, M., and Fink, M. 2004. Imaging from one-bit correlations of wideband diffuse wave fields. *J. Appl. Phys.*, **95**, 8393–8399.

Lin, F.-C., Moschetti, M. P., and Ritzwoller, M. H. 2008. Surface wave tomography of the western United States from ambient noise: Rayleigh and Love wave phase velocity maps. *Geophys. J. Int.*, **173**, 281–298.

Lin, F.-C., Tsai, V. C., Schmandt, B., Duputel, Z., and Zhan, Z. 2013. Extracting seismic core phases with array interferometry. *Geophys. Res. Lett.*, **40**, 1049–1053.

Lobkis, O. I., and Weaver, R. L. 2001. On the emergence of the Green's function in the correlations of a diffuse field. *J. Acoust. Soc. Am.*, **110**, 3011–3017.

Luo, Y., and Schuster, G. T. 1991. Wave-equation traveltime inversion. *Geophysics*, **56**, 645–653.

Malcolm, A. E., Scales, J., and van Tiggelen, B. A. 2004. Extracting the Green function from diffuse, equipartitioned waves. *Phys. Rev. E*, **70**, doi:10.1103/PhysRevE.70.015601.

McNamara, D., Boaz, R., and Buland, R. 2018. Visualization of the Ambient Seismic Noise Spectrum. Pages – of: Nakata, N., Gualtieri, L., and Fichtner, A. (eds.), *Seismic Ambient Noise*. Cambridge University Press, Cambridge, UK.

Mordret, A., Landès, M., Shapiro, N., Singh, S., and Roux, P. 2014. Ambient noise surface wave tomography to determine the shallow shear velocity structure at Valhall: depth inversion with a Neighbourhood Algorithm. *Geophysical Journal International*, **198**(3), 1514–1525.

Mordret, A., Landès, M., Shapiro, N., Singh, S., Roux, P., and Barkved, O. 2013. Near-surface study at the Valhall oil field from ambient noise surface wave tomography. *Geophysical Journal International*, ggt061.

Morse, P. M., and Feshbach, H. 1953. *Methods of Theoretical Physics*. McGraw-Hill, New York.

Nakata, N., Chang, J. P., Lawrence, J. F., and Boué, P. 2015. Body wave extraction and tomography at Long Beach, California, with ambient-noise interferometry. *J. Geophys. Res.*, **120**, 1159–1173.

Nishida, K., and Montagner, J.-P. 2009. Global surface wave tomography using seismic hum. *Science*, **326**, 5949.

Nooghabi, A. H., Boschi, L., Roux, P., and de Rosny, J. 2017. Coda reconstruction from cross-correlation of a diffuse field on thin elastic plates. *Phys. Rev. E*, under review.

Obermann, A., Froment, B., Campillo, M., Larose, E., Planes, T., Valette, B., Chen, J.-H., and Liu, Q. Y. 2014. Seismic noise correlations to image structural and mechanical changes associated with the Mw 7.9 2008 Wenchuan earthquake. *J. Geophys. Res.*, **119**, doi:10.1002/2013JB010932.

Obermann, A., Kraft, T., Larose, E., and Wiemer, S. 2015. Potential of ambient seismic noise techniques to monitor the St. Gallen geothermal site (Switzerland). *J. Geophys. Res.*, **120**, doi:10.1002/2014JB011817.

Obermann, A., Planes, T., Larose, E., and Campillo, M. 2013. Imaging preeruptive and coeruptive structural and mechanical changes of a volcano with ambient seismic noise. *J. Geophys. Res.*, **118**, 1–10.

Poli, P., Campillo, M., Pedersen, H., and LAPNET Working Group. 2012. Body-wave imaging of Earth's mantle discontinuities from ambient seismic noise. *Science*, **38**, 1063–1065.

Poli, P., Thomas, C., Campillo, M., and Pedersen, H. 2015. Imaging the D" reflector with noise correlations. *Geophys. Res. Lett.*, **42**, 60–65.

Popper, K. 1935. *Logik der Forschung (The logic of scientific discovery)*. Verlag von Julius Springer, Vienna, Austria.

Pratt, R. G. 1999. Seismic waveform inversion in the frequency domain, Part 1: Theory and verification in a physical scale model. *Geophysics*, **64**, 888–901.

Rickers, F., Fichtner, A., and Trampert, J. 2012. Imaging mantle plumes with instantaneous phase measurements of diffracted waves. *Geophys. J. Int.*, **190**, 650–664.

Ritsema, J., vanHeijst, H., and Woodhouse, J. H. 1999. Complex shear wave velocity structure imaged beneath Africa and Iceland. *Science*, **286**, 1925–1928.

Ritzwoller, M. H., and Feng, L. 2018. Overview of pre- and post-processing of ambient noise correlations. Pages – of: Nakata, N., Gualtieri, L., and Fichtner, A. (eds.), *Seismic Ambient Noise*. Cambridge University Press, Cambridge, UK.

Sabra, K. G., Gerstoft, P., Roux, P., and Kuperman, W. A. 2005. Surface wave tomography from microseisms in Southern California. *Geophys. Res. Lett.*, **32**, doi:10.1029/2005GL023155.

Sager, K., Ermert, L., Boehm, C., and Fichtner, A. 2018. Towards full waveform ambient noise inversion. *Geophys. J. Int.*, **212**, 566–590.

Saygin, E., and Kennett, B. L. N. 2012. Crustal structure of Australia from ambient seismic noise tomography. *J. Geophys. Res.*, **117**, doi:10.1029/2011JB008403.

Schimmel, M., and Paulssen, H. 1997. Noise reduction and detection of weak, coherent signals through phase-weighted stacks. *Geophys. J. Int.*, **130**, 497–505.

Schimmel, M., Stutzmann, E., and Gallart, J. 2011. Using instantaneous phase coherence for signal extraction from ambient noise data at a local to a global scale. *Geophys. J. Int.*, **184**, 494–506.

Seats, K. J., Lawrence, J. F., and Prieto, G. A. 2012. Improved ambient noise correlation functions using Welch's method. *Geophys. J. Int.*, **188**, 513–523.

Sens-Schönfelder, C., and Brenguier, F. 2018. Noise-based monitoring. Pages – of: Nakata, N., Gualtieri, L., and Fichtner, A. (eds.), *Seismic Ambient Noise*. Cambridge University Press, Cambridge, UK.

Shapiro, N. 2018. Applications with surface waves extracted from ambient seismic noise. Pages — of: Nakata, N., Gualtieri, L., and Fichtner, A. (eds.), *Seismic Ambient Noise*. Cambridge University Press, Cambridge, UK.

Shapiro, N. M., and Campillo, M. 2004. Emergence of broadband Rayleigh waves from correlations of the ambient seismic noise. *Geophys. Res. Lett.*, **31**, doi:10.1029/2004GL019491.

Shapiro, N. M., Campillo, M., Stehly, L., and Ritzwoller, M. 2005. High resolution surface wave tomography from ambient seismic noise. *Science*, **307**, 1615–1618.

Snieder, R. 2004a. Extracting the Green's function from the correlation of coda waves: A derivation based on stationary phase. *Phys. Rev. E*, **69**, doi:10.1103/PhysRevE.69.046610.

Snieder, R. 2004b. *Mathematical Methods for the Physical Sciences*. Cambridge University Press, Cambridge.

Snieder, R., and Şafak, E. 2006. Extracting the building response using seismic interferometry: Theory and application to the Millikan Library in Pasadena, California. *Bull. Seis. Soc. Am.*, **96**, 586–598.

Snieder, R., Duran, A., and Obermann, A. 2018. Locating velocity changes in elastic media with coda wave interferometry. Pages – of: Nakata, N., Gualtieri, L., and Fichtner, A. (eds.), *Seismic Ambient Noise*. Cambridge University Press, Cambridge, UK.

Stehly, L., and Boué, P. 2017. On the interpretation of the amplitude decay of noise correlations computed along a line of receivers. *Geophys. J. Int.*, **209**, 358–372.

Stehly, L., Campillo, M., Froment, B., and Weaver, R. L. 2008. Reconstructing Greens function by correlation of the coda of the correlation (C3) of ambient seismic noise. *J. Geophys. Res.*, **113**, doi:10.1029/2008JB005693.

Stehly, L., Cupillard, P., and Romanowicz, B. 2011. Towards improving ambient noise tomography using simultaneously curvelet denoising filters and SEM simulations of seismic ambient noise. *Com. Rend. Geosc.*, **343**, 591–599.

Stehly, L., Fry, B., Campillo, M., Shapiro, N. M., Guilbert, J., Boschi, L., and Giardini, D. 2009. Tomography of the Alpine region from observations of seismic ambient noise. *Geophys. J. Int.*, **178**, 338–350.

Tape, C., Liu, Q., Maggi, A., and Tromp, J. 2010. Seismic tomography of the southern California crust based upon spectral-element and adjoint methods. *Geophys. J. Int.*, **180**, 433–462.

Tarantola, A. 2006. Popper, Bayes and the inverse problem. *Nature Physics*, **2**, 492–494.

Tromp, J., Luo, Y., Hanasoge, S., and Peter, D. 2010. Noise cross-correlation sensitivity kernels. *Geophys. J. Int.*, **183**, 791–819.

Tsai, V. C. 2009. On establishing the accuracy of noise tomography traveltime measurements in a realistic medium. *Geophys. J. Int.*, **178**, 1555–1564.

———. 2010. The Relationship Between Noise Correlation and the Green's Function in the Presence of Degeneracy and the Absence of Equipartition. *Geophys. J. Int.*, **182**, 1509–1514.

———. 2011. Understanding the amplitudes of noise correlation measurements. *J. Geophys. Res.*, **116**, doi:10.1029/2011JB008483.

Valentine, A. P., and Woodhouse, J. H. 2010. Reducing errors in seismic tomography: Combined inversion for sources and structure. *Geophys. J. Int.*, **180**, 847–857.

van Leeuwen, T., and Mulder, W. A. 2010. A correlation-based misfit criterion for wave-equation traveltime tomography. *Geophys. J. Int.*, **182**, 1383–1394.

van Wijk, K. 2006. On estimating the impulse response between receivers in a controlled ultrasonic experiment. *Geophysics*, **71**, SI79–SI84.

Vasconcelos, I., and Snieder, R. 2008a. Interferometry by deconvolution, Part 1 – Theory for acoustic waves and numerical examples. *Geophysics*, **73**, S129–S141.

Vasconcelos, I., and Snieder, R. 2008b. Interferometry by deconvolution, Part 2 – Theory for elastic waves and application to drill-bit seismic imaging. *Geophysics*, **73**, S115–S128.

Verbeke, J., Boschi, L., Stehly, L., Kissling, E., and Michelini, A. 2012. High-resolution Rayleigh-wave velocity maps of central Europe from a dense ambient-noise data set. *Geophys. J. Int.*, **188**, 1173–1187.

Wapenaar, K. 2004. Retrieving the elastodynamic Green's function of an arbitrary inhomogeneous medium by cross correlation. *Phys. Rev. Lett.*, **93**, 254301.

Wapenaar, K., and Fokkema, J. 2006. Green's function representations for seismic interferometry. *Geophysics*, **71**, SI33–SI46.

Wapenaar, K., and van der Neut, J. 2010. A representation for Greens function retrieval by multidimensional deconvolution. *J. Acoust. Soc. Am.*, **128**, 366–371.

Wapenaar, K., Ruigrok, E., van der Neut, J., and Draganov, D. 2011. Improved surface-wave retrieval from ambient seismic noise by multi-dimensional deconvolution. *Geophys. Res. Lett.*, **38**, doi:10.1029/2010GL045523.

Wapenaar, K., van der Neut, J., and Ruigrok, E. 2008. Passive seismic interferometry by multidimensional deconvolution. *Geophysics*, **73**, A51–A56.

Watson, G. N. 2008. *A Treatise on the Theory of Bessel Functions*. Merchant Books, LaVergne.

Weaver, R. L., and Lobkis, O. I. 2004. Diffuse fields in open systems and the emergence of Green's function. *J. Acoust. Soc. Am.*, **116**, 2731–2734.

Weaver, R. L., Hadziioannou, C., Larose, E., and Campillo, M. 2011. On the precision of noise correlation interferometry. *Geophys. J. Int.*, **185**, 1384–1392.

Welch, P. D. 1967. The use of fast Fourier transform for the estimation of power spectra: A method based on time-averaging over short, modified periodograms. *IEEE Trans. Audio Electroacoust.*, **15**, 70–73.

Woodard, M. F. 1997. Implications of localized, acoustic absorption for heliotomographic analysis of sunspots. *Astrophys. J.*, **485**, 890–894.

Yoshizawa, K., and Kennett, B. L. N. 2004. Multi-mode surface wave tomography for the Australian region using a 3-stage approach incorporating finite-frequency effects. *J. Geophys. Res.*, **109**, doi:10.1029/2002JB002254.

———. 2005. Sensitivity kernels for finite-frequency surface waves. *Geophys. J. Int.*, **162**, 910–926.

Zhou, Y., Dahlen, F. A., and Nolet, G. 2004. Three-dimensional sensitivity kernels for surface wave observables. *Geophys. J. Int.*, **158**, 142–168.

Zhou, Y., Nolet, G., Dahlen, F. A., and Laske, G. 2006. Global upper-mantle structure from finite-frequency surface-wave tomography. *J. Geophys. Res.*, **111**, doi:10.1029/2005JB003677.

Zheng, Y., Shen, W., Zhou, L., Yang, Y., Xie, Z., and Ritzwoller, M. 2011. Crust and uppermost mantle beneath the North China Craton, northeastern China, and the Sea of Japan from ambient noise tomography. *J. Geophys. Res.*, **116**.

5

Overview of Pre- and Post-Processing of Ambient-Noise Correlations

MICHAEL H. RITZWOLLER AND LILI FENG

Abstract

All applications of ambient seismic noise, whether to study the source of the noise or the medium of propagation (static or time variable), are based fundamentally on a single observational challenge. This challenge is to process raw seismograms in a way that promotes the emergence of the signals of interest while suppressing the signals of disinterest. Here we summarize methods designed to achieve this delicate task in both continental and oceanic settings and review evidence that the observational challenge is met successfully.

5.1 Introduction

The purpose of this chapter is to discuss the problem of how to prepare seismic recordings for ambient noise data processing, and to assess how well cross-correlations of such records summarize information about the Earth. Our focus will be on broadband seismic data at relatively long periods (5–100 s) observed over relatively long distances (a few tens of kilometers to a few thousands of kilometers). Such data, recorded on recent-generation broadband seismic arrays such as USArray, provide information about the Earth with lateral resolutions on a regional scale (i.e., from a few tens to a few hundreds of kilometers). Signals of this nature derived from ambient noise are enriched in fundamental modes, so that most information derived about the Earth relates to the structure of the crust and uppermost mantle. Although our focus is fairly tight, ignoring as it does problems that may be encountered in exploration seismology and the extraction of body waves, which is discussed in Chapter 8 (Nakata and Nishida, 2018), the issues we consider are universal. The problem we address is how to construct reliable information about the Earth from observations that most of us would think of as noise.

Our approach is practical, heuristic, and non-rigorous. It is not historical and we do not attempt to provide a comprehensive summary of ambient noise pre-and post-processing procedures that have developed over the somewhat more than decade-long history of ambient noise seismology. Rather, we aim to provide a discussion that researchers who are interested in pursuing the subject may find useful as a starting point or as a guide to pursue their own creative work in ambient noise seismology.

It is remarkable that seismologists can derive deterministic structural information from recordings of seismic noise. Snieder (2004) provides a lucid and rigorous proof that the cross-correlation of seismograms recorded at a pair of stations will yield the Green's function between the stations, at least in the idealized setting underlying the proof. The Green's function is the impulse response of the medium, and provides structural information about the medium of transport between the stations.

The reader is referred to Chapter 4 (Fichtner and Tsai, 2018) for a more comprehensive discussion of the theory behind the emergence of signals from ambient noise. However, a significant aspect of the idealized setting of the proof referred to in the previous paragraph is that the sources of ambient noise are non-correlated and randomly and homogeneously distributed in space and time. Observing conditions in the earth do not meet these conditions; thus, the best that any data processing procedure can achieve is a plausible approximation to the Green's function.

Figure 5.1 motivates a heuristic explanation of why Green's functions might emerge approximately in practice from cross-correlations of ambient noise. A hyperbola is defined as the set of points in a plane in which the difference between the distances to two fixed points is a constant k. Figure 5.1 shows a set of hyperbolas where the fixed points are a pair of seismic stations (the two triangles in the figure). Each hyperbola is characterized by a single constant, k, which differs from its nearest neighbors by $\pm\pi$. Thus, the hyperbolas in Figure 5.1 are a set of level curves for the constants k. The figure illustrates that the level curves become increasingly closely spaced as they diverge from the horizontal; i.e., from the interstation direction. As a consequence, over most of the plane two closely spaced seismic events will produce waves that arrive at the two stations at different times (or phase, which is time divided by period) unless they lie nearly exactly along one of the hyperbolas, and the phase difference will depend strongly on the relative locations of the events. If such events have similar amplitudes they will have very different expressions in the seismograms recorded at the two stations, and they will therefore destructively interfere in the cross-correlation between these seismograms. There are two exceptions to the destructive interference described in the previous paragraph, the first good and the second bad for our purposes.

(1) In the first case, events that are nearly aligned with the two stations (in the so called "end-fire" direction) are more likely to arrive with similar phases on the

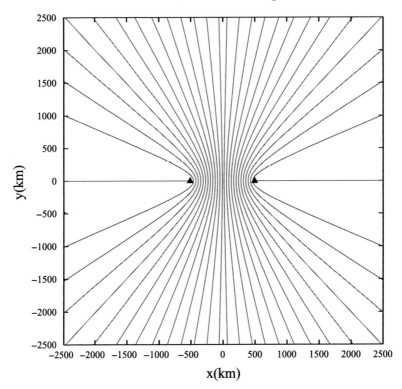

Figure 5.1. Examples of iso-phase hyperbolas (gray lines) in the plane with foci at two stations (triangles) separated from nearest neighbors by $\pm\pi$. A phase speed of 3 km/s at 50 s period is used to compute these hyperbolas. Only sources found on the same hyperbola will constructively interfere on cross-correlations of signals recorded at the two stations. Sources on different hyperbolas will destructively interfere. Sources found near or along the "end-fire" hyperbolas are least likely to be affected by destructively interfering sources, and are therefore most likely to appear on cross-correlations of long time sequences of ambient noise. This figure is adapted from Figure 16 of Lin et al. (2008).

two recordings because the level curves are more separated from one another. Such events will tend to interfere constructively on cross-correlations. It is this constructive interference between the effects of events nearly aligned with the stations in concert with the destructive interference for events in other directions that results in the recovery of waves propagating between the two stations in the cross-correlation, as long as sufficient numbers of events have taken place. As discussed in section 5.2, in this case the cross-correlation will be related to the Green's function between the pair of stations as long as there are events near the end-fire direction and if the duration of observation is long enough.

(2) The second case is a pernicious one, however, in which either an enormous event (with a much larger amplitude than other interfering sources) has occurred

or a series of smaller events occur persistently at a single location. In this case the arrivals will not destructively interfere upon cross-correlation between recordings from the two stations and an interfering signal will remain that is not related to the Green's function between the stations. Examples of such events are the Gulf of Guinea microseism (e.g., Shapiro et al., 2006), also referred to as the 26 s microseism, and the Kyushu microseism (Zeng and Ni, 2010) that results from Aso Volcano, the effects of which have been observed clearly in ambient noise correlations for stations in East Asia (Zheng et al., 2011).

This chapter discusses observational techniques that have emerged to help extract clear estimated (or empirical) Green's functions (EGFs), while discriminating against sources of noise such as earthquakes or persistent localized events. Applications of the estimated Green's functions are discussed in Chapter 7 (Shapiro, 2018). An example of one such estimated Green's function produced from a year of continuous recordings observed at broadband stations HRV and ANMO is shown in Figure 5.2 separated into different frequency bands. Records such as this are sought to form the basis for ambient noise tomography.

There are intermediate circumstances between those that deliver clear estimated Green's functions and those that generate clear artificial arrivals. These are characterized by an azimuthally inhomogeneous source distribution, which is stronger in some directions than others. Many studies have considered whether such situations, which are common in nature, will deliver reliable estimated Green's functions and find that as long as there are events in the end-fire direction, then a reliable estimated Green's functions will emerge eventually (e.g., Lin et al., 2008; Yang and Ritzwoller, 2008); it is just a matter of observing long enough. But, how does an observer know if enough time has elapsed so that a meaningful estimate of the Green's function has emerged? This is one of the principal practical concerns in ambient noise seismology. Another way to ask the same question is: Has enough signal emerged from the noise to be useful?

There is no definitive answer to these questions, but there are many relevant indicators as discussed later in section 5.6. We mention now one line of evidence that is particularly important, related to so-called precursory noise. The time of arrival of a signal on a cross-correlation will be the difference in arrival times recorded at the two stations. Events on the perpendicular bisector of the line linking the two stations will produce arrivals at the same time on the two stations, and thus will arrive at zero lag time on the cross-correlation. In contrast, events in the end-fire configuration will generate arrivals that are separated by the phase propagation time between the two stations. Thus, end-fire events, which constructively interfere to produce the estimated Green's functions, produce the latest arrivals on the cross-correlation. All other events will produce signals that arrive sooner and will superpose in the cross-correlation to produce precursory noise. The level

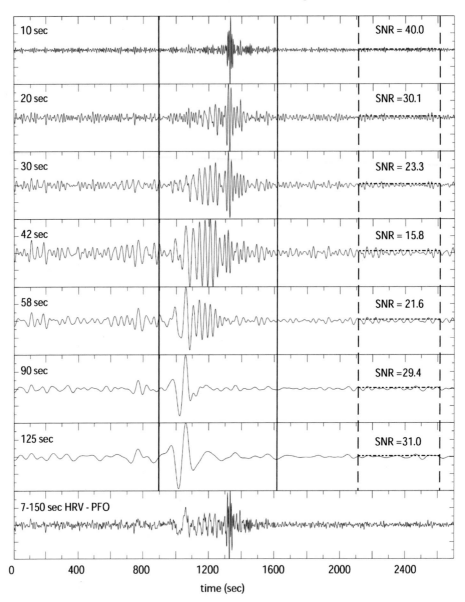

Figure 5.2. Example of a broadband symmetric-component cross-correlation of 12 months of ambient noise from stations ANMO (Albuquerque, NM, USA) and HRV (Harvard, MA, USA). The broadband signal is at the bottom, and successively longer-period passbands are presented from top to bottom in the figure, centered on the period shown at left in each panel. Vertical solid lines indicate the signal window and the vertical dashed lines the noise windows. SNR is defined as the peak amplitude in the signal window divided by the RMS amplitude in the trailing noise window (indicated with the horizontal dashed line on each panel). SNR is labeled on the right-hand side of each panel. The symmetric-component is the average of the cross-correlations at positive and negative lags, given by equation (5.8). (This figure is taken from Figure 11 of Bensen et al. (2007).)

of this precursory noise compared to the amplitude of the signal of interest is a good indicator of the convergence of the method toward a reliable estimated Green's function. Some researchers have used the observation of precursors to locate locally persistent sources (e.g., Tian and Ritzwoller, 2015). We take up the issue of precursory noise again in section 5.6.

This chapter is structured as follows. In section 5.3 we present the notation and terminology used throughout the chapter and discuss ambient noise data processing in a generalized form. In the end, we seek observational methods to yield broadband, low variance, and unbiased information about the Earth. In addition, we seek methods that will speed convergence and thus reduce observation time. With these goals in mind, sections 5.3 and 5.4 discuss the practical and specific application of ambient noise data processing in a continental setting for Rayleigh and Love waves, and section 5.5 presents a discussion of the data processing in ocean bottom environments. Section 5.6 discusses what might be referred to as post-processing assessment: How do we select some recordings to accept and others to reject (because not all will reflect Earth structure accurately), how do we quantify uncertainties in surface wave dispersion measurements and dispersion maps, and how do we know if our results, in the end, are right? The chapter closes with section 5.7, in which we present a few examples of new methods that show promise to improve the output from ambient noise data processing.

5.2 Idealized Background

Let the vector $\mathbf{u} = (u, v, w)$ be ground motion recorded at an unspecified location, such that u is the vertical component (Z) and v and w are the two horizontal components (east, E, and north, N, respectively). Then let $u_i(t)$ and $u_j(t)$ denote vertical component seismograms at stations i and j separated by distance r, recorded on a finite time interval $t \in [0, T]$. The cross-spectrum between these recordings (or the cross-correlation in the frequency domain) is as follows:

$$\gamma_{ij}(\omega) \equiv u_i(\omega)u_j^*(\omega), \tag{5.1}$$

where $u(\omega)$ is the Fourier transform of $u(t)$, ω is frequency, and $*$ denotes the complex conjugate. The cross-correlation in the time domain is the inverse Fourier transform of $\gamma_{ij}(\omega)$, denoted $\gamma_{ij}(\tau)$, where τ here is the cross-correlation lag time.

It is often of practical interest (see section 5.3) to normalize the spectrum of ground motion in some way such that

$$\tilde{u}(\omega) \equiv \frac{u(\omega)}{N(\omega)}, \tag{5.2}$$

where $N(\omega)$, usually a real-valued function, is the spectral normalization function. We use '~' to denote frequency normalization. In this case we end up with the spectrally normalized cross-spectrum:

$$\tilde{\gamma}_{ij}(\omega) \equiv \tilde{u}_i(\omega)\tilde{u}_j^*(\omega) = \frac{u_i(\omega)u_j^*(\omega)}{N_i(\omega)N_j(\omega)}. \qquad (5.3)$$

An example is $N(\omega) = |u(\omega)|$, where $|\cdot|$ denotes the modulus. With this normalization, $\tilde{\gamma}_{ij}(\omega)$ would be the complex coherency and $\tilde{\gamma}(\tau)$ would be coherency in the time domain, the inverse Fourier transform of $\tilde{\gamma}(\omega)$.

Snieder (2004) and others have argued that under idealized conditions the time derivative of the cross-correlation will be proportional to the Green's function between the two stations where the frequency-dependent proportionality constant will depend on the source spectrum of the ground motion. Aki (1957) and others have argued that these conditions may be satisfied approximately if the ensemble average of the cross-spectrum is taken,

$$\Gamma_{ij}(\omega) \equiv < \gamma_{ij}(\omega) >, \qquad (5.4)$$

where $< \cdot >$ denotes the ensemble average. We use the upper case to represent ensemble averaging. In this case,

$$\frac{d\Gamma_{ij}(\tau)}{d\tau} \propto \begin{cases} -G_{ij}(\tau) & \tau \geq 0 \\ G_{ji}(-\tau) & \tau < 0, \end{cases} \qquad (5.5)$$

where G_{ij} is the Green's function between stations i and j, G_{ji} is the reciprocal Green's function between stations j and i, and $\propto$ denotes proportionality. This proportionality also holds for the ensemble average of the frequency-normalized cross-correlation $\tilde{\Gamma}_{ij}(\tau)$.

Equation (5.5) illustrates the basis of ambient noise tomography, which is, at least in this idealized case, that the cross-correlation may be used to estimate the Green's function between a pair of stations.

We have not discussed yet the meaning of the ensemble average $< \cdot >$. It is convenient to follow Aki (1957) and define the ensemble average as an average over time or time intervals. Let $u_i^k(t)$ and $u_j^k(t)$ be the vertical components of ground motion measured at a pair of stations i and j on a finite sequence of time intervals denoted by index $k = 1, \ldots, K$. The ensemble average of the cross-correlation in the time domain, therefore, is as follows:

$$\Gamma_{ij}(\tau) = < \gamma_{ij}(\tau) > \approx \sum_k W_k(\tau)\gamma_{ij}^k(\tau), \qquad (5.6)$$

where the functions $W_k(\tau)$ compose a set of (possibly time-dependent) time-domain weights. Thus, the ensemble average can be approximated as the weighted average of the cross-correlations taken over a discrete set of time intervals. The process of ensemble averaging described by equation (5.6) is commonly referred to as "stacking." Stacked signals are sometimes called "estimated Green's functions" or EGFs, which is somewhat confusing because it is their time derivative that is related to the Green's functions. We note that signals on a cross-correlation and the time derivative of the cross-correlation will have the same group velocity, which is one of the reasons for calling the cross-correlation the EGF. However, this terminology is sloppy because the phase velocities will differ.

Time τ in equation (5.6) is the cross-correlation lag time, which can be positive or negative. It is often useful to separate the cross-correlation at positive and negative lag times as follows:

$$\Gamma_{ij}^{+}(\tau) = \Gamma_{ij}(\tau) \quad \text{and} \quad \Gamma_{ij}^{-}(\tau) = \Gamma_{ij}(-\tau), \qquad \tau \geq 0. \qquad (5.7)$$

A related concept is the so-called "symmetric component" of the cross-correlation, which is defined as follows:

$$\Gamma_{ij}^{sym}(\tau) = \frac{1}{2}\left(\Gamma_{ij}^{+}(\tau) + \Gamma_{ij}^{-}(-\tau)\right), \qquad \tau \geq 0. \qquad (5.8)$$

This is simply the average of the cross-correlation reflected symmetrically about lag time $\tau = 0$. With this definition, equation (5.5) can be rewritten as follows:

$$\frac{d\Gamma_{ij}^{sym}(\tau)}{d\tau} \propto -G_{ij}(\tau), \qquad \tau \geq 0. \qquad (5.9)$$

Equations similar to (5.4)–(5.9) also hold for the ensemble average of the frequency normalized cross-correlation, $\tilde{\Gamma}_{ij}(\tau)$, and its symmetric component, $\tilde{\Gamma}_{ij}^{sym}(\tau)$.

5.3 Practical Implementation: Continental Rayleigh Waves

In section 5.2 we noted that if certain "idealized conditions" are manifest, then the ensemble average of the cross-correlation of ambient ground motion recorded at two observing stations may be proportional to the Green's function between these stations. Such a function is sometimes referred to as an "estimated Green's function." Such idealized conditions include the azimuthal homogeneity of the ambient seismic wave field and the equipartition of energy among the normal modes of the medium. These conditions do not hold in the real Earth in essentially all applications. At the length scales and frequency bands of consideration in this chapter, surface waves dominate the cross-correlations; thus modal equipartition is unsatisfied. In addition, the ambient noise that propagates over long distances typically originates in the oceans and is excited at certain azimuths more strongly than others. Other strong isolated noise sources (e.g., earthquakes and certain other persistent sources) further deviate reality from ideality.

The practical challenge that faces the observational seismologist is to attempt to extract meaningful approximate or estimated Green's functions from cross-correlations of ambient noise without the theoretical guarantee that such attempts will be successful. As mentioned in the Introduction, the aim is to minimize bias, to estimate uncertainty, and to maximize the bandwidth of observation. To do this, there are several variables referred to in section 5.2 that can be tuned, and much of the practical work in ambient noise methodology has been dedicated to the systematic exploration of this parameter space. These variables include the tempo or the cadence rate of the cross-correlations in the time domain (T), the time-domain weights ($W(t)$), and the frequency-domain normalization ($N(\omega)$). A great many papers have been written by researchers who have varied these (and other) quantities in the attempt to extract reliable Green's functions for the medium of their study. We will make no attempt to summarize these studies, but will present a general discussion of the issues involved in the choice of these variables.

Practically speaking again, the key issues that interest most researchers, especially for crustal imaging, are to expand the frequency band of observation outside the microseism band (8–20 s), to attempt to homogenize the azimuthal content of ambient noise as much as possible, and to identify and eliminate any instrumental irregularities. In terms of optimizing azimuthal homogeneity, a major focus is the attempt to minimize the effect of earthquakes as well as persistent localized sources of noise, both of which may be azimuthally limited.

One of the key indicators of success is that precursory signals, signals that arrive prior to the signal of interest, which is typically the fundamental mode surface wave, are small in comparison to the signal of interest.

5.3.1 The Cadence Rate of Cross-Correlation

In most of the examples presented here, the length of the time series for cross-correlation, T, is one day. After selecting one day of data, we typically first remove the instrument response, the mean and trend, and the band-pass filter.

The choice of a one-day data length is ad hoc, and a shorter time series length is advocated by some researchers (e.g., Prieto et al., 2011). One advantage of a faster cadence rate for cross-correlation relates to the treatment of earthquake sources. With a time series length of 1 hour, for example, if an earthquake or short-term instrumental irregularity is identified, the 1-hour time period can simply be discarded. One is less likely to be willing to sacrifice a whole day to eliminate such effects, so with a slower rate to cross-correlation the tendency is to introduce some sort of time-dependent weights to continuously down-weight the time periods where such effects are identified, as described in the following paragraph. The downside to introducing such weights, however, is the loss of meaningful

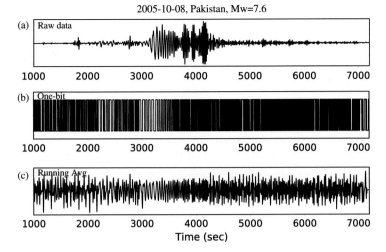

Figure 5.3. Waveforms displaying three different examples of time-domain normalization, $W(t)$, band-passed between 20 and 100 s period to clarify the contamination by the earthquake signal. Data are from station ANMO. (a) Raw seismogram showing about 6200 seconds of data around the earthquake (Mw = 7.6, Pakistan region). (b) One-bit normalized waveform in which the signal is set to ± 1 depending on the sign of the original waveform. (c) Running absolute mean normalization in which the waveform is normalized by a running average of its absolute value.

amplitude information. Thus, researchers who desire reliable amplitude information may be interested in increasing the rate of cross-correlation and applying a binary time-domain weighting scheme: 0 for time periods where earthquakes or instrumental irregularities are identified and 1 for other time periods.

5.3.2 Time-Domain Weighting

Most practitioners of ambient noise seismology employ a more detailed time-domain weighting scheme than the binary choice mentioned in section 3.1. Some of these alternatives are designed to homogenize the ambient noise signals in azimuth by down-weighting azimuthally limited signals that are either exceptionally strong (e.g., earthquakes) or are persistent over time (e.g., 26 s microseism, Kyushu microseism). Figure 5.3 presents three example alternatives for the time-domain weights, $W(t)$: no weights, one-bit normalization, and running absolute mean weights.

The running absolute mean time-domain weighting scheme is defined as follows:

$$W(t_n) = \frac{1}{N+1} \sum_{j=n-N/2}^{n+N/2} |u(t_j)|, \tag{5.10}$$

where time is presented on a discrete grid. Thus, at time grid point t_j, the weight is the absolute mean of the seismogram in a time window of length N spanning the time point. Typically, N is chosen to be some multiple of the maximum period in the band-pass filter applied to the data. To ensure that the longer periods are not filtered out, N should be chosen to be at least the length of the maximum period. Note that if $N = 0$, the approach is equivalent to one-bit normalization, in which signal amplitudes become ± 1. One down-side of the method is that it does not surgically remove data spikes, unlike one-bit normalization, but spreads them out in time. This weighting scheme is applied before cross-correlation.

Figures 5.4a–c show the effect of applying these three weighting schemes on cross-correlations stacked over a year of ambient noise data acquired at stations ANMO and HRV. The added effect of frequency-domain weighting, which is discussed in section 3.3, is shown in Figures 5.4d–f.

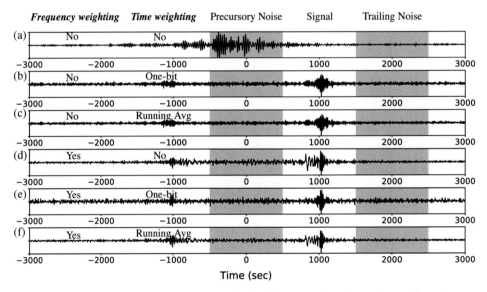

Figure 5.4. Twelve-month cross-correlation between data from the station pair ANMO-HRV (as in Figure 5.2) for various frequency and time-domain normalizations, band-passed between 20 and 100 s period. Panels in (a)–(f) are labeled with the types of normalization applied: (a) no time- or frequency-domain normalization, (b) one-bit time-domain normalization but no frequency-domain normalization, (c) running absolute average time-domain normalization but no frequency-domain normalization, (d) no time-domain normalization but frequency-domain normalization is applied, (e) one-bit time-domain normalization with frequency normalization, and (f) running absolute average time-domain normalization with frequency normalization. Signal and precursory and trailing noise windows are indicated with shading. Precursory and trailing SNR for each of these cases is presented in Table 1, whose rows are ordered corresponding to (a)–(f).

Table 5.1 *SNR using precursory (pre) and trailing (trail) noise for Figure 5.4 for different time- and frequency-domain weighting schemes in two period bands. RA stands for the running average.*

Normalization		10–20 s		20–100 s	
Freq.	Time	A_{max}/RMS_{pre}	A_{max}/RMS_{trail}	A_{max}/RMS_{pre}	A_{max}/RMS_{trail}
No	No	1.478	6.291	0.650	8.074
No	One-bit	12.273	14.928	8.189	10.586
No	RA	14.104	17.993	9.481	12.169
Yes	No	13.446	18.916	8.588	22.982
Yes	One-bit	10.555	10.854	6.323	7.319
Yes	RA	16.626	20.217	9.274	14.032

Figure 5.4a illustrates the motivation for applying time-domain weights. The near zero lag-time response illustrates that the recovered signals are not homogeneously distributed in azimuth, but are precursory noise, presumably from earthquakes in this case. The relative arrival time between the artifact and the real signal provides information about the azimuth of the interfering source relative to the interstation path. The near-zero lag time for the cross-correlation in Figure 5.4a reveals that the source of that signal lies predominantly in an azimuthal band nearly perpendicular to the line linking these stations.

Weighting schemes (b) one-bit normalization, and (c) running absolute average normalization, from Figure 5.3 produce correlations that reveal more realistic Rayleigh wave signals at both positive and negative correlation lags. The one-bit normalization scheme is not only simple (Figure 5.3b), but as shown in Figure 5.4b it is remarkably effective. Table 5.1 presents signal-to-noise ratios (SNR) computed in two period bands for the recordings in Figure 5.4a–c. Signal is the maximum amplitude in the time domain in the signal window (identified in Figure 5.4b on the positive lag in this case), which is filtered into the bands between 10–20 s and 20–100 s period in the table. Noise is the root mean square (RMS) in the time domain measured either in the precursory or trailing window. In the absence of frequency weights, the signal-to-noise ratio is largest for the running absolute mean normalization, but the one-bit normalization method also performs well.

As discussed in section 5.4, in extracting Love waves we seek to apply the same time-domain normalization to both horizontal components. The one-bit filter is so intimately related to each component that defining a single filter to apply to both components is impractical. In addition, as we discuss below, the one-bit normalization methods tends not to perform well with spectral weighting designed to increase the bandwidth of the cross-correlations. For these reasons we tend to prefer the running absolute mean weighting scheme (c) of Figure 5.3, irrespective of the simplicity of one-bit normalization. Results shown in the remainder of this chapter are based on the running absolute mean time-domain weighting scheme unless explicitly noted otherwise.

5.3.3 Frequency-Domain Weighting

Frequency-domain weighting is applied for two main reasons: to broaden the bandwidth of the estimated Green's functions and to diminish the effect of band-limited spatially localized sources such as the 26 s microseism (e.g., Shapiro et al., 2006) or the Kyushu microseism (e.g., Zeng and Ni, 2010). Like time-domain normalization, it is applied to seismic records prior to the cross-correlation. The effect on cross-correlations of applying frequency-domain normalization is exemplified in Figures 5.4d–f.

Ambient noise is not spectrally white; that is, its spectrum is not flat. Rather, in the frequency band of regional to global-scale tomography (e.g., 5–200 s period), it is peaked near the primary (~15 s period) and secondary (~7.5 s period) microseisms and then rises again at periods above 50 s to form Earth "hum" (see Chapter 1 [McNamara and Boaz, 2018]). Figure 5.5a shows an example spectrum for one summer day of vertical component data from station HRV, which illustrates

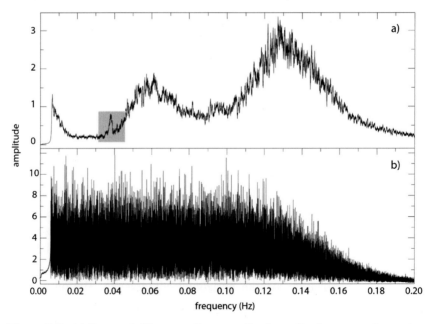

Figure 5.5. (a) Raw and (b) spectrally normalized amplitude spectra for station HRV from July 5, 2004. The shaded box in (a) shows the 26 s Gulf of Guinea microseism. The tapering at the ends reflects the 7–150 s period band-pass filter. This figure is taken from Figure 7 of Bensen et al. (2007).

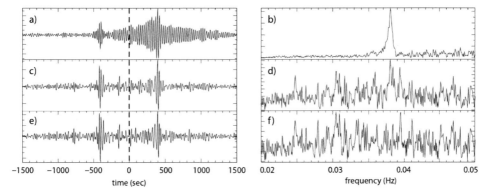

Figure 5.6. Illustration of the effect of the 26 s microseism on cross-correlations of ambient noise and its removal by spectral whitening. (a) Cross-correlation between 12 months of data from stations ANMO and CCM (Cathedral Cave, MO, USA). The broad monochromatic cigar-shaped arrival is the effect of the 26 s microseism. (b) Amplitude spectrum of the cross-correlation in (a) showing the peak near 26 s period. (c–d) Similar to (a) and (b), but for data that were spatially whitened prior to cross-correlation, such that the effect of the 26 s microseism has been largely eliminated. (e–f) Similar to (c) and (d), but the data were notch-filtered around the 26 s period prior to cross-correlation. The notch filter delivers little improvement in eliminating the 26 s microseism signal at the expense of losing dispersion information in the period band of the notch filter. This figure is taken from Figure 8 of Bensen et al. (2007).

the two microseismic humps and the rise of the noise level at low frequencies. In addition, isolated peaks are also sometimes apparent in ambient noise, as the gray shaded box in Figure 5.5a illustrates around the 26 s microseism. Because the 26 s microseism propagates coherently over long distances, it appears even more strongly on cross-correlations between distant stations, as illustrated by Figures 5.6a, b. Similarly, the dual-band nature of the microseismic component of ambient noise is also accentuated by cross-correlation as Figures 5.7a, c illustrates.

One means of frequency-domain weighting is spectral whitening, which flattens the observed spectrum in some way. Whitening acts to broaden the bandwidth of ambient noise and also reduces the impact of spatially isolated persistent noise sources. Whitening can take several forms. Some researchers advocate for dividing the observed spectrum by its modulus in some finite band, so that $N(\omega) = |u(\omega)|$ in that band. This results in an absolutely flat spectrum and the construction of the coherence after cross-correlation, as discussed in section 5.2. One of several other alternatives is to normalize by a smoothed version of the amplitude spectrum. In this case, $N(\omega) = S(\omega) \star |u(\omega)|$, where $S(\omega)$ is the smoothing filter and $\star$ represents convolution. An example showing the effect of this operation is presented in Figure 5.5b. Example results of such spectral smoothing on cross-correlations are presented in Figures 5.4, 5.6, and 5.7. A whitened spectrum typically does not

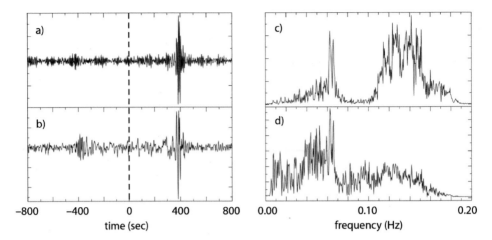

Figure 5.7. Comparison of cross-correlations and their amplitude spectra without (a, c) and with (b,d) spectral whitening, computed for one month of data (April 2004) using stations CCM and SSPA (Standing Stone, PA, USA). Primary and secondary microseisms are more dominant without spectral whitening (c) and low frequencies are accentuated with spectral whitening (d). This figure is taken from Figure 9 of Bensen et al. (2007).

remain white after cross-correlation, presumably because the geographical distribution of microseismic sources is more favorable to recovering some signals than others. However, as illustrated in Figures 5.7b, d the result is broader band and more continuous than without spectral whitening (Figure 5.7a, c), and is particularly effective at improving the quality of signals at frequencies lower than the microseismic band. Whitening is particularly successful at eliminating the effect of the 26 s microseism as Figures 5.6c–f shows. The application of a notch filter around the 26 s microseism provides little improvement over spectral whitening without the notch.

Spectral whitening appears to be less successful at removing some local persistent noise sources, such as the Kyushu signal generated from Aso volcano, from cross-correlations between stations in China or locally isolated wave–wave interactions. Zheng et al. (2011) found that they needed to use the opposite correlation time lag from the direction to Aso volcano. Tian and Ritzwoller (2017) identified persistent signals in their cross-correlations between OBS stations that they argued were related to local persistent microseismic sources in shallow waters of the Juan de Fuca plate.

Figures 5.4d–f illustrate how the application of frequency weights brings out the longer periods, which arrive earlier than the shorter periods in the Rayleigh wave train. Table 5.1 quantifies this effect. Two things are noteworthy. First, one-bit normalization does not work well in conjunction with frequency-domain weights, and cross-correlations tend to be more band limited than when running absolute mean normalization is applied. Second, the highest SNR at long periods results without time-domain weighting, which is an indication that the length of the time window we used to measure the absolute mean (N in equation 5.10) was probably too short to optimize results at the longer periods.

Results shown here use smoothed spectral whitening unless explicitly indicated otherwise. Shen et al. (2012) present a revised method of spectral normalization in which spectral whitening is performed on shorter time windows along with time-domain normalization that they argue further extends the spectral band of observation to longer periods.

5.3.4 Cross-Correlation and Stacking

After the data have been prepared for each day (or for a shorter or longer duration T), including spectral and temporal normalization if desired, we cross-correlate and stack over time as shown in equation (5.6). Example results for different total time series lengths are shown in Figure 5.8. The SNR typically grows with stacking length. We define the SNR as the peak amplitude in the signal window divided by the RMS amplitude in the noise window, as illustrated in Figure 5.4. Typically, signal to trailing noise grows approximately as the square root of the time, series length. This is because signal amplitude tends to grow approximately linearly with time, and trailing noise grows approximately as the square root of time as shown

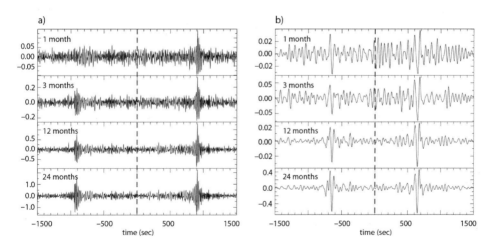

Figure 5.8.   Example cross-correlation results with different total time series lengths (labeled on each panel) and in different pass bands: (a) 5–40 s period and (b) 40–100 s period. Both positive and negative correlation lags are shown. Data are from the station pair ANMO and DWPF (Disney Wilderness Preserve, FL, USA). This figure is adapted from Figure 10 of Bensen et al. (2007).

by Lin et al. (2011). There can be deviations from the square root of time rule that depend on time variations in the strength of ambient noise, but it is a useful rule-of-thumb. Bensen et al. (2007) reported a deviation from the square root of time rule, but much of this was related to how they measured time series length.

The SNR as we define it in the previous paragraph is a useful indicator of data quality and is used regularly as an automated data selection metric, as discussed in section 5.6. However, because precursory noise is indicative of incomplete destructive inference for off-interstation axis events, the ratio of peak signal to the RMS of precursory noise level may be a better metric to indicate the extent of convergence of the algorithm.

5.3.5 Measurement of Surface Wave Dispersion

After the daily cross-correlations have been computed and stacked, group and phase speeds can be measured as functions of the period on the resulting waveform, $s(t)$, using a variety of different time-domain and frequency-domain methods. Here we describe a time-domain method, which is traditional frequency-time analysis (e.g., Levshin and Ritzwoller, 2001). Other researchers prefer frequency-domain methods (e.g., Ekström et al., 2009).

Roughly following the terminology and notation of Bracewell (1978), the Fourier transform of $s(t)$ is defined with a positive exponent as:

$$S(\omega) = \int_{-\infty}^{\infty} s(t) \exp(i\omega t) dt. \tag{5.11}$$

We obtain group and phase time measurements by considering the "analytic signal," which is defined in the frequency domain as

$$S_a(\omega) = S(\omega)(1 + \text{sgn}(\omega)), \tag{5.12}$$

where "sgn" is the sign function. The inverse Fourier transform of equation (5.12) in the time domain is the analytic time series

$$s_a(t) = s(t) + ih(t) = A(t)\exp(i\phi(t)), \tag{5.13}$$

where $h(t)$ is the Hilbert transform of $s(t)$ and the positive sign in front of the $h(t)$ is chosen to be consistent with the positive exponent in the Fourier transform, equation (5.11). The frequency-time function is constructed by applying to the analytic time series a set of narrow band-pass Gaussian filters with center frequencies ω_0:

$$S_a(\omega, \omega_0) = S(\omega)(1 + \text{sgn}(\omega))G(\omega - \omega_0), \tag{5.14}$$

where

$$G(\omega - \omega_0) = \exp\left(-\alpha \left(\frac{\omega - \omega_0}{\omega_0}\right)^2\right), \tag{5.15}$$

where α is a tunable parameter that defines the complementary resolutions in the time and frequency domains and is commonly made range dependent (e.g., Levshin et al., 1989).

Inverse Fourier transforming each band-passed function $S_a(\omega, \omega_0)$ back to the time domain yields the 2D envelope function, $A(t, \omega_0)$, and phase function, $\phi(t, \omega_0)$. Group speed is measured from $A(t, \omega_0)$ and phase speed from $\phi(t, \omega_0)$. $A(t, \omega_0)$ is sometimes called the "frequency-time analysis" or "FTAN" diagram, an example of which is shown later in the chapter. The group travel time, t_g, is measured using the peak of the envelope function at each center frequency such that group velocity $U(\omega) = r/t_g$, where r is the interstation distance. If the group speed changes rapidly with center frequency, Bracewell suggests that the center frequency be replaced by the so-called "instantaneous frequency," ω, at t_g:
$\omega = \left[\partial\phi(t, \omega_0)/\partial t\right]_{t=t_g}$.

For instantaneous frequency ω, the phase of the cross-correlation function observed at time t can be expressed as follows:

$$\phi(t, \omega) = kr - \omega t + \frac{\pi}{2} - \frac{\pi}{4} + n \cdot 2\pi, \tag{5.16}$$

where n is an integer. The negative sign in front of the ωt is chosen to be consistent with choice of the positive exponent in the Fourier transform, equation (5.11). With this definition, phase decreases with an increase in time. In equation (5.16), k is the wavenumber, $\pi/2$ is the phase shift from the negative time derivative relating the stacked cross-correlation with the Green's function (equation (5.5)), $-\pi/4$ is the asymptotic remnant of the Bessel function under the far-field approximation (Snieder, 2004), and $n \cdot 2\pi$ is the intrinsic phase ambiguity of phase measurements. The $\pi/2$ phase shift can equivalently be thought of as accounting for the phase shift between the applied force and the displacement response of the system.

From equation (5.16), the phase velocity c at instantaneous frequency ω when measured on the cross-correlation function is given by

$$c = \frac{\omega}{k} = \frac{r\omega}{\phi(\omega, t_g) + \omega t_g - \pi/4 - n \cdot 2\pi} \tag{5.17}$$

and the phase time is r/c and n is an unknown. n is determined, essentially, by using reference dispersion curves and iterating. Its determination is discussed at length by Lin et al. (2008). We only note that n is more easily resolved at longer periods, and once it is known for a station-pair at any frequency it is known for all frequencies.

Figure 5.9 presents a few examples of Rayleigh wave phase and group speed dispersion curves measured between a station in New Mexico and a set of closely

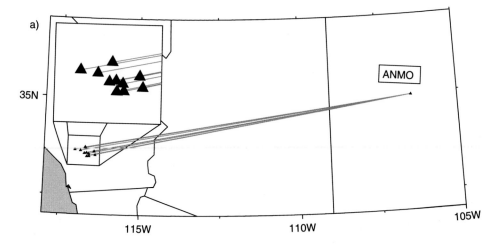

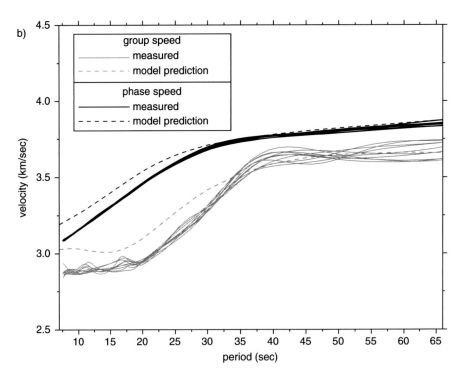

Figure 5.9. Group and phase speed measurements obtained from year-long correlations between station ANMO and 10 stations in southern California. (a) The cluster of 10 paths. (b) Measurements are shown with solid lines, and the prediction from the global 3D model of Shapiro and Ritzwoller (2002) is shown with the dashed lines. This figure is modified from Figure 16 of Bensen et al. (2007).

located stations in Southern California. It illustrates an important point. Phase speed is typically determined more reliably than group speed.

Researchers often apply an interstation distance criterion such that dispersion measurements are discarded unless the interstation spacing is greater than two or perhaps more wavelengths. One reason for this is that time-domain dispersion measurements reflected in equations (5.16) and (5.17) are based on a far-field approximation, and the far-field is typically defined as setting on after two or more wavelengths. For this reason, frequency-domain dispersion measurements may be preferable (e.g., Ekström et al., 2009) for short interstation distances. Luo et al. (2015) present evidence, however, that time-domain methods are reliable at least down to an interstation spacing of one wavelength. A more compelling reason to apply a multiple wavelength selection criterion is that at longer interstation distances, signals of interest become more separated from precursory noise and the interference from precursory noise lessens appreciably. This is true for both the time- and frequency-domain measurements of phase speed.

5.3.6 Closing Remarks

The methods presented here are devised to recover meaningful and reliable surface wave dispersion information. Optimal methods probably depend on the nature, time period, and location of the experiment, and the reader is encouraged to be creative and try their own ideas. It should be remembered, however, that both the time-domain normalization and frequency whitening methods presented here, and presumably most others, are nonlinear. This means that the order of application of the time-domain and frequency-domain filters matters.

5.4 Practical Implementation: Continental Love Waves

In early research, ambient noise tomography was applied predominantly to vertical component seismic records to recover fundamental mode Rayleigh waves. This was partially because the generation of Love waves in ambient noise was (and remains) more poorly understood than the generation of Rayleigh waves (see Chapter 3 [Ardhuin et al., 2018]). Thus, it was not clear initially if ambient noise cross-correlations would recover Love waves, and less attention was paid to horizontal components. Nevertheless, it is now understood that Love waves are well represented in ambient noise and they provide important information about both crustal and uppermost mantle anisotropy through a growing number of studies (e.g., Moschetti et al., 2010; Lin et al., 2011; Xie et al., 2013, 2017).

There are three principal differences in processing ambient noise data for Love waves compared to Rayleigh waves. (1) First, to retain meaningful geometrical information upon rotation transverse to the interstation direction, the East and

North component records for each station must be normalized identically (in both time and frequency) prior to cross-correlation. This can be done in a variety of ways, but Lin et al. (2008) suggest normalizing by the maximum of the values from the two components (time-domain weight, frequency-domain amplitude at each frequency). (2) Second, cross-correlations are computed between all components. Given the three components of the seismograms at two stations (Z - vertical, E - East, N - North), nine cross-correlations can be performed, ZZ, EE, NN, EN, NE, ZN, NZ, ZE, and EZ, where the first letter represents the component of the first station and the second letter the component of the second station. (3) Finally, the cross-correlations are rotated into the Radial (R) and Transverse (T) components, each of which point in the same direction at both stations. Rotating after cross-correlation is computationally more efficient than rotating before cross-correlation. The resulting four horizontal cross-correlations are TT, RR, TR, and RT, which are computed by rotating the four components EE, EN, NN, and NE.

An example of cross-correlations for several rotated components is presented in Figure 5.10. Rayleigh waves are observed at both positive and negative lags on the ZZ and RR components. Love waves, which are faster than Rayleigh waves in this period band, appear on the TT component. If Love and Rayleigh waves are confined predominantly to the T and R components, respectively, the RT and TR components will not show coherent arrivals, as exemplified here. The relative amplitudes for Rayleigh and Love waves on the two correlation lags differ, such that Rayleigh waves are stronger on the positive lag and Love waves on the negative lag, which means that the azimuthal content of Rayleigh and Love waves differ. Contrary to initial expectations, Love wave amplitudes on the TT components are often higher than Rayleigh wave amplitudes on either the ZZ or RR components. Love waves tend to be narrower band than Rayleigh waves, however, and Love wave dispersion measurements commonly do not extend to periods as long as those of Rayleigh waves. This may reflect a narrower bandwidth of excitation for Love waves, but the bandwidth of Love wave dispersion measurements on TT components is similar to Rayleigh waves on RR components; thus, the relative lack of longer-period Love wave dispersion measurements probably results mostly from the higher noise levels on horizontal components.

5.5 Practical Implementation: Ocean Bottom Rayleigh Waves

The primary assumption that underlies ambient noise tomography is that seismic waves generated by noise sources propagate coherently between the pairs of stations from which recordings are cross-correlated. Highly localized seismic waves that are generated near one station but do not propagate to the second station will corrupt ambient noise cross-correlations and compromise the effectiveness

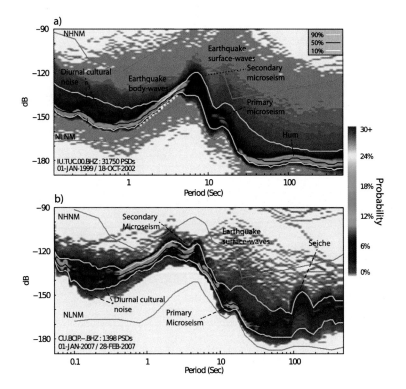

Figure 1.1.

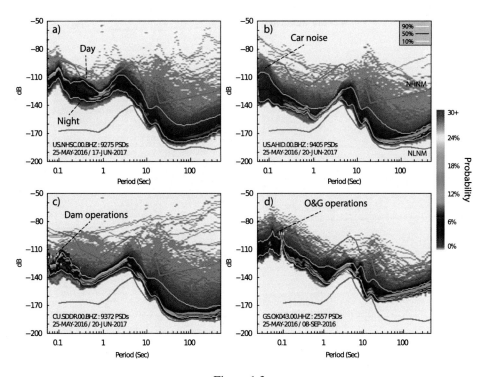

Figure 1.3.

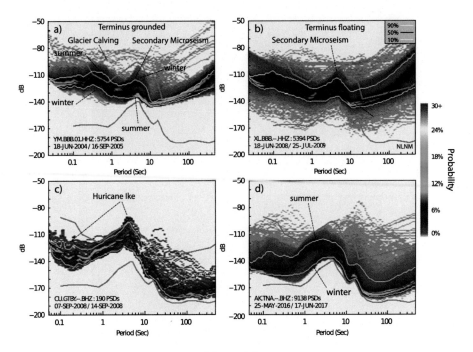

Figure 1.4.

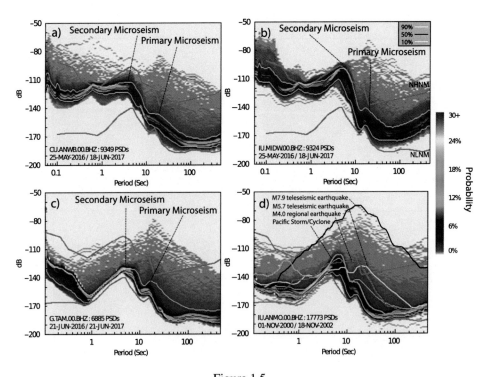

Figure 1.5.

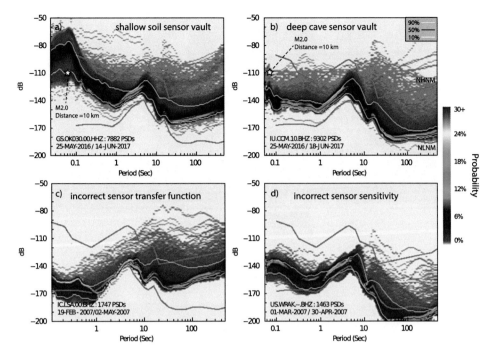

Figure 1.6.

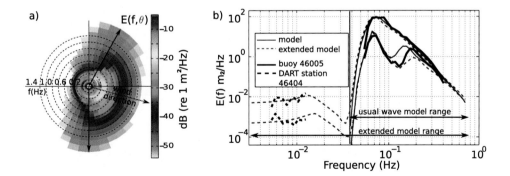

Figure 3.4.

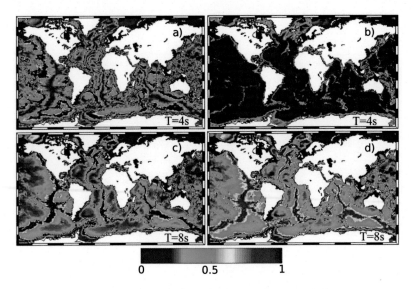

Figure 3.12.

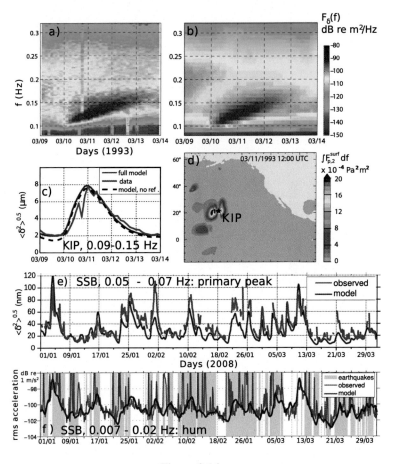

Figure 3.14.

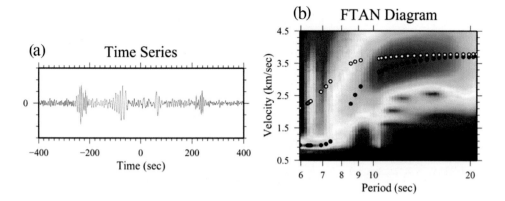

(a) Time Series

(b) FTAN Diagram

Figure 5.11.

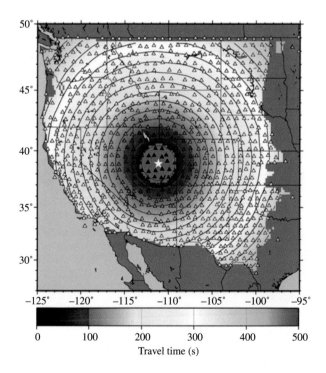

Travel time (s)

Figure 5.13.

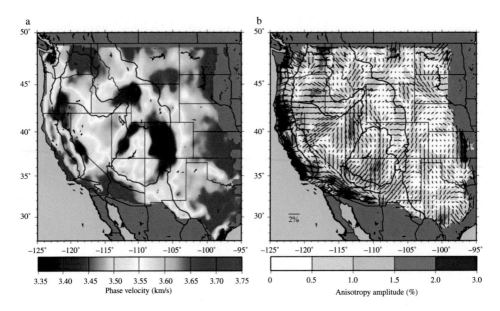

Figure 5.16.

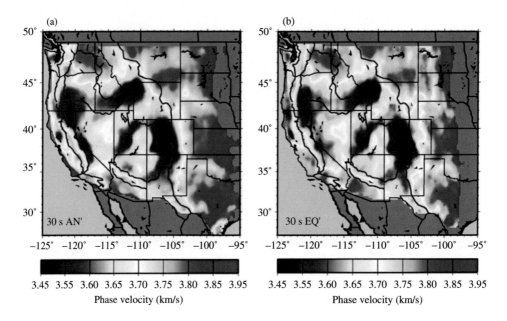

Figure 5.18.

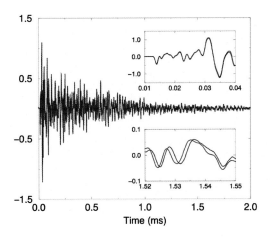

Figure 6.1.

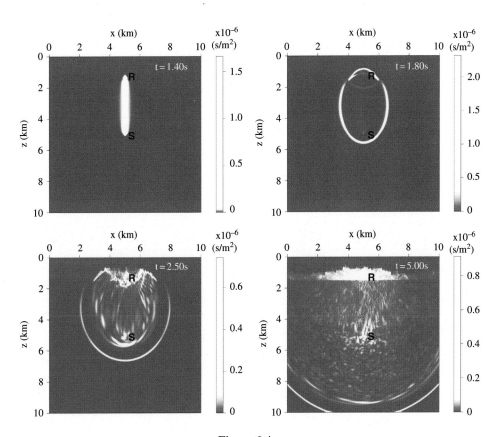

Figure 6.4.

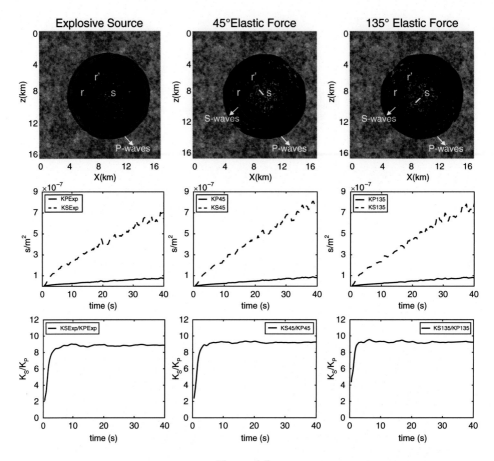

Figure 6.6.

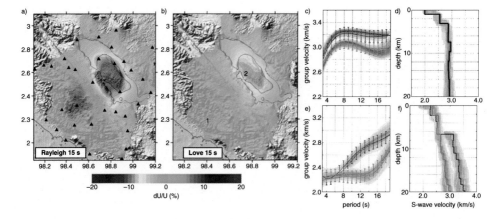

Figure 7.3.

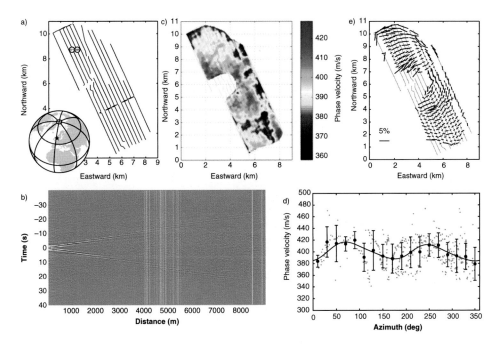

Figure 7.4.

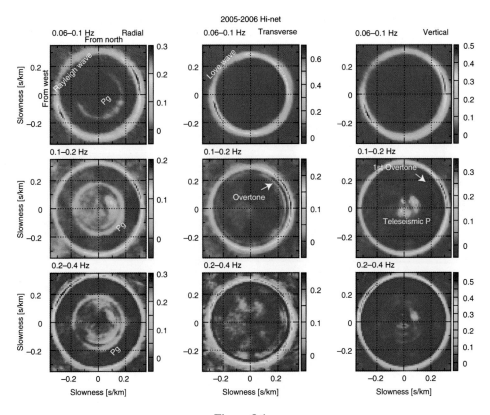

Figure 8.1.

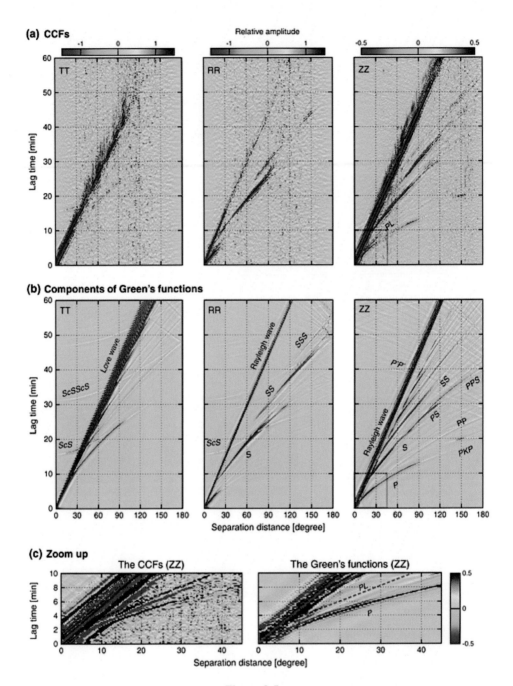

Figure 8.5.

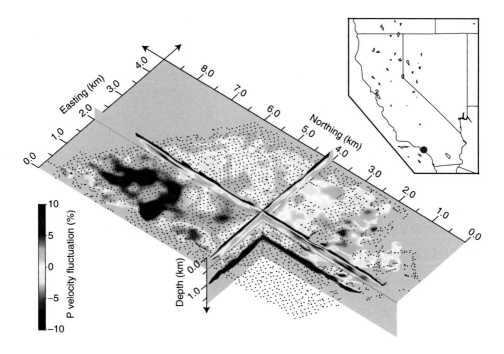

Figure 8.10.

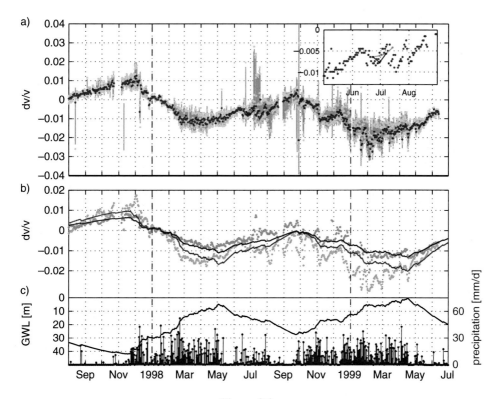

Figure 9.3.

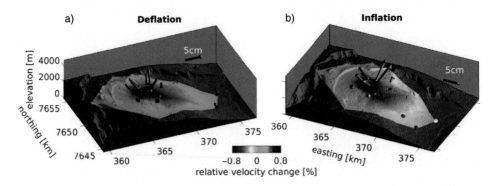

Figure 9.4.

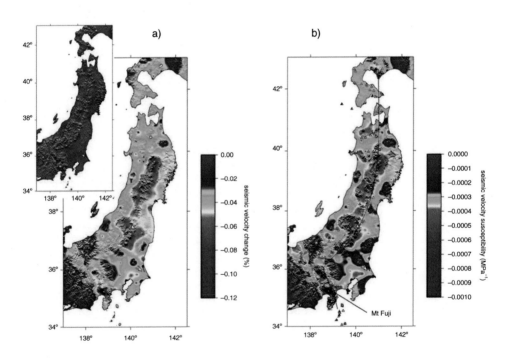

Figure 9.5.

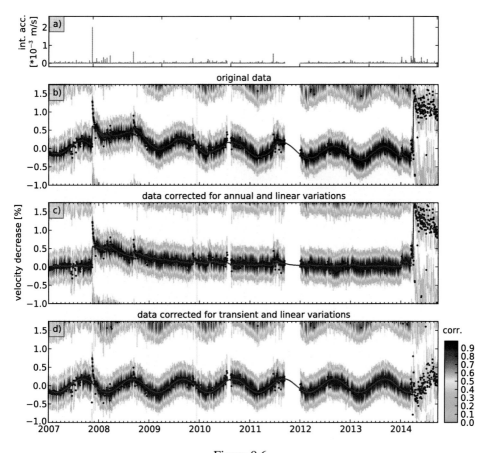

Figure 9.6.

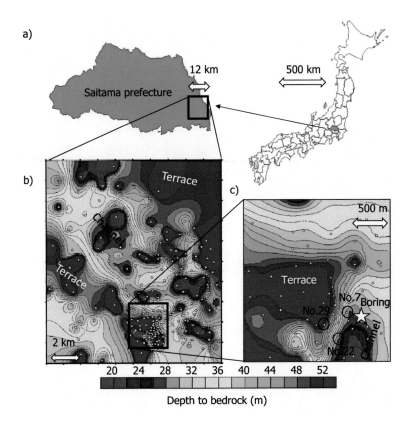

Figure 10.11.

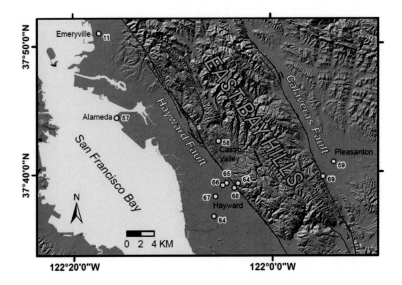

Figure 10.14.

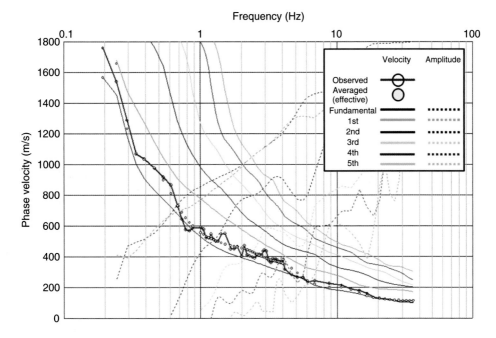

Figure 10.16.

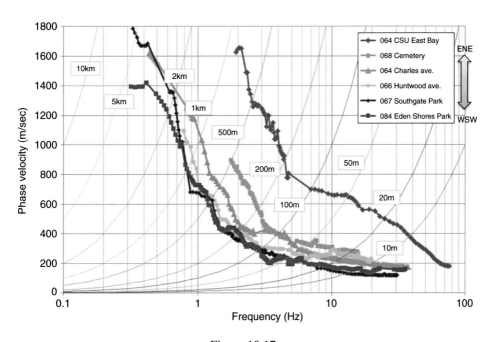

Figure 10.17.

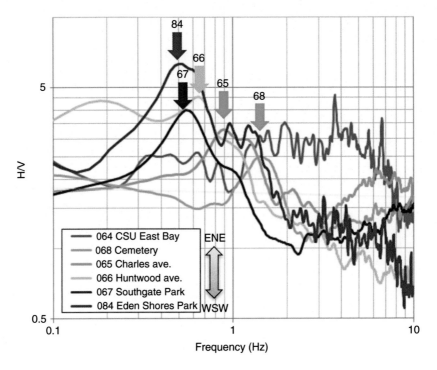

Figure 10.18.

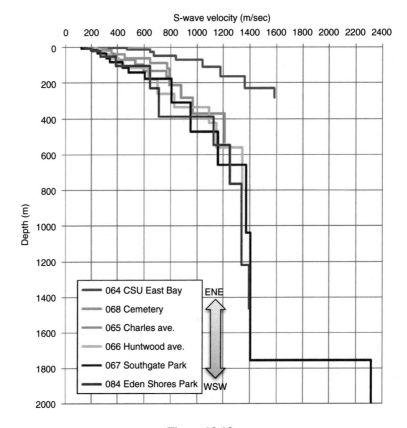

Figure 10.19.

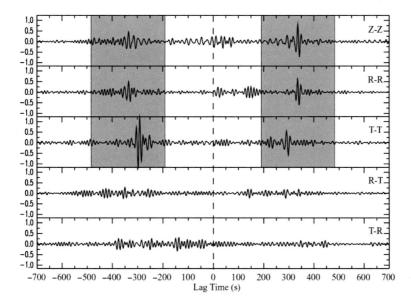

Figure 5.10. Normalized cross-correlations observed between two USArray stations, 116A (Elroy, Arizona) and R06C (Coleville, CA). The gray boxes indicate a group speed window between 2 km/s and 5 km/s. Vertical, radial, and transverse are indicated with Z, R, and T, respectively. Results are band-pass filtered between 10 and 25 s period. This figure is taken from Figure 3 of Lin et al. (2008).

of ambient noise tomography. Such local noise sources are particularly strong in ocean bottom environments and are reflected in higher ambient noise levels, particularly on horizontal components of ocean bottom seismometers (OBS). The data processing procedures discussed in sections 5.3 and 5.4 have been developed for application to land-based stations and are not calibrated for ocean-bottom environments. Nevertheless, several researchers have used data processing schemes similar to these to produce the first studies of ambient noise tomography using OBSs to image crustal and mantle structures (e.g., Harmon et al., 2007; Gao and Shen, 2015; Tian et al., 2013; Yao et al., 2011) and to determine the directional dependence of ambient noise (e.g., Tian and Ritzwoller, 2015).

An example cross-correlation and associated FTAN diagram are presented in Figure 5.11 for two deep-water stations near the Juan de Fuca Ridge. Useful cross-correlations based on OBS data using the land-based data processing procedures described above result only between the relatively quiet deep-water stations or between deep-water and land stations. Such recordings are typically narrow band, extending from less than about 5 s period up to about 20 s, although longer periods can be recovered if one of the stations is on land. Love waves are much more difficult to recover. Local noise levels are high at stations in all but the deepest water,

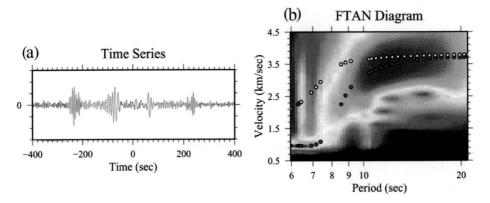

Figure 5.11. (a) Example 6-month vertical component cross-correlation for deep-water stations J29A and J47A near the Juan de Fuca Ridge recorded as part of the Cascadia Initiative. (b) Frequency-time analysis (FTAN) diagram showing Rayleigh wave dispersion from the symmetric component in (a): grey dots (red dots in the plate section) are group speed and white dots are phase speed. The background color is the envelope amplitude, $A(t, \omega)$, discussed in section 3.5. This figure is adapted from Figure 1 of Tian et al. (2013). (A black-and-white version of this figure appears in some formats. For the color version, please refer to the plate section.)

for example near the Juan de Fuca Ridge in the Cascadia Initiative experiment, and make even the vertical components of these stations difficult to use in ambient noise tomography.

As seen in Figure 5.11, Rayleigh waves observed on OBSs recovered from ambient noise bifurcate into slow ocean propagating waves (< 8 s period) and fast solid earth propagating waves (>10 s period). It is the latter that provide information about the solid earth. We would like data processing methods targeting ocean bottom environments to help extend the observations to longer periods in order to improve constraints on mantle structures. We would also like to make observations in shallower water more useful for ambient noise tomography.

The nature of local noise on ocean bottom seismic recordings has been well studied (e.g., Webb, 1988; Duennebier and Sutton, 1995), and two types of noise are considered most important in degrading local conditions: tilt noise and compliance noise. Tilt noise is produced by seafloor currents rocking unstably situated seismometers, and is most significant where bottom currents are strongest. Compliance noise is produced by pressure variations induced by ocean gravity waves that deform the solid earth below the seismometer, and is most significant where the ocean is shallow enough for waves on the ocean's surface to couple to the solid earth. Tilt and compliance noise are local sources of noise, which are distinct from the coherently propagating long range noise that is the basis for ambient noise tomography. Both types of local noise are strongest in shallow water, although the depth extent of compliance noise is frequency dependent. Pressure variations induced by surface gravity waves decay with depth faster at higher frequencies, thus for a given water depth there is a cut-off frequency above which compliance noise is less important. Compliance noise can extend to quite deep waters, but only at very long periods.

Crawford and Webb (2000) and Webb and Crawford (1999) showed that both types of local noise are greatly reduced by predicting the effect of tilt and compliance on vertical components based on the horizontal components and a local pressure gauge, respectively. In their methods, time dependent transfer functions are found that convert horizontal noise and local pressure to the vertical component record, and then these predictions are subtracted from the vertical component. These so-called "de-noising" techniques have been shown to improve the SNR of earthquake data recorded on the seafloor and to reduce distortions (e.g., Ball et al., 2014; Bell et al., 2015; Dolenc et al., 2007). Additionally, Bowden et al. (2016) and Tian and Ritzwoller (2017) have shown that these methods also improve vertical (ZZ) component cross-correlations of ambient noise by improving the SNR of the first-overtone and the fundamental mode Rayleigh waves, particularly for shallower water stations, and, importantly, by extending the measurements to longer periods. Tian and Ritzwoller (2017) also showed that the methods produce records that provide more information about the origin of the coherently propagating ambient noise that is the basis for ambient noise tomography.

Figure 5.12 presents example record sections of vertical component ambient noise cross-correlations in two pass-bands observed using Cascadia Initiative data from the Juan de Fuca Plate. The common station in the record section is in shallow water, station J49A at a depth of 123 m. The reduction of tilt and compliance noise greatly improves the SNR of the recovered Rayleigh waves, particularly at periods above 20 s that are difficult to observe without de-noising. Improvements extend up to about 40 s period.

These de-noising steps do not reduce the effect of local noise on the horizontal components, and therefore do not improve the ability to observe Love waves in ocean bottom settings. This remains a technical challenge for the future.

5.6 Reliability

As discussed earlier, there is no theoretical guarantee that the computed cross-correlations obtained using the data processing methods we describe here or others will reliably reproduce Green's functions (see also Chapter 4 [Fichtner and Tsai, 2018]). The inhomogeneous distribution of ambient noise caused by temporally persistent and spatially localized sources (e.g., 26 s Gulf of Guinea microseism, Kyushu microseism), by strong temporally and spatially localized sources (e.g.,

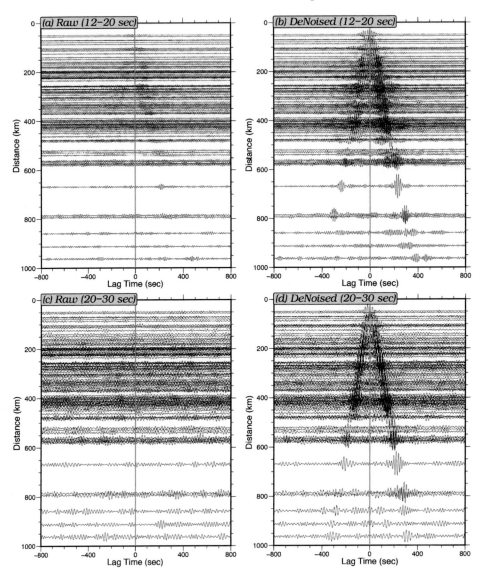

Figure 5.12. Record section of vertical component ambient noise cross-correlations for shallow water station J49A (123 m water depth) from the Cascadia Initiative on the Juan de Fuca plate, in which between 160 and 270 daily cross-correlations are stacked. (a,c) So-called "Raw" cross-correlations are shown in which only land-based data processing has been applied, including temporal normalization and spectral whitening. (b,d) "De-Noised" cross-correlations in which each record has had tilt and compliance noise removed prior to application of the same temporal normalization and spectral whitening procedures applied in the raw cross-correlations. Two period bands are shown: (top row) 12–20 s period, (bottom row) 20–30 s period. In both bands Rayleigh waves emerge more clearly with the local noise sources, tilt and compliance, removed. This figure is adapted from Figure 8 of Tian and Ritzwoller (2017).

earthquakes), by highly localized incoherently propagating noise sources near one station, by a short duration of observation, and so forth, can vitiate the cross-correlation and limit its utility. Given the fact that such circumstances do arise in practice, we now consider three questions, the first two of which are:

Question 1. What tools can the observer bring to bear to determine whether a given cross-correlation is meaningfully related to the Green's function between the pair of stations from which it is computed?

Question 2. How can the observer quantify the extent of this fidelity, if it is deduced that the cross-correlation does provide a reasonable approximation to the Green's function?

To address these questions, we find it useful to think about the conditions that necessarily relate observed cross-correlations with Green's functions. The first of these questions is discussed in section 6.1. The second, which is related to the estimation of uncertainties, is discussed in section 6.2.

There is a third and final question:

Question 3. How do we know if the results are right?

This is different from asking if a measurement is acceptable. As we will see, acceptability depends to a substantial degree on quantifying the variability of the observations such that acceptable measurements exhibit low variability. There are, however, systematic errors that can bias results without introducing variability. Question 3, therefore, relates to an assessment of the nature and extent of systematic errors, which is discussed in section 6.3. We base this analysis in terms of conditions that are sufficient to establish the correspondence of a cross-correlation with the Green's function.

5.6.1 Acceptance or Rejection of a Cross-Correlation

We define a necessary condition X as follows.

Necessary Condition: If an ensemble-averaged (temporally stacked) cross-correlation $\Gamma_{ij}(t)$ computed from recordings observed at stations i and j is a good approximation to the Green's function $G_{ij}(t)$ between the stations, consistent with equation (5.5), then criterion X will be satisfied.

Necessary conditions are desired characteristics that would be satisfied if a cross-correlation in fact is a reasonable approximation to the Green's function. Together they define a set of criteria to test the hypothesis that a given cross-correlation acceptably approximates a Green's function and therefore is worthy of acceptance in an experiment. If a necessary condition X is not satisfied for a given cross-correlation, then there is reason to consider rejecting the cross-correlation from further consideration. This data acceptance and rejection stage is a critical part of the data processing at the heart of ambient noise tomography.

It is up to the observer to define the necessary conditions that form the basis for the acceptance or rejection of cross-correlations in a given experiment. We list several criteria that have played useful roles as necessary conditions in a number of studies. These criteria, a subset of them, or in conjunction with others can be thought of as a set of filters applied to cross-correlations of ambient noise before they are welcome as a final component of an experiment.

(1) Geological coherence. Surface waves are sensitive to relatively shallow earth structures about which other information often exists. We, therefore, expect qualitative similarity of dispersion measurements obtained from ambient noise cross-correlations with geological structures. If measurements disagree with considered judgments about the structure of the Earth, it is either a very interesting observation or a cause for concern about the measurement. For example, at periods below about 15 s group and phase speeds tend to be depressed in well-understood ways by sedimentary basins. An observation of high group or phase speeds at short periods in sedimentary basins is either an error or high-speed material (perhaps of volcanic origin) must reside near the surface of the basin, as it does in the Pasco Basin in southern Washington.

(2) High SNR. As discussed above, the ratio of peak signal amplitude to the RMS of trailing noise tends to increase as the square root of observing time, as illustrated in Figure 5.8. If this proves not to be the case, then the reason should be resolved by the observer. SNR affects the variance in group and phase speed measurements. SNR values of 10–20 are commonly used acceptance criteria, but in data-starved circumstances, an SNR as low as 5–8 has been used in some circumstances.

(3) Stability. This condition relates to how cross-correlations or measurements made on them are modified when subjected to small changes in observing conditions, such as station location or time period of observation (seasonal variability, for example). Figure 5.9 presents an example of how group and phase speed vary as a function of observing station. Observations of the variability of surface wave dispersion measurements are often used as conservative estimates of the uncertainty of the measurement.

(4) Reduction in precursory noise. As the total time series length of the stacked cross-correlations increases, it is expected that the relative amplitude of the arrivals precursory to the signal of interest will decrease. In this case, longer observing times will yield better surface wave dispersion measurements. The rate of increase of the signal to precursory noise will depend on local observing conditions, but if signal level does not increase with increasing time series length relative to precursory noise, the observer should be concerned. It may be the case that the precursory noise appears only on a particular lag related to the azimuth of a spatially localized noise source, as with the Kyushu microseism in the study by Zheng et al. (2011). In this case, a cross-correlation may be salvaged by using only the correlation lag opposite from where the noise source arrives. This illustrates that it is a good idea for the observer to subject both lags of the cross-correlation independently to the selection criteria listed here, and decide which inter-station cross-correlations and lags to accept or reject based on this full set of information.

(5) **Self-consistency.** We refer here to two characteristics of cross-correlations as "self-consistent." First, group and phase speed measurements form period-dependent dispersion curves, which for realistic earth structures typically vary smoothly with period. Dispersion measurements that perturb the smoothness of the dispersion curve are suspect. Second, if both the positive and negative lags of a cross-correlation pass other selection criteria and are believed to be of high quality, then they should provide similar group and phase speed measurements. There are limitations to this criterion, as one lag may be enriched at different frequencies than the other. However, this criterion has been used by Stehly et al. (2007) and Lin et al. (2007) to discover station timing errors, and differences between dispersion measurements at positive and negative lags may also provide useful information about measurement uncertainties.

(6) **Cross-consistency and simplicity.** The determination of cross-consistency, the agreement of a cross-correlation between one pair of stations with those between other pairs, is less straightforward to test than self-consistency. We usually do this with the group and phase speeds (or times) measured from the cross-correlations rather than the cross-correlations themselves. One method is to determine if the travel time field computed for a single central station varies smoothly spatially (e.g., Figure 5.13). Another is to determine if the measurements at a given period can be fit well with a smooth dispersion map. Figure 5.14 presents example histograms of the misfit of Rayleigh wave phase speed measurements across China at different periods to a set of smooth dispersion maps that form the basis for a 3D model (Shen et al., 2016). The standard deviations of such phase travel time misfits are typically about 1 s (but are higher for group times). Measurements that are fit much worse than this (for example, worse than about 3 standard deviations) are typically rejected. Such misfit histograms plotted for measurements in which a particular station participates are also useful to identify problems with stations such as timing errors, location errors, and instrument response errors (e.g., Zhou et al., 2012), which are either corrected or the station is removed from the experiment. We sometimes refer to the ability of a smooth dispersion map to fit the dispersion measurements as a "simplicity" criterion.

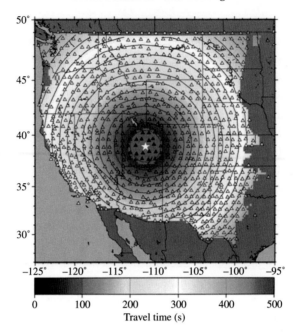

Figure 5.13. The 24 s Rayleigh wave phase travel time surface computed from cross-correlations of ambient noise observed across the western United States based on central station (USArray, Transportable Array) Q16A in Utah. Travel time lines are presented in increments of wave period. The map is truncated within two wavelengths of the central station and where the travel times are not well determined. Station Q16A operated simultaneously with the 843 stations shown (triangles), but only for a short time near the western and eastern boundaries of the map. This figure is taken from Figure 2 of Ritzwoller et al. (2011). (A black-and-white version of this figure appears in some formats. For the color version, please refer to the plate section.)

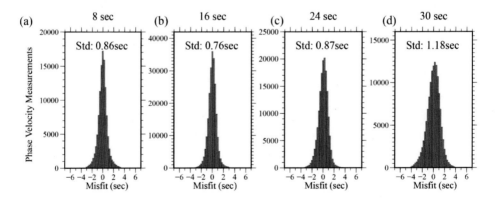

Figure 5.14. Examples of misfit histograms defined as observed Rayleigh wave phase times minus predicted phase times (in sec) computed from an estimated phase speed map at different periods across China. The standard deviation of each misfit distribution is presented in each panel. This figure is adapted from Figure 3 of Shen et al. (2016).

Information about measurement errors and in inferred quantities such as dispersion maps and earth models is contained in the tests represented by the above criteria, particularly tests of stability, self-consistency, and cross-consistency. As discussed in section 6.2, ideally we prefer to use cross-consistency to constrain errors in dispersion maps rather than other criteria that provide information about errors in the dispersion measurements themselves. But all are relevant both to the selection of data retained in the experiment and to the assessment of uncertainties in the inferred quantities.

5.6.2 Quantifying Uncertainty

The necessary conditions discussed in section 6.1 provide circumstantial evidence about the reliability of a given cross-correlation and measurements obtained from it (e.g., surface wave dispersion). This evidence is useful for the acceptance or rejection of data in the experiment. Once accepted in an experiment, however, one wishes to assess the data's reliability and quantify that assessment as data uncertainties. Thus, we take up Question 2 here.

There are four levels of uncertainty worth mentioning: (1) errors in the cross-correlations themselves, (2) errors in measurements obtained on the cross-correlations (e.g., group and phase speed measurements), (3) errors in surface wave dispersion maps, and (4) errors in structural models of the Earth. We will not consider here the first and last of these, but will focus on the assessment of uncertainties in dispersion measurements and dispersion maps, which are commonly produced intermediate products in the construction of structural models of the Earth. As discussed in section 6.1, the stability, repeatability, and self-consistency of measurements are useful tools to quantify the uncertainty of dispersion measurements. In particular, the spatial (e.g., Figure 5.9) and temporal (notably seasonal) variability of dispersion measurements, the variation between measurements obtained on positive and negative lags, and the misfit produced by dispersion maps (e.g., Figure 5.14) have all proven useful to quantify dispersion measurement uncertainty. SNR has also been shown to be a useful proxy so that measurement uncertainty can be inferred from it.

Dispersion maps, however, are what are most commonly used in inversions for 3D structural models rather than raw dispersion measurements. Uncertainty estimates in tomographic maps can be produced either in the context of a linearized tomographic inversion in which estimates of data uncertainties are propagated to model uncertainties (i.e., dispersion maps) in standard ways (e.g., Barmin et al., 2001) or via non-tomographic, spatially localized methods such as eikonal (Lin et al., 2009) and Helmholtz tomography (Lin and Ritzwoller, 2011) in which no forward matrix is constructed, decomposed, or inverted. As instrumental seismology has advanced, the installation of arrays at which the localized non-tomographic methods can be applied is an increasingly common practice. For this reason, we discuss here uncertainties in the context of such array-based methods, acknowledging that the methods are not friendly to spatial gaps in station coverage.

Eikonal tomography is based exclusively on observed travel time surfaces like that shown for the 24 s Rayleigh wave in Figure 5.13, although separate maps are constructed for each period and for each central station. In Helmholtz tomography, similar observed amplitude maps are used to apply a finite frequency correction. Because absolute amplitude information is commonly lost in ambient noise data processing, such amplitude-based corrections are typically not applied in ambient noise studies. As described by Lin et al. (2009), based on the eikonal equation, the local phase speed and direction of wave propagation can be inferred from the gradient of the local travel time. Using maps from many central stations allows the observation of the mean and azimuthal variation of local phase speed, as illustrated Figure 5.15. Following Smith and Dahlen (1973), these quantities can be modeled as a function of azimuthal angle, ψ, as follows:

$$c(\psi) \approx c_0 + A \cos[2(\psi - \phi)], \tag{5.18}$$

in which the c_0 represents the local average of the isotropic component of phase speed, and A and ϕ are mean estimates of the amplitude and fast-axis direction of azimuthal anisotropy. An example of such mean estimates is presented in Figure 5.16 for the 24 s Rayleigh wave.

The error bars in Figure 5.15 represent the variation of the measurements in each azimuthal bin. They are, therefore, related to stability, repeatability, and self-consistency as discussed in section 6.1. If they are interpreted as one-standard-deviation Gaussian errors, they can be transformed to uncertainties in the quantities c_0, A, and ϕ in a standard way. Figure 5.17 illustrates such error estimates on isotropic and anisotropic dispersion curves below 25 s period for Rayleigh waves. Such estimates may not capture the potential effects of systematic errors, however.

5.6.3 Assessing Systematic Error

The necessary conditions discussed in sections 6.1 and 6.2 provide indicators that measure the variance in the cross-correlations, measurements made from them, and inferences drawn from those measurements. Such indicators do not guarantee against systematic error. To assess the extent of systematic error, or bias, it is useful to consider sufficient conditions that may link cross-correlations and Green's functions.

We define a sufficient condition X as follows.

Sufficient Condition: If criterion X is satisfied, then an ensemble-averaged (temporally stacked) cross-correlation $\Gamma_{ij}(t)$ computed from recordings observed at stations i and j will be a good approximation to the Green's function $G_{ij}(t)$ between the stations, consistent with equation (5.5).

Well-formulated sufficient conditions, as distinct from necessary conditions, are powerful as they guarantee that a cross-correlation reliably approximates a Green's function. But such conditions are hard to come by. We consider only one here – the comparison of dispersion measurements and maps obtained from ambient noise to those obtained from earthquakes. Such comparisons cannot be performed at every period because earthquake results do not extend to periods as short as those of ambient noise results. This is one of the ways ambient noise tomography has provided new information about the Earth (Chapters 7 and 8 [Shapiro, 2018; Nakata and Nishida, 2018]). Conversely, earthquake results often extend to longer periods than ambient noise results. Consequently, these two sources of information are complementary and better used together than independently (e.g., Yang et al., 2008, and many others).

Both Figures 5.17 and 5.18 present comparisons between phase speeds derived from ambient noise and earthquakes. First, Figure 5.18 shows Rayleigh wave dispersion maps at 30 s period, which is in the band of overlap between the methods. There are differences in detail between these maps, but a quantitative comparison does not indicate systematic differences. The result of the similarity of these different measurements is that they can be used simultaneously, as in the isotropic and

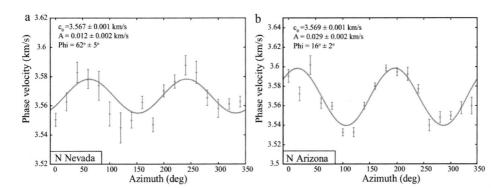

Figure 5.15. Phase speed as a function of interstation azimuth averaged in each 20° bin for the 24 s Rayleigh wave is plotted as error bars that represent the RMS variation in each bin. The example points are in (a) Nevada (242°E, 42°N) and (b) Arizona (250°E, 36°N). The best-fitting 2ψ curve (equation (5.18)) is presented as the solid line in each panel. Estimated values with 1 standard deviation errors for c_0, A, and ϕ are listed at upper left in each panel. The 2ψ component of anisotropy is clear in both panels. This figure is adapted from Figure 4 of Ritzwoller et al. (2011).

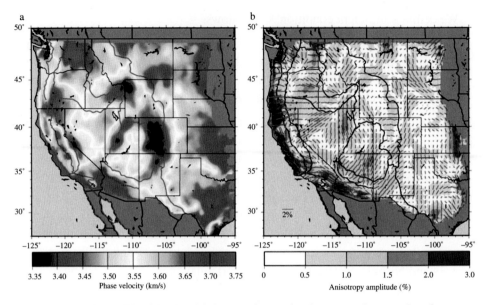

Figure 5.16. (a) The 24 s Rayleigh wave isotropic phase speed map taken from ambient noise by averaging all local phase speed measurements at each point on the map. (b) The amplitudes and fast directions of the 2ψ component of the 24 s Rayleigh wave phase velocities. The amplitude of anisotropy is identified with the length of the bars, which point in the fast-axis direction, and is color-coded in the background. At 24 s period, Rayleigh wave anisotropy reflects a combination of crust and uppermost mantle. This figure is adapted from Figure 5 of Ritzwoller et al. (2011). (A black-and-white version of this figure appears in some formats. For the color version, please refer to the plate section.)

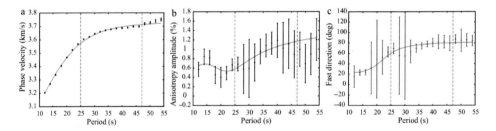

Figure 5.17. Isotropic and azimuthally anisotropic dispersion curves estimated from a location in northern Nevada. Only ambient noise measurements are used at periods below 25 s, ambient noise and earthquake measurements are averaged between 25 s and 45 s period, and only earthquake measurements are used above 45 s period. Phase velocity is presented in km/s, anisotropy amplitude in percent, and the fast direction of anisotropy in degrees east of north. Measurement uncertainties are presented with one standard deviation error bars. The best-fitting curves based on the isotropic and anisotropic inversions are presented as the continuous line in each panel. This figure is adapted from Figure 8 of Ritzwoller et al. (2011).

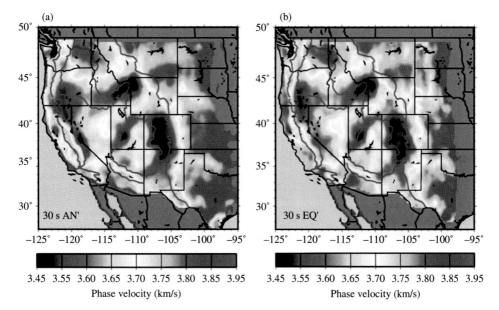

Figure 5.18. (a) Isotropic maps for the 30 s period Rayleigh wave phase speed observed via eikonal tomography applied to ambient noise data. (b) A similar map for comparison constructed by eikonal tomography applied to earthquake data. This figure is adapted from Figure 7 of Ritzwoller et al. (2011). (A black-and-white version of this figure appears in some formats. For the color version, please refer to the plate section.)

anisotropic dispersion curves presented in Figure 5.17 in which in the period band between 25 and 45 s ambient noise and earthquake dispersion measurements are averaged.

Sufficient condition comparisons such as those we present here provide great confidence in ambient noise cross-correlations to deliver unbiased information about the crust and uppermost mantle. However, such comparisons involve a highly processed product of ambient noise data processing and tomography. Consequently, unlike the necessary conditions discussed earlier in this section, sufficient conditions tend to cause the cross-correlations to stand or fall together and do not test a single or small subset of the cross-correlations or measurements.

5.7 Recent Developments in Ambient Noise Data Processing

In this section, we briefly review some recently developed methods designed to amplify or extend the methods described in previous sections. It is impossible to cover all the studies related to this topic and we, therefore, only list a few in several different categories that we consider to be representative. We divide these methods into four categories:

(1) Pre-processing procedures applied before cross-correlation,
(2) De-noising techniques applied to cross-correlation waveforms and advanced stacking schemes,
(3) Post-processing methods applied to the stacked cross-correlation data, and
(4) Advanced seismic interferometry methods.

Categories (1)–(3) are the techniques that are designed to be applied in the context of conventional seismic interferometry for seismic tomography, in which surface wave arrivals emitted from one virtual source are recorded at one station. Category (4), however, includes sophisticated interferometric methods that go beyond conventional approaches.

5.7.1 Pre-Processing Techniques Applied Before Cross-Correlation

Although the time- and frequency-domain weighting or normalization techniques described above have proven to be quite useful, they have some drawbacks, and researchers have proposed new methods to enhance their performance. We present two examples here.

Prieto et al. (2009) used complex coherency (see section 5.2 for a definition), which preserves the amplitude information in ambient noise. They applied this technique to data from 154 broadband seismic stations in southern California and received a similar Q model as earlier work based on earthquakes (Yang and Forsyth, 2008). Following this work, Seats et al. (2012) demonstrated that a faster convergence to empirical Green's functions can be achieved by using short-duration overlapped time windows rather than longer, non-overlapped time windows for cross-correlation, the so-called Welch's method (Welch, 1967).

Inspired by Gallot et al. (2012) and Carriére et al. (2014), Seydoux et al. (2017) developed an array-based technique to equalize noise energy coming from different azimuths. The cross-spectrum matrix defined by equation (5.1) earlier in the chapter defines a positive semi-definite Hermitian matrix, $\mathbf{G}$, termed the array covariance matrix by Seydoux et al. (2017). It can be diagonalized as $\mathbf{G} = \mathbf{\Psi} \mathbf{\Lambda} \mathbf{\Psi}^\dagger$, where $\mathbf{\Lambda} = diag(\lambda_1, \lambda_2, \ldots)$ is the diagonal matrix formed by the non-negative eigenvalues λ_i and where $\mathbf{\Psi}$ is the matrix of the eigenvectors. The authors construct an equalized covariance (cross-spectrum) matrix, $\hat{\mathbf{G}}$, such that $\hat{\mathbf{G}} = \mathbf{\Psi} \hat{\mathbf{\Lambda}} \mathbf{\Psi}^\dagger$. $\hat{\mathbf{\Lambda}}$ contains the equalized eigenvalues, obtained by setting all the eigenvalues $\lambda_i = 1$ when $i \leq L$ and $\lambda_i = 0$ otherwise. The index L is the cut-off for eigenvectors that are considered as noise-related and should be discarded. By applying this technique to both synthetic and real datasets, the authors demonstrated that the symmetry of cross-correlation waveforms can be enhanced and precursory noise suppressed. The beam energy of the processed array data also indicates a more isotropic distribution of sources.

5.7.2 De-Noising Techniques Applied to Cross-Correlation Waveforms and Advanced Stacking Schemes

Another branch of data processing techniques for ambient noise is designed to be applied after the cross-correlation computation. Typically, finalized cross-correlation waveforms for seismic tomography are obtained by stacking all the daily (or hourly, depending on the length of the windows chosen for cross-correlation) correlation data. The stacking scheme can be improved by selecting signals wisely, and we summarize several ideas here to improve this selection. All the methods mentioned in this section have one thing in common: They all assume that coherent signals are contained in each daily cross-correlation and attempt to extract that signal while discarding the incoherent noise using an intelligent selection criterion.

By transforming the daily records into the frequency-time domain based on the S-transform (Stockwell et al., 1996), Baig et al. (2009) proposed to construct a weight function for each octave and time range. This weight function can be designed to optimize the coherence in phase/amplitude for daily cross-correlation data, and thus could serve as a filter to de-noise the stacked cross-correlation waveform. The authors applied the technique to real data recorded in California and observed that more meaningful Rayleigh and Love wave travel time measurements were obtained. The method of Baig et al. (2009) was later called time-frequency phase-weighted stacking (tf-PWS) by Schimmel et al. (2011), who proposed to use a different version of S-transform to perform the weighted stacking scheme. Schimmel et al. (2011) also defined a new type of cross-correlation, the phase cross-correlation (PCC). Different from the conventional cross-correlation, the PCC is more sensitive to small amplitude signals. Schimmel et al. (2011) applied both the tf-PWS and PCC methods to synthetic and real data and observed enhanced SNR in the cross-correlation waveforms.

Liu et al. (2016) proposed another advanced stacking scheme based on a bootstrap resampling approach. They performed a statistical analysis of the daily (or hourly/monthly, depending on the time window length) cross-spectra to identify outliers to discard during stacking. The analysis also yields the variance of the cross-spectra and provides a probabilistic determination of the reliability of the stacked cross-spectrum. Figure 10 in Liu et al. (2016) compares a quality-controlled stacked cross-spectrum with raw stacked cross-spectra, along with estimates of uncertainties for each data point. The noise in the finalized cross-spectrum can be significantly reduced via this method to discard outliers.

Another way of "selecting" data is to de-noise the cross-correlation waveform before stacking. Stehly et al. (2011) proposed the use of a curvelet de-noising filter to be applied to the daily cross-correlation before stacking. The method is based on the idea that the wavefront can be sparsely represented with curvelet coefficients. Given the 2D image of the daily cross-correlation, the sum of all the daily records yields an empirical Green's function with a low SNR. To perform the de-noising, the 2D image is transformed into the 2D curvelet domain. Because the signals in the original 2D image construct a 2D planar wavefront, which can be sparsely represented by a subset of the curvelet coefficients with relatively large values, the large number of small curvelet coefficients can be discarded. Then, by performing an inverse curvelet transform, the authors received de-noised cross-correlation data shown in Figure 4c in their paper. The stacked cross-correlation data from the de-noised 2D image has an improved SNR. The authors also proposed a more sophisticated filtering scheme by using the original stacked cross-correlation waveform as a reference for the selection of the curvelet coefficients. They applied this technique to get more reliable empirical Green's functions whose coda waves were used to monitor seismic wave velocity changes associated with the Mw 7.9 Wenchuan earthquake (Stehly et al., 2015).

5.7.3 Post-Processing Methods Applied to the Stacked Cross-Correlation Data

After the stacked cross-correlation waveforms are generated, researchers typically perform seismic tomography using the travel time measurements obtained on the correlation waveforms. However, some additional information can also be retrieved from the cross-correlation data using various post-processing approaches, of which we mention a couple here.

van Wijk et al. (2011) and Takagi et al. (2014) proposed to use cross-terms of the ambient noise Green's tensor to separate body waves from Rayleigh waves. Their idea is based on the fact that, theoretically, the RZ and ZR components of the cross-correlation tensors are time symmetric for Rayleigh waves while time antisymmetric for P-waves for isotropic media. The use of (ZR-RZ)/2 may suppress the P-wave and reduce noise precursory to the Rayleigh wave. Similarly, (ZR+RZ)/2 will accentuate the P-waves at the expense of the Rayleigh waves.

Another interesting post-processing procedure is to compute the correlation of the coda of the cross-correlation, which is referred to as C3 (Stehly et al., 2008), in contrast with traditional cross-correlations of ambient noise (C1). Ideally, the empirical Green's function between a station-pair should include all the propagation effects between the two stations. Thus, not only the ballistic waves but also the coda should be included in the original cross-correlation. Because the coda wavefield is typically diffuse, empirical Green's functions can be extracted by correlating the coda window of the noise correlation. In fact, before the era of ambient noise tomography, Campillo and Paul (2003) discovered that the cross-correlation of earthquake coda can yield empirical Green's functions. Compared with the ambient noise cross-correlation function, the C3 function may be more symmetric if the stations used for the computation are more evenly distributed geographically. Moreover, Zhang and Yang (2013) proved that C3 is more suitable for the extraction of attenuation information compared with the traditional noise cross-correlation.

A potentially much more significant advantage of C3 over traditional ambient noise correlation is that it can be used to extract empirical Green's functions between station-pairs even when the two stations are not operating simultaneously. This requires that there are some other stations that have operation times that overlap those two asynchronous ones. Ma and Beroza (2012) demonstrated the feasibility of this using data from southern California. A potential downside of the use of C3 is that the coda tends to be more band-limited than the Rayleigh wave observed with C1 and the recovery of meaningful empirical Green's functions is more difficult than with the use of C1.

5.7.4 Advanced Seismic Interferometric Theories and Methods

Several researchers have proposed more advanced seismic interferometry theories that go beyond approaches that have been applied previously. We briefly introduce two of them.

As mentioned previously in this chapter, precursory noise can emerge in ambient noise cross-correlation waveforms when the distribution of noise sources differs significantly from azimuthal homogeneity. Conventionally, observers attempt to eliminate or suppress these spurious arrivals. In fact, traditional ambient noise tomographic inversions have been performed by assuming the only meaningful arrival is the ballistic wave emitted from one station (or virtual source) and recorded by another station. Recently, Fichtner et al. (2017) developed a more general seismic interferometry theory for cross-correlations that can use the precursory noise to constrain Earth structure (see also section 4.5 of Fichtner and Tsai, 2018). Figure 13 of Fichtner et al. (2017) illustrates that the sensitivity kernels for different waveform windows, including precursory arrivals, can be used to constrain the Earth. Fichtner's theory is sophisticated and it is not trivial to apply the method to real data. However, at the very least, it demonstrates the feasibility of using precursory arrivals from cross-correlation data for seismic imaging.

Another advanced seismic interferometry theory that goes beyond conventional cross-correlation approaches has been developed by Kees Wapenaar and his collaborators. These methods are designed to improve the performance of ambient noise cross-correlation in the presence of highly inhomogeneously distributed noise sources. Wapenaar and his coworkers developed a new type of computation called multidimensional deconvolution (MDD) and proved theoretically that this procedure can guarantee the retrieval of accurate empirical Green's functions even when the source distribution is one-sided (Wapenaar and van der Neut, 2010; Wapenaar et al., 2011). Figures 11 in Wapenaar et al. (2011) demonstrates the improvement that MDD can achieve compared with cross-correlation. An application of this technique to real data for surface waveform retrieval was presented by van Dalen et al. (2015). Although the method appears to be promising, MDD also has some limitations. To retrieve an empirical Green's function for a station pair, a regular array of receivers is required to construct the so-called point spread function (PSF) for the deconvolution computation (see Wapenaar and van der Neut (2010) for more detail). The MDD is computationally more expensive than cross-correlation, and the matrix inversion in the deconvolution may be unstable. Despite these caveats, the MDD theory provides a possible new path to retrieve more reliable Green's functions from ambient noise.

5.8 Conclusions

This chapter presents a discussion of some of the practical issues involved with processing ambient noise recorded at two stations and cross-correlating such recordings to recover reliable estimated Green's functions between the stations, at least for the surface wave parts of the Green's function. Our focus is on regional-scale broadband ambient noise seismology, which delivers information at lateral resolutions from tens to hundreds of kilometers about the crust and uppermost mantle.

A number of data processing variables affect the resulting cross-correlations, including the length of the records in the time domain that are cross-correlated, and the nature of the normalization of records in the time and frequency domains. We discuss how to obtain surface wave dispersion (group and phase speed) measurements in the time domain. We also discuss the reduction of tilt and compliance noise, which is required in an ocean bottom setting.

A significant part of ambient noise data processing is the selection and rejection of particular cross-correlations. We list a number of useful criteria on which to base such decisions, which we present as necessary conditions to be satisfied for cross-correlations to provide reliable approximations to Green's functions. These conditions include the SNR of the cross-correlations, their temporal stability (particularly over seasons), the level of precursory noise, the self-similarity of measurements obtained on the cross-correlations over time and frequency, and their correspondence to other cross-correlations. The stability, repeatability, and self-consistency of dispersion measurements are useful tools to quantify the uncertainty of dispersion measurements. Array-based methods are particularly useful to estimate uncertainties in dispersion maps (e.g., eikonal tomography).

The ultimate question is: Why should we believe that ambient noise provides reliable information about the Earth? Part of this question is answered with the uncertainty estimates discussed in the previous paragraph, but only part. Such uncertainty estimates primarily quantify the variability of the measurements, which does not include an assessment of systematic errors. The principal reason we believe that ambient noise does not provide a biased estimator, at least in many circumstances, is that ambient noise results can be compared with earthquake results. For example, phase velocity maps from ambient noise and earthquake tomography can be compared in detail. When this has been done carefully, bias has been seen to be small and reduces with the length of ambient noise observation and the number of earthquakes.

Further improvements in ambient noise data processing are needed, and we summarize a few recent amplifications and extensions of the methods presented in this chapter. Particular needs include the fact that Love waves are narrower band than Rayleigh waves on continents and are very hard to observe in ocean bottom settings using OBS data. The reader is encouraged to remember that what we seek are observational methods to yield broadband, low variance, and unbiased information about the Earth and, preferably, methods that will speed convergence and thus reduce observation time. Ambient noise seismology remains a young discipline, and readers are encouraged to view the current state of the art as merely the starting point to supersede with their own creative work in ambient noise seismology.

Acknowledgments

The authors are grateful to two anonymous reviewers and Nori Nakata for criticisms that helped to improve this chapter. Much of the work on which this chapter rests has been supported by grants from the US National Science Foundation and the US Department of Defense. The facilities of the IRIS Data Management System were used to access all of the data used in this study. The IRIS DMS is funded through the US National Science Foundation under Cooperative Agreement EAR-0552316. The authors are also grateful to the Cascadia Initiative Expedition Team for acquiring the Amphibious Array Ocean Bottom Seismograph data and appreciate the open data policy that makes these data available. This work utilized the Janus supercomputer, which is supported by the National Science Foundation (award number CNS-0821794), the University of Colorado Boulder, the University of Colorado Denver, and the National Center for Atmospheric Research. The Janus supercomputer is operated by the University of Colorado at Boulder.

References

Aki, K. 1957. Space and Time Spectra of Stationary Stochastic Waves, with Special Reference to Microtremors. *Bull. Earthq. Res. Inst.*, **35**, 415–456.

Ardhuin, F., Gualtieri, L., and Stutzmann, E. 2018. Physics of ambient noise generation by ocean waves. Pages – of: Nakata, N., Gualtieri, L., and Fichtner, A. (eds.), *Seismic Ambient Noise*. Cambridge University Press, Cambridge, UK.

Baig, A. M., Campillog, M., and Brenguier, F. 2009. Denoising seismic noise cross correlations. *J. Geophys. Res.*, **114**, B08310.

Ball, J. S., Sheehan, A. F., Stachnik, J. C., Lin, F.-C., and Collins, J. A. 2014. A joint Monte Carlo analysis of seafloor compliance, Rayleigh wave dispersion and receiver functions at ocean bottom seismic stations offshore New Zealand. *Geochem. Geophys. Geosyst.*, **15**, 5051–5068.

Barmin, M. P., Ritzwoller, M. H., and Levshin, A. L. 2001. A fast and reliable method for surface wave tomography. *Pure Appl. Geophys.*, **158**, 1351–1375.

Bell, S. W., Forsyth, D. W., and Ruan, Y. 2015. Removing noise from the vertical component records of ocean-bottom seismometers: Results from year one of the Cascadia Initiative. *Bull. Seismol. Soc. Am.*, **105**(1), 300–313.

Bensen, G. D., Ritzwoller, M. H., Barmin, M. P., Levshin, A. L., Lin, F., Moschetti, M. P., Shapiro, N. M., and Yang, Y. 2007. Processing seismic ambient noise data to obtain reliable boroad-band surface wave dispersion measurements. *Geophys. J. Int.*, **169**, 1239–1260.

Bowden, D. C., Kohler, M. D., Tsai, V. C., and Weeraratne, D. S. 2016. Offshore southern California lithospheric velocity structure from noise cross-correlation functions. *J. Geophys. Res.*, **121**(5), 3415–3427.

Bracewell, R. N. 1978. *The Fourier transform and its applications*. 2nd edn. McGraw-Hill.

Campillo, M., and Paul, A. 2003. Long-range correlations in the diffuse seismic coda. *Science*, **299**, 547–549.

Carriére, O., Gerstoft, P., and Hodgkiss, W. S. 2014. Spatial filtering in ambient noise interferometry. *J. Acoust. Soc. Am.*, **135**(3), 1186–1196.

Crawford, W. C., and Webb, S. C. 2000. Identifying and removing tilt noise from low-frequency (<0.1 Hz) seafloor vertical seismic data. *Bull. Seismol. Soc. Am.*, **90**(4), 952–963.

Dolenc, D., Romanowicz, B., Uhrhammer, R., McGill, P., Neuhauser, D., and Stakes, D. 2007. Identifying and removing noise from the Monterey ocean bottom broadband seismic station (MOBB) data. *Geochem. Geophys. Geosyst.*, **8**(2), Q02005.

Duennebier, F. K., and Sutton, G. H. 1995. Fidelity of ocean bottom seismic observations. *Marine Geophys. Res.*, **17**, 535–555.

Ekström, G., Abers, G. A., and Webb, S. C. 2009. Determination of surface-wave phase velocities across USArray from noise and Aki's spectral formulation. *Geophys. Res. Lett.*, **36**, L18301.

Fichtner, A., and Tsai, V. 2018. Theoretical foundations of noise interferometry. Pages – of: Nakata, N., Gualtieri, L., and Fichtner, A. (eds.), *Seismic Ambient Noise*. Cambridge University Press, Cambridge, UK.

Fichtner, A., Stehly, L., Ermert, L., and Boehm, C. 2017. Generalized interferometry – I: theory for interstation correlations. *Geophys. J. Int.*, **208**, 603–638.

Gallot, T., Catheline, S., Roux, P., and Campillo, M. 2012. A passive inverse filter for Green's function retrieval. *J. Acoust. Soc. Am.*, **131**(1), EL21–EL27.

Gao, H., and Shen, Y. 2015. A preliminary full-wave ambient-noise tomography model spanning from the Juan de Fuca and Gorda spreading centers to the Cascadia volcanic arc. *Seismol. Res. Lett.*, **86**(5), 1253–1260.

Harmon, N., Forsyth, D., and Webb, S. 2007. Using ambient seismic noise to determine short-period phase velocities and shallow shear velocities in young oceanic lithosphere. *Bull. Seismol. Soc. Am.*, **97**(6), 2009–2023.

Levshin, A. L., and Ritzwoller, M. H. 2001. Automated detection, extraction, and measurement of regional surface waves. *Pure Appl. Geoophys.*, **158**, 1531–1545.

Levshin, A. L., Yanovskaya, T. B., Lander, A. V., Bukchin, B. G., Barmin, M. P., Ratnikova, L. I., and Its, E. N. 1989. *Seismic surface waves in laterally inhomogeneous Earth.* Kluwer Publ.

Liu, X., Ben-Zion, Y., and Zigone, D. 2016. Frequency domain analysis of errors in cross-correlations of ambient seismic noise. *Geophys. J. Int.*, **207**, 1630–1652.

Lin, F.-C., Moschetti, M. P., and Ritzwoller, M. H. 2008. Surface wave tomography of the western United States from ambient seismic noise: Rayleigh and Love wave phase velocity maps. *Geophys. J. Int.*, **173**, 281–298.

Lin, F.-C., and Ritzwoller, M. H. 2011. Helmholtz surface wave tomography for isotropic and azimuthally anisotropic structure. *Geophys. J. Int.*, **186**, 1104–1120.

Lin, F.-C., Ritzwoller, M. H., and Snieder, R. 2009. Eikonal tomography: surface wave tomography by phase front tracking across a regional broad-band seismic array. *Geophys. J. Int.*, **177**, 1091–1110.

Lin, F.-C., Ritzwoller, M. H., Townend, J., Bannister, S., and Savage, M. K. 2007. Ambient noise Rayleigh wave tomography of New Zealand. *Geophys. J. Int.*, **170**, 649–666.

Lin, F.-C., Ritzwoller, M. H., Yang, Y., Moschetti, M. P., and Fouch, M. J. 2011. Complex and variable crustal and uppermost mantle seismic anisotropy in the western United States. *Nature Geosci.*, **4**, 55–61.

Luo, Y., Yang, Y., Xu, Y., Xu, H., Zhao, K., and Wang, K. 2015. On the limitations of interstation distances in ambient noise tomography. *Geophys. J. Int.*, **201**, 652–661.

Ma, S., and Beroza, G. C. 2012. Ambient-field Green's functions from asynchronous seismic observations. *Geophys. Res. Lett.*, **39**, L06301.

McNamara, D., and Boaz, R. 2018. Visualization of the Seismic Ambient Noise Spectrum. Pages — of: Nakata, N., Gualtieri, L., and Fichtner, A. (eds.), *Seismic Ambient Noise.* Cambridge University Press, Cambridge, UK.

Moschetti, M. P., Ritzwoller, M. H., Lin, F., and Yang, Y. 2010. Seismic evidence for widespread western-US deep-crustal deformation caused by extension. *Nature*, **464**, 885–890.

Nakata, N., and Nishida, K. 2018. Body wave exploration. Pages – of: Nakata, N., Gualtieri, L., and Fichtner, A. (eds.), *Seismic Ambient Noise.* Cambridge University Press, Cambridge, UK.

Prieto, G. A., Denolle, M., Lawrence, J. F., and Beroza, G. C. 2011. On amplitude information carried by the ambient seismic field. *C.R. Geosci.*, **343**, 600–614.

Prieto, G. A., Lawrence, J. F., and Beroza, G. C. 2009. Anelastic earth structure from the coherency of the ambient seismic field. *J. Geophys. Res.*, **114**, B07303.

Ritzwoller, M. H., Lin, F.-C., and Shen, W. 2011. Ambient noise tomography with a large seismic array. *C. R. Geoscience*, **343**, 558–570.

Schimmel, M., Stutzmann, E., and Gallart, J. 2011. Using instantaneous phase coherence for signal extraction from ambient noise data at a local to a global scale. *Geophys. J. Int.*, **184**(1), 494–506.

Seats, K. J., Lawrence, J. F., and Prieto, G. A. 2012. Improved ambient noise correlation functions using Welch's method. *Geophys. J. Int.*, **188**, 513–523.

Seydoux, L., de Rosny, J., and Shapiro, N. M. 2017. Pre-processing ambient noise cross-correlations with equalizing the covariance matrix eigenspectrum. *Geophys. J. Int.*, **210**, 1432–1449.

Shapiro, N. 2018. Applications with surface waves extracted from ambient seismic noise. Pages — of: Nakata, N., Gualtieri, L., and Fichtner, A. (eds.), *Seismic Ambient Noise*. Cambridge University Press, Cambridge, UK.

Shapiro, N. M., and Ritzwoller, M. H. 2002. Monte-Carlo inversion for a global shear-velocity model of the crust and upper mantle. *Geophys. J. Int.*, **151**, 88–105.

Shapiro, N. M., Ritzwoller, M. H., and Bensen, G. D. 2006. Source location of the 26 sec microseism from cross-correlations of ambient seismic noise. *Geophys. Res. Lett.*, **33**, L18310.

Shen, W., Ritzwoller, M. H., Kang, D., Kim, Y., Lin, F.-C., Ning, J., Wang, W., Zheng, Y., and Zhou, L. 2016. A seismic reference model for the crust and uppermost mantle beneath China from surface wave dispersion. *Geophys. J. Int.*, **206**, 954–979.

Shen, Y., Ren, Y., Gao, H., and Savage, B. 2012. An improved method to extract very-broadband empirical Green's functions from ambient seismic noise. *Bull. Seismol. Soc. Am.*, **102**(4), 1872–1877.

Smith, M. L., and Dahlen, F. A. 1973. The azimuthal dependence of Love and Rayleigh wave propagation in a slightly anisotropic medium. *J. Geophys. Res.*, **78**(17), 3321–3333.

Snieder, R. 2004. Extracting the Green's function from the correlation of coda waves: A derivation based on stationary phase. *Phys. Rev. E*, **69**, 046610.

Stehly, L., Campillo, M., and Shapiro, N. M. 2007. Traveltime measurements from noise correlation: stability and detection of instrumental time-shifts. *Geophys. J. Int.*, **171**, 223–230.

Stehly, L., Campillo, M., Froment, B., and Weaver, R. L. 2008. Reconstruction Green's function by correlation of the coda of the correlation (C^3) of ambient seismic noise. *J. Geophys. Res.*, **113**, B11306.

Stehly, L., Cupillard, P., and Romanowicz, B. 2011. Towards improving ambient noise tomography using simultaneously curvelet denoising filters and SEM simulations of seismic ambient noise. *C. R. Geoscience*, **343**, 591–599.

Stehly, L., Froment, B., Campillo, M., Liu, Q. Y., and Chen, J. H. 2015. Monitoring seismic wave velocity changes associated with the M_W 7.9 Wenchuan earthquake: increasing the temporal resolution using curvelet filters. *Geophys. J. Int.*, **201**, 1939–1949.

Stockwell, R. G., Mansinha, L., and Lowe, R. P. 1996. Localization of the complex spectrum: The S transform. *IEEE Trans Signal Process*, **44**(4), 998–1001.

Takagi, R., Nakahara, H., Kono, T., and Okada, T. 2014. Separating body and Rayleigh waves with cross terms of the cross-correlation tensor of ambient noise. *J. Geophys. Res.*, **119**, 2005–2018.

Tian, Y., and Ritzwoller, M. H. 2015. Directionality of ambient noise on the Juan de Fuca plate: implications for source locations of the primary and secondary microseisms. *Geophys. J. Int.*, **201**, 429–443.

———. 2017. Improving ambient noise cross-correlations in the noisy ocean bottom environment of the Juan de Fuca plate. *Geophys. J. Int.*, **210**, 1787–1805.

Tian, Y., Shen, W., and Ritzwoller, M. H. 2013. Crustal and uppermost mantle shear velocity structure adjacent to the Juan de Fuca Ridge from ambient seismic noise. *Geochem. Geophys. Geosyst.*, **14**(8), 3221–3233.

van Dalen, K. N., Mikesell, T. D., Ruigrok, E. N., and Wapenaar, K. 2015. Retrieving surface waves from ambient seismic noise using seismic interferometry by multidimensional deconvolution. *J. Geophys. Res.*, **120**, 994–961.

van Wijk, K., Mikesell, T. D., Schulte-Pelkum, V., and Stachnik, J. 2011. Estimating the Rayleigh-wave impulse response between seismic stations with the cross terms of the Green tensor. *Geophys. Res. Lett.*, **38**, L16301.

Wapenaar, K., and van der Neut, J. 2010. A representation for Green's function retrieval by multidimensional deconvolution. *J. Acoust. Soc. Am.*, **128**(6), EL366–EL371.

Wapenaar, K., van der Neut, J., Ruigrok, E., Draganov, D., Hunziker, J., Slob, E., Thorbecke, J., and Snieder, R. 2011. Seismic interferometry by crosscorrelation and by multidimensional deconvolution: a systematic comparison. *Geophys. J. Int.*, **185**, 1335–1364.

Webb, S. C. 1988. Long-period acoustic and seismic measurements and ocean floor currents. *IEEE J Ocean Eng*, **13**(4), 263–270.

Webb, S. C., and Crawford, W. C. 1999. Long-period seafloor seismology and deformation under ocean waves. *Bull. Seismol. Soc. Am.*, **89**(6), 1535–1542.

Welch, P. D. 1967. The use of fast Fourier transform for the estimation of power spectra: A method based on time averaging over short, modified periodograms. *IEEE Trans. Audio Electroacoust.*, **AU-15**(2), 70–73.

Xie, J., Ritzwoller, M. H., Shen, W., and Wang, W. 2017. Crustal anisotropy across eastern Tibet and surroundings modeled as a depth-dependent tilted hexagonally symmetric medium. *Geophys. J. Int.*, **209**, 466–491.

Xie, J., Ritzwoller, M. H., Shen, W., Yang, Y., Zheng, Y., and Zhou, L. 2013. Crustal radial anisotropy across Eastern Tibet and the Western Yangtze Craton. *J. Geophys. Res.*, **118**, 4226–4252.

Yang, Y., and Forsyth, D. W. 2008. Attenuation in the upper mantle beneath Southern California: Physical state of the lithosphere and asthenosphere. *J. Geophys. Res.*, **113**, B03308.

Yang, Y., and Ritzwoller, M. H. 2008. Teleseismic surface wave tomography in the western U.S. using the Transportable Array component of USArray. *Geophys. Res. Lett.*, **35**, L04308.

Yang, Y., Ritzwoller, M. H., Lin, F.-C., Moschetti, M. P., and Shapiro, N. M. 2008. Structure of the crust and uppermost mantle beneath the western United States revealed by ambient noise and earthquake tomography. *J. Geophys. Res.*, **113**, B12310.

Yao, H., Gouédard, P., Collins, J. A., McGuire, J. J., and van der Hilst, R. D. 2011. Structure of young East Pacific Rise lithosphere from ambient noise correlation analysis of fundamental- and higher-mode Scholte-Rayleigh waves. *C. R. Geoscience*, **343**, 571–583.

Zeng, X., and Ni, S. 2010. A persistent localized microseismic source near the Kyushu Island, Japan. *Geophys. Res. Lett.*, **37**, L24307.

Zhang, J., and Yang, X. 2013. Extracting surface wave attenuation from seismic noise using correlation of the coda of correlation. *J. Geophys. Res.*, **118**, 2191–2205.

Zheng, Y., Shen, W., Zhou, L., Yang, Y., Xie, Z., and Ritzwoller, M. H. 2011. Crust and uppermost mantle beneath the North China Craton, northeastern China, and the Sea of Japan from ambient noise tomography. *J. Geophys. Res.*, **116**, B12312.

Zhou, L., Xie, J., Shen, W., Zheng, Y., Yang, Y., Shi, H., and Ritzwoller, M. H. 2012. The structure of the crust and uppermost mantle beneath South China from ambient noise and earthquake tomography. *Geophys. J. Int.*, **189**, 1565–1583.

6

Locating Velocity Changes in Elastic Media with Coda Wave Interferometry

ROEL SNIEDER, ALEJANDRO DURAN, AND ANNE
OBERMANN

Abstract

In coda wave interferometry the long wave paths of coda waves are used to detect minute changes in the velocity. When the relative velocity perturbation is constant in space, it is related to the travel time change δt by $\delta v / v = -\delta t / t$. But when the velocity change depends on space, the relation between the measured travel time change and the velocity change is more complicated. We show that in that case the estimation of velocity change can be formulated as a standard linear inverse problem. The sensitivity kernel that relates the travel time change to the velocity depends on the energy density of the coda waves in space. We derive these kernels for (1) diffusive acoustic waves, (2) acoustic waves that obey radiative transfer, and (3) diffusive elastic waves, and illustrate the theory with numerical examples for acoustic and elastic waves.

6.1 Introduction

We introduce coda waves in Figure 6.1 with ultrasound measurements of waves that propagate through a granite cylinder (Snieder et al., 2002; Grêt et al., 2006b). Let us first consider the red waveforms. The waves are chaotic looking; one cannot identify isolated arrivals in the waveform. The reason is that these waves reverberate and are scattered in the granite sample along a multitude of wave paths. As a result, it is difficult to use these waves to create an image of the heterogeneity in the sample.

The tail of reverberating and scattered waves in Figure 6.1 is called *coda waves*, in music the word *coda* refers to the closing of a piece. The temporal decay of these waves is due to attenuation, and hence the decay of scattered waves can be used to estimate attenuation (Sato and Fehler, 2012). Changes in the decay of coda waves have been used to estimate changes in the attenuation in the Earth (Chouet,

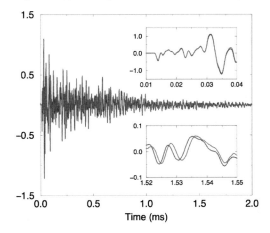

Figure 6.1. Ultrasonic waves recorded in a granite cylinder at a temperature of 45°C (dark grey line (blue in the plate section)) and 50°C (light grey line (red in the plate section)). (A black-and-white version of this figure appears in some formats. For the color version, please refer to the plate section.)

1979; Sato, 1986, 1988; Fehler et al., 1998). The decay of the coda waves, however, does not utilize the phase information of the waveforms, and hence is insensitive to changes in the velocity.

The red waves in Figure 6.1 were recorded when the sample was at a temperature of 50° C. When the wavefield is measured with the same source and receiver in the same sample, but now at a temperature of 45°C, the blue waveforms result (Snieder et al., 2002; Grêt et al., 2006b). Again, the waveforms are chaotic looking, and these waves are also difficult to interpret. But it is interesting to consider the *change* in the waveforms as the temperature is changed. The inset in the top right shows the waves at a time around the first-arriving waves. The red and blue waveforms are almost identical. The first-arriving waves have propagated through the sample for a short time, hence the change in the arrival time of the early arriving waves due to the temperature change of 5°C is too small to lead to an appreciable change in these early arriving waves. A blowup of the later-arriving waves in a time window around 1.54 s is shown in the lower right panel of Figure 6.1. For these later-arriving waves the red waves are a time-shifted version of the blue waves. The later-arriving waves have spent more time propagating through the sample, hence they are more sensitive to the velocity change caused by the 5°C change in temperature.

Since the coda waves have spent a longer time propagating through the medium than the direct waves, the coda waves are more sensitive to changes in the velocity than the direct wave, which makes coda waves useful for detecting small time-lapse changes in the velocity. This concept was originally proposed and applied to Earthquake doublets (Poupinet et al., 1984) and ultrasound data (Roberts et al., 1992). As shown in the bottom-right inset of Figure 6.1, a velocity change corresponds to a change in the arrival time of time-windowed coda waves. That change in arrival time can be measured using a cross-correlation (Snieder et al., 2002; Snieder, 2006), which essentially is an interferometric measurement. For this reason the extraction of changes in media from changes in recorded coda waves has been called *coda wave interferometry*. An alternative, and more robust, way to extract the velocity change is the stretching method where one seismogram is stretched to match the other seismogram (Sens-Schönfelder and Wegler, 2006; Hadziioannou et al., 2009).

Coda wave interferometry has been applied to a large number of problems that include coseismic and postseismic changes in seismic velocity (Schaff and Beroza, 2004; Brenguier et al., 2008; Wegler et al., 2009; Nakata and Snieder, 2011; Hobiger et al., 2012; Takagi et al., 2012; Obermann et al., 2014; Gassenmeier et al., 2016), volcano monitoring (Nishimura et al., 2000; Yamawaki et al., 2004; Grêt et al., 2005; Brenguier et al., 2011), monitoring changes in the near surface (Sens-Schönfelder and Wegler, 2006; Mainsant et al., 2012; Larose et al., 2015a; Gassenmeier et al., 2015) and in concrete (Tremblay et al., 2010; Zhang et al., 2016), stress changes in a mining environment (Grêt et al., 2006a), geo-engineering (Hillers et al., 2015; Planès et al., 2016; Obermann et al., 2015), structural health monitoring (Lu and Michaels, 2005; Nakata et al., 2013; Larose et al., 2015b; Salvermoser et al., 2015), and even the detection of velocity changes on the moon (Sens-Schönfelder and Larose, 2008). Coda wave interferometry has not only been used to detect velocity changes, the principle can also be used to estimate the relative distance between repeat Earthquakes (Snieder and Vrijlandt, 2005; Robinson et al., 2011, 2013), measuring the motion of scatterers in fluid flow (Cowan et al., 2000; Page et al., 2000), and detecting changes in the strength of scatterers (Larose et al., 2010; Rossetto et al., 2011; Planès et al., 2014; Margerin et al., 2016). To a large extent, these different perturbations of the wavefield can be distinguished because they leave a different imprint on the change in the coda waves (Snieder, 2006). A great boost was given to monitoring by the development of seismic interferometry when one retrieves the waves that propagate between two sensors by cross-correlating the noise recorded at these sensors (Lobkis and Weaver, 2001; Campillo and Paul, 2003; Curtis et al., 2006; Larose et al., 2006; Snieder and Larose, 2013; Fichtner and Tsai, 2018, Chapter 4). Since the noise is always present, one can measure waves in a quasi-continuous way. Chapter 9 by Sens-Schönfelder and Brenguier (2018) provides a review of recent coda wave interferometry applications and measurement techniques.

Many studies report changes in the velocity as a function of time, but do not report *where* in space the velocity is changed. When one is only interested in the temporal behavior of the velocity, that is not needed. But there are situations where localizing the velocity change is desirable. To what distance from an Earthquake does the velocity change? What is the depth extent of the velocity change? One approach to compute the imprint of a velocity change on coda waves is to numerically compute waveforms before and after the velocity change. This approach is most useful when one considers a prescribed spatial pattern of the velocity change, such as a horizontal layer in depth or a slab (Obermann et al., 2013, 2016). This approach is, however, not practical when one seeks to find a velocity change that can have any spatial distribution. In that case it is useful to define sensitivity kernels that relate the travel time change τ in a given time window to the relative slowness perturbation $\delta s / s$ as a function of space:

$$\tau = \int K(\mathbf{r}) \frac{\delta s}{s}(\mathbf{r}) dV. \tag{6.1}$$

This relation has been derived for single scattered waves (Pacheco and Snieder, 2006) and for multiple scattered waves (Pacheco and Snieder, 2005). The sensitivity kernel $K(\mathbf{r})$ depends on the source and receiver used, as well as on the time window in which the travel time change is measured. Equation (6.1) constitutes a linear inverse problem for the relative velocity change from travel time changes measured with coda wave interferometry (Menke, 1984; Aster et al., 2004). Kanu and Snieder (2015b) show how one can invert equation (6.1) for a space-dependent slowness perturbation given a set of measured changes in the arrival time of coda waves. Obermann et al. (2013, 2014) have used measurements to locate time-lapse velocity changes related to volcanic eruptions and Earthquakes.

In this work we present derivations of the kernel $K(\mathbf{r})$ for a number of situations. We first consider acoustic waves that are strongly scattered. The energy density of such waves behaves as a diffusion process (van Rossum and Nieuwenhuizen, 1999; Tourin et al., 2000), which can intuitively be understood from the fact that strongly scattered waves follow a random walk. We also analyze acoustic waves whose intensity follows the equations of radiative transfer (Chandrasekhar, 1960; Özisik, 1973). For long propagation times this produces diffusive wave propagation, but the equations of radiative transfer also hold for the direct wave, and scattered waves for early times (Paasschens, 1997). Our derivation is similar to recent derivations (Mayor et al., 2014; Margerin et al., 2016), but we elucidate some steps, in particular the role of the Chapman-Kolmogorov equation, in more detail. We also derive the kernels for diffusive elastic waves, which is important for seismological applications since the Earth is elastic. Note that we make no assumptions about the nature of the scatterers; they may scatter isotropically, or they may have an arbitrary radiation pattern. The information of the scatterers is encoded in the energy or specific intensity of the propagating waves; this is all the information that is needed to compute the sensitivity kernels.

This chapter consists of the following sections. In section 6.2 we derive the sensitivity kernels for strongly scattered acoustic waves. We generalize the derivation to acoustic waves that follow radiative transfer in section 6.3. In section 6.4 we present numerical examples of sensitivity kernels assuming diffusive wave propagation. We show an example that an inappropriate application of the diffusion approximation leads to erroneous kernels. We derive the kernels for diffuse elastic waves in section 6.5. The equations of radiative transfer for elastic waves have been developed (Ryzhik et al., 1996), but because of its complexity we refrain from deriving the sensitivity kernels for this case. Section 6.6 features numerical examples of the sensitivity kernels for elastic waves. In appendices 6.A and 6.B we derive the Chapman-Kolmogorov equation for the diffusion equation and the equation of radiative transfer, respectively, because this theorem plays a central role in the derivation.

6.2 The Travel Time Change for Diffusive Acoustic Waves

Since the Earth is elastic, it may seem strange to treat seismological waves as acoustic waves. The coda, however, is mostly comprised of shear waves (Aki and Chouet, 1975). The ratio of S-wave energy density to P-wave energy density in strongly scattering 3D elastic media is given by

$$\frac{I_S}{I_P} = 2 \left(\frac{v_P}{v_S} \right)^3, \tag{6.2}$$

where v_P is the P-wave velocity and v_S the S-wave velocity (Weaver, 1982; Trégourès and van Tiggelen, 2002; Snieder, 2002). For reasons of brevity we refer to the *energy density*, defined as the energy per unit volume, also as *energy*. For a Poisson medium, where $v_P/v_S = \sqrt{3}$, the energy ratio (6.2) satisfies $I_S/I_P \approx 10$; hence, most of the energy resides in the shear waves. For this reason, treating the waves as scalar acoustic waves can be a reasonable approach. In this approach one retrieves the perturbation in the shear wave velocity.

The sensitivity kernels for velocity changes for scalar waves in strongly scattering media have been derived by Pacheco and Snieder (2005). Their derivation included some ad hoc steps, notably the insertion of the velocity change in their equation (18). We rederive in this section the sensitivity kernels of the travel time of acoustic waves in strongly scattering media with the purpose of (a) avoiding some of the ad hoc steps in the derivation of Pacheco and Snieder (2005), and (b) presenting a derivation that can be extended to elastic media.

In coda wave interferometry one measures the effective travel time changes by cross-correlating the unperturbed and perturbed coda waves over a time window centered at a central time t (Snieder, 2006). The time-windowed cross-correlation between unperturbed waves $u(t)$ and perturbed waves $\tilde{u}(t)$ is defined as

$$R(t_s) = \frac{\int_{t-t_w}^{t+t_w} u(t')\tilde{u}(t' + t_s)dt'}{\sqrt{\int_{t-t_w}^{t+t_w} u^2(t')dt' \int_{t-t_w}^{t+t_w} \tilde{u}^2(t')dt'}}, \tag{6.3}$$

where the time window has center time t and width $2t_w$, and where t_s is the time shift. The time shift $t_{s,max}$ for which this cross-correlation is a maximum is given by Snieder (2006)

$$t_{s,max} = \frac{\sum_T I_T \tau_T}{\sum_T I_T}. \tag{6.4}$$

This expression is based on the path summation where the scattered waves are written as a sum of the waves that propagate along all possible scattering trajectories T. The travel time change for a wave that travels along trajectory T is denoted by τ_T, and the energy density of that wave is given by I_T. Equation (6.4) thus states that the travel time change obtained from coda wave interferometry is the energy-weighted average of the travel time perturbation of all waves that arrive within the time window used for the cross-correlation (Snieder, 2006). Since the last term in equation (6.4) is the energy-weighted average of the travel time change we denote this quantity also as an average:

$$\langle \tau \rangle \equiv \frac{\sum_T I_T \tau_T}{\sum_T I_T}. \tag{6.5}$$

The travel time change caused by a slowness perturbation δs for a wave propagating along trajectory T is given by

$$\tau_T = \int_T \delta s \, dl, \tag{6.6}$$

where the integration is along trajectory T. This is a linear approximation of the travel time change that is based on Fermat's theorem (Aldridge, 1994; Nolet, 2008), but since the velocity changes inferred from coda wave interferometry are usually a fraction of a percent (Snieder et al., 2002; Sens-Schönfelder and Wegler, 2006; Brenguier et al., 2008), this linearization can be expected to work well.

Consider first the case of a constant relative slowness perturbation: $\delta s/s = A = constant$. The travel time perturbation along trajectory T in equation (6.6) is then given by $\tau_T = \int A s dl = A \int dt = At_T$, where t_T is the travel time along trajectory T. In this case $\tau_T/t_T = A = \delta s/s$. Thus, for a constant relative slowness perturbation the relative travel time perturbation is equal to the relative slowness perturbation:

$$\frac{\tau_T}{t_T} = \frac{\delta s}{s} \approx -\frac{\delta v}{v}. \tag{6.7}$$

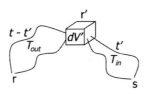

Figure 6.2. Definition of geometric variables for the contribution of a velocity change in volume dV' at $\mathbf{r}'$, a source at $\mathbf{s}$, and a receiver at $\mathbf{r}$ in a diffusive treatment. Only two incoming and outgoing trajectories are shown. Incoming trajectories T_{in} and outgoing trajectories T_{out} can be combined in any way to form the total trajectory from $\mathbf{s}$ to $\mathbf{r}$ through dV'.

The last identity is only true to first order in the velocity perturbation, but since the perturbation is typically less than 1%, the last identity is very accurate. This chapter is concerned with the situation that the relative slowness perturbation is not constant in space. How can the velocity change be localized in space? Specifically, what is the contribution of the slowness change in a volume element dV' to the observed change in arrival time of the coda waves in different time windows?

In the following we consider all trajectories that travel through a volume dV' as shown in Figure 6.2. These waves intersect dV' in unknown directions, which may be different for each trajectory, hence the corresponding length of intersection dl of a trajectory with the volume dV' is unknown. This complication can be avoided by expressing the line-integral (6.6) as a time-integral using $dl = s^{-1}dt'$, where s is the slowness:

$$\tau_T = \int_0^t \frac{\delta s}{s}(\mathbf{r}_T)\, dt', \tag{6.8}$$

where the integration is along the path $\mathbf{r}_T(t')$ of trajectory T as a function of t'. The unperturbed travel time is given by t, hence the time integral is limited to the interval $0 < t' < t$ taken for the waves to propagate from the source $\mathbf{s}$ to the receiver at $\mathbf{r}$. Inserting this expression in equation (6.5) gives

$$\langle \tau \rangle = \frac{\sum_T I_T \int_0^t \frac{\delta s}{s}(\mathbf{r}_T)\, dt'}{\sum_T I_T}. \tag{6.9}$$

Equation (6.9) gives the change in arrival of the coda waves in terms of the slowness perturbation along all possible paths. We next reorder this sum over paths to find the contribution of the slowness change in the volume dV' in Figure 6.2. As shown in that figure we consider incoming trajectories T_{in} that propagate from the source at $\mathbf{s}$ to dV' in a time t', and outgoing trajectories T_{out} that propagate from

dV' to the receive $\mathbf{r}$ in the remaining time $t - t'$. For notational simplicity we first consider the contribution to the energy density of all paths that propagate through dV' at a fixed time t'; we indicate this quantity by $\sum_{T \text{ through } dV'} I_T$. We assume for the moment that the energy of the waves that propagate from the source at $\mathbf{s}$ through dV' to the receiver at $\mathbf{r}$ is the product of the energy of the waves that propagate from $\mathbf{s}$ to dV' and from dV' to $\mathbf{r}$. We substantiate this assumption below in equation (6.12), but first provide a heuristic explanation. As shown in Figure 6.2, every path T from $\mathbf{s}$ to $\mathbf{r}$ through dV' consists of an incoming segment T_{in} from $\mathbf{s}$ to dV' and an outgoing segment T_{out} from dV' to $\mathbf{r}$. Therefore the sum of all paths from $\mathbf{s}$ to $\mathbf{r}$ that traverse dV' can be written as a double sum over paths T_{in} and T_{out}: $\sum_T = \sum_{T_{in}} \sum_{T_{out}}$. This means that when we consider all the paths that intersect dV' at time t', we can write the contribution of the energy of the paths that intersect dV' as

$$\sum_{T \text{ through } dV'} I_T(\mathbf{r}, \mathbf{s}, t) = \sum_{T_{out}} I_{T_{out}}(\mathbf{r}, \mathbf{r}', t - t') \sum_{T_{in}} I_{T_{in}}(\mathbf{r}', \mathbf{s}, t') dV', \qquad (6.10)$$

where $I_{T_{in}}(\mathbf{r}', \mathbf{s}, t')$ is the energy of the wave that propagates along a trajectory T_{in} from $\mathbf{s}$ to $\mathbf{r}'$ in time t'. The total energy of these waves in the volume dV' is given by $\sum_{T_{in}} I_{T_{in}}(\mathbf{r}', \mathbf{s}, t') dV'$. This wave energy then propagates in a time $t - t'$ to $\mathbf{r}$; the propagation of the energy is accounted for by the term $\sum_{T_{out}} I_{T_{out}}(\mathbf{r}, \mathbf{r}', t - t')$. Denoting the total energy of the waves along all incoming trajectories by $I(\mathbf{r}', \mathbf{s}, t')$, and the energy of the waves that propagate along all outgoing trajectories by $I(\mathbf{r}, \mathbf{r}', t - t')$, the contribution to the energy of the waves that propagate through dV' is given by

$$\sum_{T \text{ through } dV'} I_T(\mathbf{r}, \mathbf{s}, t) = I(\mathbf{r}, \mathbf{r}', t - t') I(\mathbf{r}', \mathbf{s}, t') dV'. \qquad (6.11)$$

This expression is a consequence of the Chapman-Kolmogorov theorem for diffusion that is derived in equation (6.46) of appendix 6.A. Equation (6.11) is, however, dimensionally not correct. In fact, following equation (6.46), it is more precise to write equation (6.11) as

$$\sum_{T \text{ through } dV'} I_T(\mathbf{r}, \mathbf{s}, t) = G^D(\mathbf{r}, \mathbf{r}', t - t') I(\mathbf{r}', \mathbf{s}, t') dV', \qquad (6.12)$$

where $G^D(\mathbf{r}, \mathbf{r}', t - t')$ is the Green's function for the diffuse energy that propagates from $\mathbf{r}'$ to $\mathbf{r}$ in a time $t - t'$. This Green's function satisfies the diffusion equation for an impulsive source

$$\frac{\partial G^D(\mathbf{r}, \mathbf{r}', t)}{\partial t} - \nabla \cdot (D(\mathbf{r}) \nabla G^D(\mathbf{r}, \mathbf{r}', t)) = \delta(\mathbf{r} - \mathbf{r}') \delta(t), \qquad (6.13)$$

where $D(\mathbf{r})$ is the diffusion constant of the energy density.

Integrating equation (6.12) over dV' shows that the energy density at location $\mathbf{r}$ is given by

$$I(\mathbf{r}, \mathbf{s}, t) = \int G^D(\mathbf{r}, \mathbf{r}', t - t') I(\mathbf{r}', \mathbf{s}, t') dV'. \qquad (6.14)$$

This expression states that the energy at a time t follows from the energy at an arbitrary earlier time t' if the diffusive Green's function is known. Margerin et al. (2016) derive this result heuristically from Bayes' theorem, which presumes that the energy can be treated as a probability. Our treatment does not invoke a probabilistic interpretation, but both treatments give the same result. Applying the reasoning of section 18.4 of Snieder and van Wijk (2015) to equation (6.13), it follows that the Green's function G^D has dimension $1/volume$, hence equation (6.14) is dimensionally correct.

Using equation (6.12) in the numerator of equation (6.9), and integrating over all volume elements dV' gives

$$\langle \tau \rangle = \frac{\int \int_0^t G^D(\mathbf{r}, \mathbf{r}', t - t') I(\mathbf{r}', \mathbf{s}, t') \frac{\delta s}{s}(\mathbf{r}') \, dt' dV'}{I(\mathbf{r}, \mathbf{s}, t)}, \qquad (6.15)$$

where we replaced the denominator of equation (6.9) by $I(\mathbf{r}, \mathbf{s}, t)$, the energy density of the waves that propagate from the source at $\mathbf{s}$ to the receiver at $\mathbf{r}$ in time t. Equation (6.15) can also be written as

$$\langle \tau \rangle = \int K(\mathbf{r}') \frac{\delta s}{s}(\mathbf{r}') \, dV', \qquad (6.16)$$

with

$$K(\mathbf{r}') = \frac{\int_0^t G^D(\mathbf{r}, \mathbf{r}', t - t') I(\mathbf{r}', \mathbf{s}, t') \, dt'}{I(\mathbf{r}, \mathbf{s}, t)}. \qquad (6.17)$$

Reciprocity applies to the diffusion equation (Morse and Feshbach, 1953a); hence $G^D(\mathbf{r}, \mathbf{r}', t) = G^D(\mathbf{r}', \mathbf{r}, t)$, and as a result

$$K(\mathbf{r}') = \frac{\int_0^t G^D(\mathbf{r}', \mathbf{r}, t - t') I(\mathbf{r}', \mathbf{s}, t') \, dt'}{I(\mathbf{r}, \mathbf{s}, t)}. \qquad (6.18)$$

In this expression, the properties of the source – for example an explosion or a point force, as well as the source strength – are contained in the energies $I(\mathbf{r}', \mathbf{s}, t')$ and $I(\mathbf{r}, \mathbf{s}, t)$, because these energies depend on the wave field, and hence on the source that excites the waves.

Equation (6.16) poses the determination of the velocity change as a standard linear inverse problem with sensitivity kernel $K(\mathbf{r}')$ (Kanu and Snieder, 2015b). By combining measurements of the travel time change for different sources, receivers, and travel times t one can estimate $\delta s/s$ as a function of location. The sensitivity kernel in equation (6.17) is the same as derived earlier (Pacheco and Snieder, 2005). In order to compute this kernel, one needs to (1) compute the energy density of the waves that are excited by the source at $\mathbf{s}$, (2) compute the Green's function $G^D(\mathbf{r}', \mathbf{r}, t)$ that accounts for the diffusive energy generated by a unit source at the receiver location $\mathbf{r}$ that propagates to $\mathbf{r}'$, and (3) convolve these energies (the time-integral in equation (6.17)).

Note that the kernel in equation (6.18) has similar properties as the gradient computed by adjoint methods in full waveform inversion for earthquake- and active-source data (Tarantola, 1984a,b; Tromp et al., 2005; Fichtner et al., 2006), and for noise-correlation waveforms (Fichtner and Tsai, 2018, Chapter 4). In these adjoint methods one can compute updates to an Earth model by convolving the field propagated forward in time from the source with the waveform residual propagating backward in time from the receivers. In equation (6.18) one does the same, except that one backpropagates the energy from the receiver instead of the waveform residual.

There are different ways to compute the sensitivity kernels. As shown in equation (6.18) one needs to know the energies, hence the computation of the kernels reduces to the computation of the energies. The first way to do this is to model the energies by solving the diffusion equation for the energy density. This approach presumes that one knows the diffusion constant for a given model of the random medium that generates the wave scattering. One also needs to ensure that at interfaces the diffusion equation satisfies boundary conditions that agree with the boundary conditions of the underlying wave propagation problem. In the presence of a free surface – the Earth has a free surface – one also needs to account for the energy carried by surface wave modes compared to the energy of body waves. These complexities imply that using the diffusion equation is mostly practical for random media whose statistical properties are constant in space. Second, one can use a Monte-Carlo simulation that simulates a random walk that corresponds to a diffusive process. A third, and simple, alternative is to numerically model the wavefield instead of the energy density, and to compute the energy from these wavefield simulations. Since for a given realization of a random medium the wavefield has statistical fluctuations, one may have to average the energy density computed for several realizations of the random medium. This approach was taken by Kanu and Snieder (2015a), who show examples for scattering media whose statistical properties are not constant in space. We show in section 6.4 examples of sensitivity kernels that are computed in this way.

6.3 The Travel Time Change from Radiative Transfer of Acoustic Waves

Radiative transfer accounts for the distribution of energy in a scattering medium as a function of space, time, and the direction $\hat{\mathbf{n}}$ of wave propagation (Chandrasekhar, 1960; Özisik, 1973). Diffusive wave transport follows from the radiative transfer equations for late times when the energy propagation is almost independent of direction (van Rossum and Nieuwenhuizen, 1999). The equation of radiative transfer, however, holds for early times as well and describes the transition from ballistic wave propagation to weak scattering to strong multiple scattering (Paasschens, 1997). The price one pays for this refinement of the description of energy transfer is that the equations of radiative transfer depend not only on space and time but also on the direction of wave propagation. As a consequence, the radiative transfer solution depends on six variables (time, three space variables, and two angles). For elastic waves, there are different wave modes that need to be taken into account, which makes the treatment more involved (Ryzhik et al., 1996).

The equation of radiative transfer for scalar waves is given by (Chandrasekhar, 1960; Özisik, 1973)

$$\frac{\partial I(\mathbf{r}, \hat{\mathbf{n}}, t)}{\partial t} + v(\mathbf{r}) \hat{\mathbf{n}} \cdot \nabla I(\mathbf{r}, \hat{\mathbf{n}}, t) - \oint S(\mathbf{r}, \hat{\mathbf{n}}, \hat{\mathbf{n}}_0) I(\mathbf{r}, \hat{\mathbf{n}}_0, t) d^2 n_0 + q(\mathbf{r}) I(\mathbf{r}, \hat{\mathbf{n}}, t) = 0.$$

$$(6.19)$$

In this expression $I(\mathbf{r}, \hat{\mathbf{n}}, t)$ is the intensity of waves at location $\mathbf{r}$ at time t that propagate in the direction $\hat{\mathbf{n}}$; this quantity is called the *specific intensity*. The advection of energy with wave velocity $v(\mathbf{r})$ is described by the second term. The integral $\oint S(\mathbf{r}, \hat{\mathbf{n}}, \hat{\mathbf{n}}_0) I(\mathbf{r}, \hat{\mathbf{n}}_0, t) d^2 n_0$ accounts for the energy gain from energy propagating in other directions $\hat{\mathbf{n}}_0$, while the last term $q(\mathbf{r}) I(\mathbf{r}, \hat{\mathbf{n}}, t)$ accounts for energy lost to wave propagation in other directions and for inelastic damping.

The purpose of the derivation is to determine the change in the arrival time of the coda waves caused by a slowness perturbation in a volume dV'. This change in the arrival time is according to equation (6.6) given by the integrated slowness change along the wave path. In the diffusive regime, the waves propagate with near-equal intensity in all directions (van Rossum and Nieuwenhuizen, 1999); as a result we don't need to keep track of the direction of propagation. But in radiative transfer, we do keep track of the direction of wave propagation. Since we only consider the impact of the slowness perturbation in dV' on the arrival time, we don't consider the scattering caused by the slowness perturbation, and as a consequence of equation (6.6) we assume that the direction of wave propagation does not change in dV'. The waves propagate, in general, in all possible directions $\hat{\mathbf{n}}'$ through dV', but this direction is the same for the incoming and outgoing waves. We thus replace equation (6.11) by

$$\sum_{T \text{ through } dV'} I_T(\mathbf{r}, \mathbf{s}, t) = \left(\oint I(\mathbf{r}, \mathbf{r}', \hat{\mathbf{n}}', t - t') I(\mathbf{r}', \mathbf{s}, \hat{\mathbf{n}}', t') d^2 n' \right) dV', \quad (6.20)$$

where $I(\mathbf{r}', \mathbf{s}, \hat{\mathbf{n}}', t')$ is the specific intensity of the wave that travels from the source $\mathbf{s}$ to $\mathbf{r}'$ and that propagates at $\mathbf{r}'$ in the $\hat{\mathbf{n}}'$ direction, while $I(\mathbf{r}, \mathbf{r}', \hat{\mathbf{n}}', t - t')$ is the wave that leaves $\mathbf{r}'$ in the $\hat{\mathbf{n}}'$ direction, and then propagates to the receiver $\mathbf{r}$. The integration over $d^2 n'$ accounts for the waves that propagate through dV' in all possible directions of propagation.

A more precise derivation, based on the Chapman-Kolmogorov theorem for radiative transfer that is presented in appendix 6.B, shows that

$$\sum_{T \text{ through } dV'} I_T(\mathbf{r}, \mathbf{s}, t) = \left(\oint \oint G^{RT}(\mathbf{r}, \hat{\mathbf{n}}, \mathbf{r}', \hat{\mathbf{n}}', t - t') I(\mathbf{r}', \mathbf{s}, \hat{\mathbf{n}}', t') d^2 n' d^2 n \right) dV',$$

$$(6.21)$$

where the radiative transfer Green's function is defined by

$$\frac{\partial G^{RT}(\mathbf{r}, \hat{\mathbf{n}}, \mathbf{r}', \hat{\mathbf{n}}', t)}{\partial t} + v(\mathbf{r}) \hat{\mathbf{n}} \cdot \nabla G^{RT}(\mathbf{r}, \hat{\mathbf{n}}, \mathbf{r}', \hat{\mathbf{n}}', t)$$

$$- \oint S(\mathbf{r}, \hat{\mathbf{n}}, \hat{\mathbf{n}}_0) G^{RT}(\mathbf{r}, \hat{\mathbf{n}}_0, \mathbf{r}', \hat{\mathbf{n}}', t) d^2 n_0 + q(\mathbf{r}) G^{RT}(\mathbf{r}, \hat{\mathbf{n}}, \mathbf{r}', \hat{\mathbf{n}}', t)$$

$$= \delta(\mathbf{r} - \mathbf{r}') \delta(\hat{\mathbf{n}} - \hat{\mathbf{n}}') \delta(t). \quad (6.22)$$

This Green's function gives the radiative transfer solution for a source that injects a unit pulse of energy propagating in the $\hat{\mathbf{n}}'$ direction at position $\mathbf{r}'$. Using reciprocity for radiative transfer (Case, 1957)

$$G^{RT}(\mathbf{r}, \hat{\mathbf{n}}, \mathbf{r}', \hat{\mathbf{n}}', t) = G^{RT}(\mathbf{r}', -\hat{\mathbf{n}}', \mathbf{r}, -\hat{\mathbf{n}}, t). \quad (6.23)$$

Physically, the minus signs in the right-hand side are caused by the fact that when we interchange $\mathbf{r}$ and $\mathbf{r}'$, we must interchange the direction of energy propagation as well. Using this in equation (6.21) gives

$$\sum_{T \text{ through } dV'} I_T = \left(\oint \oint G^{RT}(\mathbf{r}', -\hat{\mathbf{n}}', \mathbf{r}, -\hat{\mathbf{n}}, t - t') I(\mathbf{r}', \mathbf{s}, \hat{\mathbf{n}}', t') d^2 n' d^2 n \right) dV'.$$

$$(6.24)$$

By analogy with equation (6.15) we obtain for the travel time perturbation

$$\langle \tau \rangle = \frac{\int \int_0^t \oint \oint G^{RT}(\mathbf{r}', -\hat{\mathbf{n}}', \mathbf{r}', -\hat{\mathbf{n}}, t - t') I(\mathbf{r}', \hat{\mathbf{n}}', \mathbf{s}, t') d^2 n' d^2 n dt' \frac{\delta s}{s}(\mathbf{r}') dV'}{\oint I(\mathbf{r}, \hat{\mathbf{n}}, \mathbf{s}, t) d^2 n}.$$

$$(6.25)$$

Writing this equation in the form of equation (6.1), the sensitivity kernel derived from radiative transfer is given by

$$K(\mathbf{r}') = \frac{\int_0^t \oint \oint G^{RT}(\mathbf{r}', -\hat{\mathbf{n}}', \mathbf{r}, -\hat{\mathbf{n}}, t - t')I(\mathbf{r}', \hat{\mathbf{n}}', \mathbf{s}, t')d^2n'd^2ndt'}{\oint I(\mathbf{r}, \hat{\mathbf{n}}, \mathbf{s}, t)d^2n}. \tag{6.26}$$

A comparison of the radiative scattering kernel above and the kernel (6.18) for diffuse waves is that in radiative transfer theory one uses waves traversing dV' that propagate in the same direction $\hat{\mathbf{n}}'$, while in the kernel (6.18) there is no accounting for the direction of waves that enter dV' and those that leave dV'. Physically this corresponds to the fact that diffuse waves propagate with nearly equal intensity in all directions, while in radiative transport the energy propagation may depend on direction. The radiative transfer kernel (6.26) stipulates that the incoming and outgoing waves in dV' propagate in the same direction. This corresponds to the fact that we consider the imprint of velocity variations in dV' on the arrival time scattered waves, but not the scattering of waves by inhomogeneities in dV'.

One only needs the specific intensity to compute the sensitivity kernel (6.26) for radiative transfer. There are several ways to compute the specific intensity (Wegler et al., 2006). One way to achieve this is to solve the equation of radiative transfer directly. Since the radiative transfer equation depends on six variables, and it is an integro-differential equation, this can be an involved and numerically demanding process. (Not to mention the intellectual demands.) An alternative is to use Monte-Carlo simulations where one shoots rays into the random medium that are scattered in statistically the same way as the wave scattering (Gusev and Abubakirov, 1996; Yoshimoto, 2000; Margerin et al., 2000; Sens-Schönfelder et al., 2009). A third alternative is to first compute the wavefield numerically, and derive the specific intensity from this wavefield. This involves locally decomposing the wavefield into the different directions of propagation, followed by squaring to convert the wave-field into specific intensity. The directional decomposition can be carried out by a local Fourier transform or by local beamforming.

Since radiative transfer keeps track of the direction of wave propagation, one can, in principle, use this theory also to retrieve anisotropic slowness perturbations that depend on the direction of wave propagation $\hat{\mathbf{n}}'$ by rewriting equations (6.25) and (6.26) as

$$\langle \tau \rangle = \int \oint K(\mathbf{r}', \hat{\mathbf{n}}') \frac{\delta s}{s}(\mathbf{r}', \hat{\mathbf{n}}') d^2n'dV', \tag{6.27}$$

and

$$K(\mathbf{r}', \hat{\mathbf{n}}') = \frac{\int_0^t \oint G^{RT}(\mathbf{r}', -\hat{\mathbf{n}}', \mathbf{r}, -\hat{\mathbf{n}}, t - t')I(\mathbf{r}', \hat{\mathbf{n}}', \mathbf{s}, t')d^2ndt'}{\oint I(\mathbf{r}, \hat{\mathbf{n}}, \mathbf{s}, t)d^2n}, \tag{6.28}$$

where $K(\mathbf{r}', \hat{\mathbf{n}}')$ measures the sensitivity to the slowness at location $\mathbf{r}'$ of waves that propagate in the $\hat{\mathbf{n}}'$-direction.

6.4 An Example of Sensitivity Kernels and of the Breakdown of Diffusion

To illustrate the kernels, and their limitations, we show in Figure 6.4 the diffuse wave kernels (6.18) computed by Kanu and Snieder (2015a) for acoustic waves. The used model consists of a random medium that is overlain by a low-velocity layer in the near surface and a free surface with a rough topography (Kanu and Snieder, 2015a). The source and receiver locations are marked with S and R, respectively. The four panels are for four different lag times, which are shown in the upper right-hand corner of each panel. These different lag times correspond to four different regimes of wave propagation. The kernels are computed by modeling the wavefield by finite difference simulations, and computing the energy density from the obtained waveforms. In order to reduce fluctuations, the kernels were averaged over a few realizations of the random medium (Kanu and Snieder, 2015a).

The upper left panel is for a time of $t = 1.4$ s, which corresponds to the travel time of the ballistic, or direct, wave that propagates from source to receiver. This wave can only be influenced by velocity perturbations on the path of the ballistic wave, and indeed the upper left panel shows a kernel that is only nonzero in the first Fresnel zone for the ballistic wave. (A description of the Fresnel zone is given by Spetzler and Snieder (2004).) The wave in this time window is in the *ballistic regime*.

The panel in the top right is for a slightly later time $t = 1.8$ s. This time is large enough that the waves have had the time to be scattered, but the propagation time is so short that multiple scattering is not yet important. The sensitivity kernel is nonzero mostly on an ellipse with the source and receiver as focal points, although a weak reflected wave generated at the bottom of the near-surface layer is visible just below the receiver. Since this kernel corresponds to single scattered waves, the waves are in the *single-scattering regime*. At a later time $t = 2.5$ s, shown in the bottom left panel, the waves have been scattered more often. As a result the single-scattering ellipse is filled in. The nonzero value of this scattering kernel within the single-scattering ellipse is caused by multiple scattering. These waves are in the *multiple-scattering regime*. In addition to the random infill of the single-scattering ellipse by multiple scattering, a secondary ellipse is present within the single-scattering ellipse. This is due to scattered waves that reflect off the bottom of the low-velocity layer near the surface before propagating to the receiver.

The bottom right panel, for travel time $t = 5.0$ s, may be least familiar. For this late time there still is a remnant of the single-scattering ellipse near the bottom, and the speckle in the interior corresponds to multiple scattering. But the most conspicuous feature is the strong value of the kernels in a horizontal band running through the receiver R. The location of this horizontal band corresponds to the low-velocity layer that is bounded by the corrugated free surface. The strong value of the kernel in this low-velocity layer implies that most of the wave energy, and hence most of the sensitivity to velocity changes, is in the low-velocity layer just below the free surface. For this late time, the waves are in the *surface-saturated regime*. Perhaps surprisingly, the near-surface layer attracts a large fraction of the wave energy, even though the source is located many wavelengths below the base of the low-velocity layer. This signifies the power of the near-surface layer to trap energy. Simulations as shown in Figure 6.4 give insight into the sensitivity of the coda wave for velocity perturbations in different regions of the model. Of particular practical importance is the depth-dependence of this sensitivity.

The single-scattering ellipse in the upper right panel of Figure 6.4 may seem perfectly natural, but something is wrong with this figure. The problem is that the kernels are for the sensitivity of the arrival time of coda waves to a velocity perturbation; these are not the kernels that account for the generation of scattered waves. The sensitivity of the single-scattered waves for velocity perturbations should be distributed over the interior of the single-scattering ellipse, because this interior covers the region of space traversed by the single-scattered waves, and hence velocity changes in this inner region are the cause of the changes in the arrival time of single-scattered waves.

So what went wrong? The sensitivity kernels in Figure 6.4 are computed with the kernel (6.18) for diffuse waves, even though the propagation regime of the waves in the upper right panel of Figure 6.4 is far from diffusive. As sketched in Figure 6.2, we allow in the diffusive regime the incoming and outgoing waves at dV' to travel in different directions. For diffuse waves, this difference in direction of wave propagation does not matter because these waves travel with near-equal

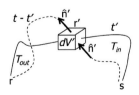

Figure 6.3. Definition of geometric variables for the contribution of a velocity change in volume dV' at $\mathbf{r}'$, a source at $\mathbf{s}$, and a receiver at $\mathbf{r}$ in a treatment based on radiative transfer. In a radiative transfer approach, incoming trajectories T_{in} and outgoing trajectories T_{out} must be in the same direction $\hat{\mathbf{n}}'$, i.e., the solid trajectories are combined with each other and the dashed trajectories are combined with each other.

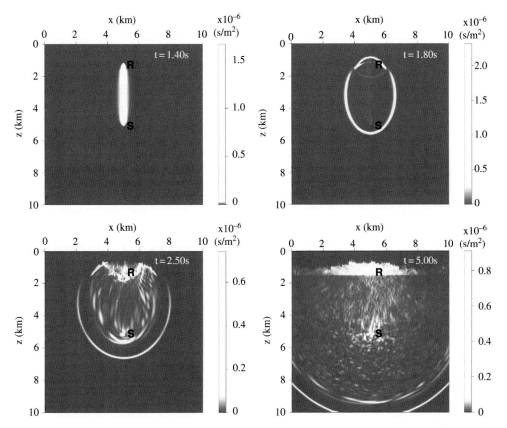

Figure 6.4. Sensitivity kernels for diffuse waves for a random medium with a near-surface layer as computed by Kanu and Snieder (2015a) for different lag times shown in the upper right corner of each panel. The source and receiver positions are marked by S and R, respectively. The model has a near-surface layer and rough topography of the free surface (Kanu and Snieder, 2015a). The four panels correspond to four different wave propagation regimes: ballistic wave propagation (top left), single scattering (top right), multiple scattering (bottom left), and the surface-saturated regime (bottom right). (A black-and-white version of this figure appears in some formats. For the color version, please refer to the plate section.)

intensity in all directions anyhow (van Rossum and Nieuwenhuizen, 1999). But the single-scattered waves are highly directional. For a volume dV' in the interior of the single-scattering ellipse, the situation sketched in Figure 6.3 is much more realistic because for a volume element dV' within the single-scattering ellipse, the waves continue on a straight line to the scattering point on the scattering ellipse. For this reason, the direction of propagation $\hat{n}$ must be preserved in the wave propagation within the single-scattering ellipse. Since this is not the case with the used diffuse wave theory for the computation of the sensitivity kernels in Figure 6.4, the sensitivity is erroneously confined to the single-scattering ellipse instead of the interior of this ellipse.

There are two ways to obtain the correct kernels for single-scattered waves. The first is to use the kernels that are designed for single-scattering (Pacheco and Snieder, 2006). The other alternative is to use the radiative transfer kernel (6.26) because these kernels stipulate that the direction of wave propagation does not change in dV', as shown in Figure 6.3. This restriction precludes the contribution from velocity changes on the single-scattering ellipse.

6.5 The Travel Time Change for Diffuse Elastic Waves

The somewhat lengthy derivation of section 6.2 makes it relatively easy to determine the sensitivity kernels for the travel time change of the coda waves caused by changes in the P-wave slowness s_P and the S-wave slowness s_S. The elastic displacement $\mathbf{u}$ can be separated into a curl-free part $\mathbf{u}_P$ and a divergence-free part $\mathbf{u}_S$ that corresponds to the P- and S-waves, respectively (Aki and Richards, 2002): $\mathbf{u} = \mathbf{u}_P + \mathbf{u}_S$. One might think that the energy can also be decomposed into a P-wave energy I_P and an S-wave energy I_S. Perhaps surprisingly, this is, strictly speaking, not the case. This can be seen for example by considering the contribution of the kinetic energy I^{KIN} to the total energy, which is given by

$$I^{KIN} = \frac{1}{2}\rho|\dot{\mathbf{u}}_P + \dot{\mathbf{u}}_S|^2 = \frac{1}{2}\rho\dot{u}_P^2 + \frac{1}{2}\rho\dot{u}_S^2 + \rho(\dot{\mathbf{u}}_P \cdot \dot{\mathbf{u}}_S), \qquad (6.29)$$

where ρ is the mass density and the overdot denotes a time-derivative. The first term in the right-hand side is the kinetic energy of the P-waves, and the second term is the kinetic energy of the S-waves. There is, however, an additional cross-term $\rho(\dot{\mathbf{u}}_P \cdot \dot{\mathbf{u}}_S)$ that corresponds neither to the P-wave energy nor to the S-wave energy. The P- and S-wave kinetic energies $\frac{1}{2}\rho\dot{u}_P^2$ and $\frac{1}{2}\rho\dot{u}_S^2$ are always positive, but the cross-term $\rho(\dot{\mathbf{u}}_P \cdot \dot{\mathbf{u}}_S)$ can be either positive or negative. This means that the cross-term can be eliminated by local averaging over space and time. Note that the presence of the cross-term is not a peculiarity of elastic waves. Consider the superposition of two acoustic waves, $u = u_1 + u_2$. The corresponding energy is quadratic in the wavefield and satisfies $I = u_1^2 + u_2^2 + 2u_1u_2$, hence the presence of cross-terms is a general consequence of superposition.

After local averaging over space and time, the waves propagating through dV' can be decomposed into P- and S-waves, with their energies I_P and I_S, respectively. The P-wave energy is for a displacement field $\mathbf{u}$ in an isotropic medium with Lamé parameters λ and μ (given by Morse and Feshbach, 1953b; Shapiro et al., 2000)

$$I_P = (\lambda + 2\mu)(\nabla \cdot \mathbf{u})^2, \qquad (6.30)$$

and the S-wave energy satisfies

$$I_S = \mu |\nabla \times \mathbf{u}|^2. \tag{6.31}$$

This is twice the potential energy, but because the kinetic and potential energy average over time is equal (Dassios, 1979), the total energy is twice the potential energy.[1]

After spatial averaging, the energy at $\mathbf{r}'$ can be decomposed in contributions from P- and S-waves:

$$I(\mathbf{r}', \mathbf{s}, t') = I_P(\mathbf{r}', \mathbf{s}, t') + I_S(\mathbf{r}', \mathbf{s}, t'), \tag{6.32}$$

where the energy for P and S-waves needs to be computed for the source at $\mathbf{s}$, for example by prescribing the double-couple of a moment-tensor source. A similar decomposition holds for $I(\mathbf{r}, \mathbf{r}', t - t')$. With this decomposition of the energies, equation (6.11) for diffuse acoustic waves for elastic waves can be generalized to

$$\sum_{T \text{ through } dV'} I_T(\mathbf{r}, \mathbf{s}, t) = I_P(\mathbf{r}, \mathbf{r}', t - t') I_P(\mathbf{r}', \mathbf{s}, t') dV' + I_S(\mathbf{r}, \mathbf{r}', t - t') I_S(\mathbf{r}', \mathbf{s}, t') dV'. \tag{6.33}$$

Note that in doing so we have ignored cross-terms such as $I_P(\mathbf{r}, \mathbf{r}', t - t') I_S(\mathbf{r}', \mathbf{s}, t')$ between P- and S-wave energies. Such cross-terms account for the mode-conversions in dV', but since we aim to retrieve the kernels for the travel time change, in contrast to the sensitivity kernels for the waveforms, such mode conversions are not relevant. This does not mean, however, there are no conversions between P- and S-waves; conversions between these different wave types can occur along the paths from the source to $\mathbf{r}'$, and between $\mathbf{r}'$ and the receiver.

Reciprocity holds for elastic waves (Aki and Richards, 2002), and it therefore also holds for the energy associated with these waves, hence

$$I_P(\mathbf{r}, \mathbf{r}', t - t') = I_P(\mathbf{r}', \mathbf{r}, t - t') \tag{6.34}$$

This energy needs to be interpreted carefully. $I_P(\mathbf{r}, \mathbf{r}', t - t')$ is the energy associated with the Green's function that propagates P-wave energy at $\mathbf{r}'$ to the recorded component at the receiver at $\mathbf{r}$. Let $G^{DP}(\mathbf{r}', \mathbf{r}, t - t')$ be the P-wave energy at location $\mathbf{r}'$ that is generated by a unit impulsive point force at the receiver location $\mathbf{r}$ in the direction of the receiver that records the wavefield. This quantity corresponds to the P-wave energy associated with the elastic wave Green's function $G_{ic}(\mathbf{r}', \mathbf{r}, t - t')$, where i denotes the orientation of this wavefield at location $\mathbf{r}'$

---

[1] In three dimensions there are two S-wave polarizations: $\mathbf{u}_S = \mathbf{u}_{S1} + \mathbf{u}_{S2}$, and the S-wave energy is given by $I_S = \mu |\nabla \times \mathbf{u}_{S1}|^2 + \mu |\nabla \times \mathbf{u}_{S2}|^2 + 2\mu (\nabla \times \mathbf{u}_{S1}) \cdot (\nabla \times \mathbf{u}_{S2})$. The cross-term on the right vanishes for two S-waves with orthogonal polarization that propagate in the same direction, but is, in general, nonzero, which can be verified for the special case $\mathbf{u}_{S1} = \hat{\mathbf{x}} f(t - z/v_S)$ and $\mathbf{u}_{S2} = \hat{\mathbf{z}} g(t - x/v_S)$. The cross-term is, however, oscillatory in space, so it vanishes after spatial averaging. This means that when one uses equation (6.31) for the S-wave energy and applies some spatial averaging, the cross-terms between the S-wave polarizations do not contribute.

and c the orientation of the receiver at $\mathbf{r}$. A similar definition holds for the S-wave energy density Green's function G^{DS}. The Green's function G_{ic} is the elastic *wave* Green's tensor (Aki and Richards, 2002), which should not be confused with Green's function $G^{D,P \, or \, S}$ for the diffusive *energy* for P- or S-waves. In the case of a pressure receiver, G^{DP} is the P-wave energy at $\mathbf{r}'$ generated by a unit explosive source at $\mathbf{r}$, and G^{DS} is the corresponding S-wave energy. Even though an explosive source does not generate S-waves, the S-wave energy at $\mathbf{r}'$ is, in general, nonzero due to mode conversions caused by the heterogeneity. We show numerical examples in section 6.6.

The rest of the derivation proceeds in the same way as in section 6.2, with the difference that we multiplied equation (6.11) with the relative slowness change $\delta s/s$. For elastic waves each of the terms of the right-hand side of equation (6.33) must be multiplied with the relative slowness change for each wave-type. Doing so generalizes equations (6.16) and (6.17) for elastic waves to

$$\langle \tau \rangle = \int K_P(\mathbf{r}') \frac{\delta s_P}{s_P}(\mathbf{r}') \, dV' + \int K_S(\mathbf{r}') \frac{\delta s_S}{s_S}(\mathbf{r}') \, dV', \tag{6.35}$$

with

$$K_P(\mathbf{r}') = \frac{\int_0^t G^{DP}(\mathbf{r}', \mathbf{r}, t - t') I_P(\mathbf{r}', \mathbf{s}, t') \, dt'}{I(\mathbf{r}, \mathbf{s}, t)}, \tag{6.36}$$

and

$$K_S(\mathbf{r}') = \frac{\int_0^t G^{DS}(\mathbf{r}', \mathbf{r}, t - t') I_S(\mathbf{r}', \mathbf{s}, t') \, dt'}{I(\mathbf{r}, \mathbf{s}, t)}. \tag{6.37}$$

Note that in equations (6.36) and (6.37) the energies I_P and I_S in the numerator are the contributions from the P- and S-waves, respectively, to the energies at $\mathbf{r}'$. By contrast, the energy in the denominator is the total energy recorded of the c component of the motion at the receiver location $\mathbf{r}$, just as it is in equation (6.17) for scalar waves. Since for strongly scattered elastic waves the S-wave energy dominates the P-wave energy (Weaver, 1982; Snieder, 2002), the sensitivity kernel K_S for perturbations in the *S*-wave slowness is much larger than the sensitivity kernel K_P for perturbations in the P-wave slowness.

6.6 Numerical Examples

In this section we present numerical examples for the sensitivity kernels K_P and K_S for P- and S-waves. Following equations (6.30) and (6.31) the kernels are computed by taking the divergence and the curl of the wavefield computed with the spectral-element method (Specfem2D) (Tromp et al., 2008). The used two-dimensional velocity medium is a superposition of a constant background with

P-velocity 6500 m/s, and random fluctuations that obey a von Karman distribution (Sato and Fehler, 2012) with exponent $\kappa = 0.5$. At every location the P- and S-waves are scaled such that $v_P(\mathbf{r})/v_S(\mathbf{r}) = \sqrt{3}$ (Poisson medium). The variance of the velocity perturbations is 20%, and the correlation lengths are given by $a_x = a_z = 325$ m, which is of the order of the wavelength for P-waves at the used dominant frequency of 20 Hz. A free surface is present at the top ($z = 0$), and absorbing boundaries are applied to the sides and bottom of the computational domain. The simulations are described by Obermann et al. (2013), who give more details on the medium characterization.

The kernels for the P- and S-waves generated by an explosive source at location $\mathbf{s}$ and a hydrophone (pressure sensor) at $\mathbf{r}$ are shown in Figure 6.5. These kernels are averaged over four realizations of the random medium; this suppresses fluctuations caused by the specific realization used. The left panels are for a time that corresponds to the travel time of the ballistic P-wave. For this early time the P-wave sensitivity is confined to the path of the ballistic wave from $\mathbf{s}$ to $\mathbf{r}$. Just as in Figure 6.4, the P-wave kernel is for a slightly later time ($t = 1.75$ s) most prominent on the single-scattering ellipse. As argued in section 6.4, the kernel should fill the interior of the ellipse. To a certain extent this happens, in particular near the source and receiver where the sensitivity is large (Pacheco and Snieder, 2005; Mayor et al., 2014), but the largest sensitivity is confined to the single-scattering ellipse. This discrepancy is caused by treating the single-scattered waves as being diffuse, which is not realistic. For a later time ($t = 3$ s) P-waves reflected off the free surface dominate the P-wave kernel in the top right panel.

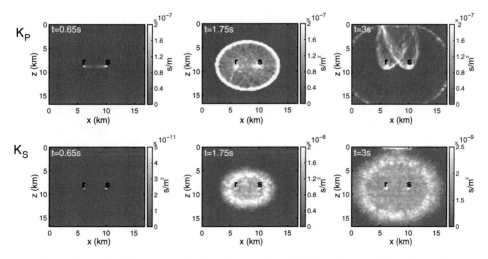

Figure 6.5. Sensitivity kernels K_P (top row) and K_S (bottom row) for three different lag times for an explosive source. The time for the left panels ($t = 0.65$ s) corresponds to the arrival time of the direct P-wave. Note that the gray scales used for the different panels are different.

The S-wave sensitivities, as shown in the bottom panels of Figure 6.5, are caused by $P \rightarrow S \rightarrow P$ conversions because the explosive source does not generate S-waves, and the used receiver, a hydrophone, does not detect S-waves. For the time of the direct P-wave, the bottom left panel of Figure 6.5, the S-wave sensitivity is confined to small regions near the source and receiver. Because the travel time is equal to the travel time for the direct P-wave, there is no possibility of the wave to propagate as an S-wave, which means that the $P \rightarrow S \rightarrow P$ conversions must take place at almost the same location. Since this is extremely unlikely, the S-wave sensitivity is about 10,000 times smaller than the P-wave sensitivity for this early time.

For later times (middle and right panels in bottom row) the S-wave kernels are nonzero, despite the fact that the explosive source does not generate shear waves. This sensitivity to the shear wave velocity is caused by $P \rightarrow S \rightarrow P$ converted waves by the fluctuations in the random medium, but for these later times the mode conversions can take place at different locations. Since these mode conversions occur throughout the random medium, the sensitivity kernels for the S-waves are distributed to the interior of an egg-like region in space centered on the source and receiver. Note that for $t = 3$ s, the bottom right panel of Figure 6.5, there is a slight sensitivity to the shear velocity near the free surface in the middle of the shown area. This sensitivity is caused by conversions from P-waves to surface waves to P-waves by near-surface heterogeneity.

In Figure 6.6 we compare the wavefields and time-dependence of the sensitivity kernels for waves excited by an explosive source (left panels) and for point forces (middle and right panels) with a direction shown in the top panels. For both wavefields the direct P-wave is clearly visible; this P-wave is isotropic for the explosive source (top left), but it is modulated by the radiation pattern of the point force in the top middle and right panels. Note that for the point force (middle and right top panels) there is a pronounced ballistic S-wave, which is absent for the explosive source (top left panel) because an explosion does not generate shear waves.

The panels in the middle row of Figure 6.6 show the P- and S-wave kernels for the location $\mathbf{r}'$ marked in the top panel for times up to 40 s. Note that the time-dependence for the kernels is similar, irrespective of the source type. For the 45° elastic force (middle panels), the point force radiates no S-waves toward $\mathbf{r}'$, yet the S-wave kernel is nonzero, as it is for the explosive source in the left panels. The scattering and mode conversion is so strong that the P- and S-wave fields equilibrate. This equilibration can be seen more clearly in the bottom panels of Figure 6.6, which show the ratio of the S-wave kernel to the P-wave kernel at location $\mathbf{r}'$ for the three source types. For $t > 4$ s, this ratio approaches an equilibrium value $K_S/K_P \approx 9$.

This ratio can be explained as follows. For a strongly scattering medium in two dimensions, the equilibrium ratio of the S-wave energy to the P-wave energy is

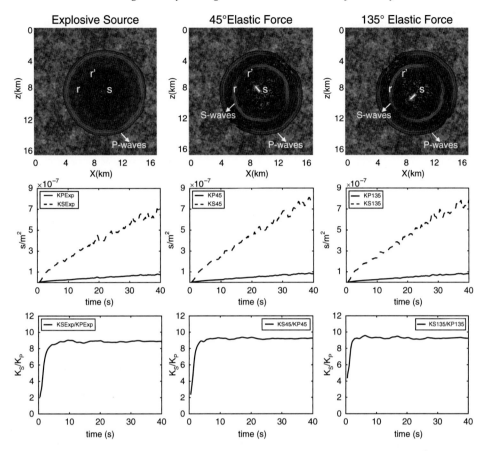

Figure 6.6. Snapshots of the wavefields (top row), sensitivity kernels $K_P(\mathbf{r}')$ and $K_S(\mathbf{r}')$ at location $\mathbf{r}'$ (middle panels), and the ratio of these kernels at $\mathbf{r}'$ (bottom panels). The examples are for a source at $\mathbf{s}$ and a seismometer at $\mathbf{r}$ as shown in the top panels. The left column is for an explosive source, while the middle and right columns are for a point force in the direction of the white arrows shown in the top panels. (A black-and-white version of this figure appears in some formats. For the color version, please refer to the plate section.)

given by $I_S/I_P = (v_P/v_S)^2$. According to equations (6.36) and (6.37), the P-wave kernel depends on the product of P-wave energies (G^{DP} and I_P), while the S-wave kernel depends on the product of S-wave energies. This means that in equilibrium $K_S/K_P \propto (I_S/I_P)^2 = (v_P/v_S)^4$. For a Poisson medium ($v_P/v_S = \sqrt{3}$), this implies that $K_S/K_P = 9$. The ratio of the kernels in the bottom panels of Figure 6.6 is slightly higher than this value. We attribute this difference to the presence of the free surface, which through the presence of surface waves slightly modifies the ratio of the P- and S-energies (Weaver, 1985; Hennino et al., 2001; Margerin et al.,

2009; Obermann et al., 2013, 2016). The large value of this ratio implies that the coda waves are mostly influenced by the S-velocity. This is even more pronounced in three dimensions where in equilibrium $I_S/I_P = 2(v_P/v_S)^3$ (Weaver, 1982), hence $K_S/K_P = 4(v_P/v_S)^6 = 98$ for a Poisson medium. The numerical examples show that in the strong scattering regime the changes in the coda depend mostly on the changes in the S-velocity, and that the sensitivity kernels do not depend strongly on the details of the seismic source.

Acknowledgments

We thank Christoph Sens-Schönfelder and two anonymous reviewers for their insightful and constructive comments.

References

Aki, K., and Chouet, L. 1975. Origin of coda waves: source, attenuation, and scattering effects. *J. Geophys. Res.*, **80**, 3322–3342.

Aki, K., and Richards, P. 2002. *Quantitative Seismology*. 2nd edn. Sausalito: Univ. Science Books.

Aldridge, D. 1994. Linearization of the Eikonal Equation. *Geophysics*, **59**, 1631–1632.

Aster, R., Borchers, B., and Thurber, C. 2004. *Parameter Estimation and Inverse Problems*. San Diego: Academic Press.

Brenguier, F., Campillo, M., Hadziioannou, C., Shapiro, N., and Larose, E. 2008. Postseismic relaxation along the San Andreas Fault at Parkfield from continuous seismological observations. *Science*, **321**, 1478–1481.

Brenguier, F., Clarke, D., Aoki, Y., Shapiro, N., Campillo, M., and Ferrazzini, V. 2011. Monitoring volcanoes using seismic noise correlations. *Comptes Rendus Geoscience*, **343**, 633–638.

Campillo, M., and Paul, A. 2003. Long-range correlations in the diffuse seismic coda. *Science*, **299**, 547–549.

Case, K. 1957. Transfer problems and the reciprocity principle. *Rev. Mod. Phys.*, **29**, 651–663.

Chandrasekhar, S. 1960. *Radiative Transfer*. New York: Dover.

Chouet, B. 1979. Temporal variation in the attenuation of earthquake coda near Stone Canyon, California. *Geophys. Res. Lett.*, **6**, 143–146.

Cowan, M., Page, J., and Weitz, D. 2000. Velocity fluctuations in fluidized suspensions probed by ultrasonic correlation spectroscopy. *Phys. Rev. Lett.*, **85**, 453–456.

Curtis, A., Gerstoft, P., Sato, H., Snieder, R., and Wapenaar, K. 2006. Seismic interferometry – turning noise into signal. *The Leading Edge*, **25**, 1082–1092.

Dassios, G. 1979. Equipartition of energy in elastic wave propagation. *Mech. Res. Comm.*, **6**, 45–50.

Fehler, M., Roberts, P., and Fairbanks, T. 1998. A temporal change in coda wave attenuation observed during an eruption of Mount St. Helens. *J. Geophys. Res.*, **93**, 4367–4373.

Feynman, R. P. and Hibbs, A. 1965. *Quantum Mechanics and Path Integrals*. New York: McGraw-Hill.

Fichtner, A., and Tsai, V. 2018. Theoretical foundations of noise interferometry. Pages – of: Nakata, N., Gualtieri, L., and Fichtner, A. (eds.), *Seismic Ambient Noise*. Cambridge University Press, Cambridge, UK.

Fichtner, A., Bunge, H.-P., and Igel, H. 2006. The adjoint method in seismology: I. Theory. *Phys. Earth. Planetary. Int.*, **157**, 86–104.

Gassenmeier, M., Sens-Schönfelder, C., Delatre, M., and Korn, M. 2015. Monitoring of environmental influences on seismic velocity at the geologic storage site for CO2 in Ketzin (Germany) with ambient seismic noise. *Geophys. J. Int.*, **200**, 524–533.

Gassenmeier, M., Sens-Schönfelder, C., Eulenfeld, T., Bartsch, M., Victor, P., Tilmann, F., and Korn, M. 2016. Field observations of seismic velocity changes caused by shaking-induced damage and healing due to mesoscopic nonlinearity. *Geophys. J. Int.*, **204**(3), 1490–1502.

Grêt, A., Snieder, R., and Özbay, U. 2006a. Monitoring in-situ stress changes in a mining environment with coda wave interferometry. *Geophys. J. Int.*, **167**, 504–508.

Grêt, A., Snieder, R., and Scales, J. 2006b. Time-lapse monitoring of rock properties with coda wave interferometry. *J. Geophys. Res.*, **111**, B03305, doi:10.1029/2004JB003354.

Grêt, A., Snieder, R., Aster, R., and Kyle, P. 2005. Monitoring rapid temporal changes in a volcano with coda wave interferometry. *Geophys. Res. Lett.*, **32**, L06304, 10.1029/2004GL021143.

Gusev, A., and Abubakirov, I. 1996. Simulated envelopes of non-isotropically scattered body waves as compared to observed ones: Another manifestation of fractal heterogeneity. *Geophys. J. Int.*, **127**, 49–60.

Hadziioannou, C., Larose, A., Coutant, O., Roux, P., and Campillo, M. 2009. Stability of monitoring weak changes in multiply scattering media with ambient noise correlation: Laboratory experiments. *J. Acoust. Soc. Am.*, **125**, 3688–3695.

Hennino, R., Trégourès, N., Shapiro, N., Margerin, L., Campillo, M., van Tiggelen, B., and Weaver, R. 2001. Observation of equipartition of seismic waves. *Phys. Rev. Lett.*, **86**, 3447–3450.

Hillers, G., Husen, S., Obermann, A., Planès, T., Larose, E., and Campillo, M. 2015. Noise-based monitoring and imaging of aseismic transient deformation induced by the 2006 Basel reservoir stimulation. *Geophysics*, **80**(4), KS51–KS68.

Hobiger, M., Wegler, U., Shiomi, K., and Nakahara, H. 2012. Coseismic and postseismic elastic wave velocity variations caused by the 2008 Iwate-Miyagi Nairiku earthquake, Japan. *J. Geophys. Res.*, **117**, 1–19.

Kanu, C., and Snieder, R. 2015a. Numerical computation of the sensitivity kernel for monitoring weak changes with multiply scattered acoustic waves. *Geophys. J. Int.*, **203**, 1923–1936.

———. 2015b. Time-lapse imaging of a localized weak change with multiply scattered waves using numerical-based senstivity kernels. *J. Geophys. Res. Solid Earth*, **119**, 5595–5605.

Larose, E., Carrieère, S., Voisin, C., Bottelin, P., Baillet, L., Guéguen, P., Walter, F., Jongmans, D., Guiller, B., Garambois, S., Gimbert, F., and Massey, C. 2015a. Environmental seismology: What can we learn on earth surface processes with ambient noise? *J. Appl. Geophys.*, **116**, 62–74.

Larose, E., Margerin, L., Derode, A., van Tiggelen, B., Campillo, M., Shapiro, N., Paul, A., Stehly, L., and Tanter, M. 2006. Correlation of random wavefields: an interdisciplinary review. *Geophysics*, **71**, SI11–SI21.

Larose, E., Obermann, A., Digulescu, A., Planès, T., Chaix, J.-F., Mazerolle, F., and Moreau, G. 2015b. Locating and characterizing a crack in concrete with diffuse ultrasound: A four-point bending test. *J. of the Acoust. Soc. of America*, **138**(1), 232–241.

Larose, E., Planès, T., Rossetto, V., and Margerin, L. 2010. Locating a small change in a multiple scattering experiment. *Appl. Phys. Lett.*, **96**, 204101.

Lobkis, O., and Weaver, R. 2001. On the emergence of the Green's function in the correlations of a diffuse field. *J. Acoust. Soc. Am.*, **110**, 3011–3017.

Lu, Y., and Michaels, J. 2005. A methodology for structural health monitoring with diffuse ultrasonic waves in the presence of temperature variations. *Ultrasonics*, **43**, 717–731.

Mainsant, G., Larose, E., Brönnimann, Jongmans, D., Michoud, C., and Jaboyedoff, M. 2012. Ambient seismic noise monitoring of a clay landslide: toward failure prediction. *J. Geophys. Res.*, **117**, F01030.

Margerin, L., Campillo, M., and van Tiggelen, B. 2000. Monte Carlo simulation of multiple scattering of waves. *J. Geophys. Res.*, **105**, 7873–7892.

Margerin, L., Campillo, M., Van Tiggelen, B., and Hennino, R. 2009. Energy partition of seismic coda waves in layered media: theory and application to Pinyon Flats Observatory. *Geophys. J. Int.*, **177**, 571–585.

Margerin, L., Planès, T., Mayor, J., and Calvet, M. 2016. Sensitivity kernels for coda-wave interferometry and scattering tomography: theory and numerical evaluation for two-dimensional anisotropically scattering media. *Geophys. J. Int.*, **204**, 650–666.

Mayor, J., Margerin, L., and Calvet, M. 2014. Sensitivity of coda waves to spatial variations of absorbtion and scattering: theory and numerical evaluation in two-dimensional anistropically scattering media. *Geophys. J. Int.*, **197**, 650–666.

Menke, W. 1984. *Geophysical Data Analysis: Discrete Inverse Theory*. San Diego: Academic Press.

Morse, P., and Feshbach, H. 1953a. *Methods of Theoretical Physics, Part 1*. New York: McGraw-Hill.

———. 1953b. *Methods of Theoretical Physics, Part 2*. New York: McGraw-Hill.

Nakata, N., and Snieder, R. 2011. Near-surface weakening in Japan after the 2011 Tohoku-Oki earthquake. *Geophys. Res. Lett.*, **38**, L17302.

Nakata, N., Snieder, R., Kuroda, S., Ito, S., Aizawa, T., and Kunimi, T. 2013. Monitoring a building using deconvolution interferometry. I: Earthquake-data analysis. *Bull. Seismol. Soc. Am.*, **103**, 1662–1678.

Nishimura, T., Uchida, N., Sato, H., Ohtake, M., Tanaka, S., and Hamaguchi, H. 2000. Temporal changes of the crustal structure associated with the M6.1 earthquake on September 3, 1998, and the volcanic activity of Mount Iwate, Japan. *Geophys. Res. Lett.*, **27**, 269–272.

Nolet, G. 2008. *A Breviary of Seismic Tomography*. Cambridge, UK: Cambridge Univ. Press.

Obermann, A., Froment, B., Campillo, M., Larose, E., Planès, T., Valette, B., Chen, J., and Liu, Q. 2014. Seismic noise correlations to image structural and mechanical changes associated with the Mw 7.9 2008 Wenchuan earthquake. *J. Geophys. Res. Solid Earth*, **119**, 3155–3168.

Obermann, A., Kraft, T., Larose, E., and Wiemer, S. 2015. Potential of ambient seismic noise techniques to monitor the St. Gallen geothermal site (Switzerland). *J. of Geophys. Res.: Solid Earth*, **120**(6), 4301–4316.

Obermann, A., Planès, T., Hadziioannou, C., and Campillo, M. 2016. Lapse-time dependent coda wave-wave depth sensitivity to local velocity perturbations in 3-D heterogeneous elastic media. *Geophys. J. Int.*, **207**, 59–66.

Obermann, A., Planès, T., Larose, E., Sens-Schönfelder, C., and Campillo, M. 2013. Depth sensitivity of seismic coda waves to velocity perturbations in an elastic heterogeneous medium. *Geophys. J. Int.*, **194**(1), 372–382.

Özisik, M. 1973. *Radiative Transfer and Interaction with Conduction and Convection.* New York: John Wiley.

Paasschens, J. 1997. Solution of the Time-Dependent Boltzmann Equation. *Phys. Rev. E*, **56**, 1135–1141.

Pacheco, C., and Snieder, R. 2005. Localizing time-lapse changes with multiply scattered waves. *J. Acoust. Soc. Am.*, **118**, 1300–1310.

———. 2006. Time-lapse traveltime change of single scattered acoustic waves. *Geophys. J. Int.*, **165**, 485–500.

Page, J., Cowan, M., and Weitz, D. 2000. Diffusing acoustic wave spectroscopy of fluidized suspensions. *Physica B*, **279**, 130–133.

Planès, T., Larose, E., Margerin, L., Rossetto, V., and Sens-Schönfelder, C. 2014. Decorrelation and phase-shift of coda waves induced by local changes: multiple scattering approach and numerical validation. *Waves in Random and Complex Media*, **24**(2), 99–125.

Planès, T., Mooney, M., Rittgers, J., Parekh, M., Behm, M., and Snieder, R. 2016. Time-lapse monitoring of internal erosion in earthen dams and levees using ambient seismic noise. *Géotechnique*, **66**, 301–312.

Poupinet, G., Ellsworth, W., and Fréchet, J. 1984. Monitoring velocity variations in the crust using earthquake doublets: an application to the Calaveras Fault, California. *J. Geophys. Res.*, **89**, 5719–5731.

Roberts, P., Scott Phillips, W., and Fehler, M. 1992. Development of the active doublet method for measuring small velocity and attenuation changes in solids. *J. Acoust. Soc. Am.*, **91**, 3291–3302.

Robinson, D., Sambridge, M., and Snieder, R. 2011. A probabilistic approach for estimating the separation between a pair of earthquakes directly from their coda waves. *J. Geophys. Res.*, **116**, B04309.

Robinson, D., Sambridge, M., Snieder, R., and Hauser, J. 2013. Relocating a cluster of earthquakes using a single station. *Bull. Seismol. Soc. Am.*, **103**, 3057–3072.

Rossetto, V., Margerin, V., Planès, T., and Larose, E. 2011. Locating a weak change using diffuse waves: Theoretical approach and inversion procedure. *J. Appl. Phys.*, **109**, 034903.

Ryzhik, L., Papanicolaou, G., and Keller, J. B. 1996. Transport equations for elastic and other waves in random media. *Wave Motion*, **24**, 327–370.

Salvermoser, J., Hadzioannou, C., and Stähler, S. 2015. Structural monitoring of a highway bridge using passive noise recordings from street traffic. *J. Acoust. Soc, Am.*, **138**, 3864–3872.

Sato, H. 1986. Temporal change in attenuation intensity before and after the eastern Yamanashi earthquake of 1983 in Central Japan. *J. Geophys. Res.*, **91**, 2049–2061.

———. 1988. Temporal change in scattering and attenuation associated with the earthquake occurence - a review of recent studies on coda waves. *Pure Appl. Geophys.*, **126**, 465–497.

Sato, H., and Fehler, M. 2012. *Seismic Wave Propagation and Scattering in the Heterogeneous Earth.* 2nd edn. New York: Springer Verlag.

Schaff, D., and Beroza, G. 2004. Coseismic and postseismic velocity changes measured by repeating earthquakes. *J. Geophys. Res.*, **109**, B10302, doi:10.1029/2004JB003011.

Sens-Schönfelder, C., and Brenguier, F. 2018. Noise-based monitoring. Pages – of: Nakata, N., Gualtieri, L., and Fichtner, A. (eds.), *Seismic Ambient Noise.* Cambridge University Press, Cambridge, UK.

Sens-Schönfelder, C., and Larose, E. 2008. Temporal changes in the lunar soil from correlation of diffuse vibrations. *Phys. Rev. E*, **78**, 045601.

Sens-Schönfelder, C., and Wegler, U. 2006. Passive image interferometry and seasonal variations at Merapi volcano, Indonesia. *Geophys. Res. Lett.*, **33**, L21302, doi:10.1029/2006GL027797.

Sens-Schönfelder, C., Margerin, L., and Campillo. 2009. Laterally heterogeneous scattering explains Lg blockage in the Pyrenees. *J. Geophys. Res.*, **114**, B07309.

Shapiro, N., Campillo, M., Margerin, L., Singh, S., Kostoglodov, V., and Pacheco, J. 2000. The energy partitioning and the diffuse character of the seismic coda. *Bull. Seism. Soc. Am.*, **90**, 655–665.

Snieder, R. 2002. Coda wave interferometry and the equilibration of energy in elastic media. *Phys. Rev. E*, **66**, 046615.

———. 2006. The theory of coda wave interferometry. *Pure and Appl. Geophys.*, **163**, 455–473.

Snieder, R., and Larose, E. 2013. Extracting Earth's elastic wave response from noise measurements. *Ann. Rev. Earth Planet. Sci.*, **41**, 183–206.

Snieder, R., and van Wijk, K. 2015. *A Guided Tour of Mathematical Methods for the Physical Sciences*. 3rd edn. Cambridge, UK: Cambridge Univ. Press.

Snieder, R., and Vrijlandt, M. 2005. Constraining relative source locations with coda wave interferometry: Theory and application to earthquake doublets in the Hayward Fault, California. *J. Geophys. Res.*, **110**, B04301, 10.1029/2004JB003317.

Snieder, R., Grêt, A., Douma, H., and Scales, J. 2002. Coda wave interferometry for estimating nonlinear behavior in seismic velocity. *Science*, **295**, 2253–2255.

Spetzler, J., and Snieder, R. 2004. The Fresnel volume and transmitted waves. *Geophysics*, **69**, 653–663.

Takagi, R., Okada, T., Nakahara, H., Umino, N., and Hesegawa, A. 2012. Coseismic velocity change in and around the focal region of the 2008 Iwate-Miyagi Nairiku earthquake. *J. Geophys. Res.*, **117**, B06315.

Tarantola, A. 1984a. Inversion of seismic reflection data in the acoustic approximation. *Geophysics*, **49**, 1259–1266.

———. 1984b. Linearized inversion of seismic reflection data. *Geophys. Prosp.*, **32**, 998–1015.

Tourin, A., Fink, M., and Derode, A. 2000. Multiple scattering of sound. *Waves Random Media*, **10**, R31–R60.

Trégourès, N., and van Tiggelen, B. 2002. Generalized diffusion equation for multiple scattered elastic waves. *Waves in Random Media*, **12**, 21–38.

Tremblay, N., Larose, E., and Rossetto, V. 2010. Probing slow dynamics of consolidated granular multicomposite materials by diffuse acoustic wave spectroscopy. *J. Acoust. Soc, Am.*, **127**, 1239–1243.

Tromp, J., Komattisch, D., and Liu, Q. 2008. Spectral-element and adjoint methods in seismology. *Communications in Computational Physics*, **3**(1), 1–32.

Tromp, J., Tape, C., and Liu, Q. 2005. Seismic tomography, adjoint methods, time reversal and banana-doughnut kernels. *Geophys. J. Int.*, **160**, 195–216.

van Rossum, M., and Nieuwenhuizen, T. 1999. Multiple scattering of classical waves: microscopy, mesoscopy and diffusion. *Rev. Mod. Phys.*, **71**, 313–371.

Weaver, R. 1982. On diffuse waves in solid media. *J. Acoust. Soc. Am.*, **71**, 1608–1609.

———. 1985. Diffuse waves at a free surface. *J. Acoust. Soc. Am.*, **78**, 131–136.

Wegler, U., Korn, M., and Przybilla, J. 2006. Modeling full seismogram envelopes using radiative transfer theory with Born scattering coefficients. *Pure Appl. Geoph.*, **163**, 503–531.

Wegler, U., Nakahara, H., Sens-Schönfelder, C., Korn, M., and Shiomi, K. 2009. Sudden drop of seismic velocity after the 2004 M_W 6.6 mid-Niigata earthquake, Japan, observed with passive image interferometry. *J. Geophys. Res. Solid Earth*, **114**, B06305.

Yamawaki, T., Nishimura, T., and Hamaguchi, H. 2004. Temporal change of seismic structure around Iwate volcano inferred from waveform correlation analysis of similar earthquakes. *Geophys. Res. Lett.*, **31**, L24616, doi:10.1029/2004GL021103.

Yoshimoto, K. 2000. Monte-Carlo simulation of seismogram envelopes in scattering media. *J. Geophys. Res.*, **105**, 6153–6161.

Zhang, Y., Planès, T., Larose, E., Obermann, A., Rospars, C., and Moreau, G. 2016. Diffuse ultrasound monitoring of stress and damage development on a 15-ton concrete beam. *J. Acoust. Soc. America*, **139**(4), 1691–1701.

6.A The Chapman-Kolmogorov Equation for Diffusion

Consider a diffusive field I that obeys the diffusion equation

$$\frac{\partial I(\mathbf{r}, t)}{\partial t} - \nabla \cdot (D(\mathbf{r})\nabla I(\mathbf{r}, t)) = 0, \tag{6.38}$$

We use the Green's function $G^D(\mathbf{r}, \mathbf{r}', t)$ for the diffusion equation that is defined in equation (6.13). This Green's function is causal is the sense that

$$G^D(\mathbf{r}, \mathbf{r}', t) = 0 \quad \text{for} \quad t < 0. \tag{6.39}$$

The Green's function at time $t = 0^+$ just after the source term $\delta(\mathbf{r} - \mathbf{r}')\delta(t)$ has acted follows by integrating equation (6.13) from $t = -\varepsilon$ to $t = \varepsilon$. This gives, using equation (6.39) and $\int_{-\varepsilon}^{\varepsilon} \delta(t)dt = 1$:

$$G^D(\mathbf{r}, \mathbf{r}', t = \varepsilon) - \nabla \cdot \left(D(\mathbf{r})\nabla \int_{-\varepsilon}^{\varepsilon} G^D(\mathbf{r}, \mathbf{r}', t)dt\right) = \delta(\mathbf{r} - \mathbf{r}'). \tag{6.40}$$

The absolute value of the integral in the left-hand side is smaller that 2ε; hence this integral vanishes in the limit $\varepsilon \to 0$. Applying this limit to equation (6.40) gives

$$G^D(\mathbf{r}, \mathbf{r}', t = 0^+) = \delta(\mathbf{r} - \mathbf{r}'). \tag{6.41}$$

We next consider the integral

$$F(\mathbf{r}, t) = \int G^D(\mathbf{r}, \mathbf{r}', t)I(\mathbf{r}', t = 0)dV'. \tag{6.42}$$

Inserting this solution into the diffusion equation (6.38), and using the fact that the Green's function satisfies equation (6.13), gives

$$\frac{\partial F(\mathbf{r}, t)}{\partial t} - \nabla \cdot (D(\mathbf{r})\nabla F(\mathbf{r}, t)) = \int \delta(\mathbf{r}-\mathbf{r}')\delta(t)I(\mathbf{r}', t = 0)dV' = \delta(t)I(\mathbf{r}, t = 0). \tag{6.43}$$

Since $\delta(t) = 0$ for $t > 0$, the right-hand side vanishes, and hence $F(\mathbf{r}, t)$ is a solution to the diffusion equation (6.38) for $t > 0$ as well. To verify that it has the same initial condition as I we set $t = 0$ in equation (6.42) and use equation (6.41); this gives

$$F(\mathbf{r}, t = 0) = \int G^D(\mathbf{r}, \mathbf{r}', t = 0^+) I(\mathbf{r}', t = 0) dV'$$

$$= \int \delta(\mathbf{r} - \mathbf{r}') I(\mathbf{r}', t = 0) dV' = I(\mathbf{r}, t = 0). \qquad (6.44)$$

This means that the integral (6.42) satisfies the same equation as the diffusive field $I(\mathbf{r}, t)$, and the same initial conditions as this field, which implies that $F(\mathbf{r}, t) = I(\mathbf{r}, t)$; hence equation (6.42) implies that

$$I(\mathbf{r}, t) = \int G^D(\mathbf{r}, \mathbf{r}', t) I(\mathbf{r}', t = 0) dV'. \qquad (6.45)$$

Since the diffusion equation is invariant for translation in time, we can replace $t = 0$ by an arbitrary time $t' < t$, so that

$$I(\mathbf{r}, t) = \int G^D(\mathbf{r}, \mathbf{r}', t - t') I(\mathbf{r}', t') dV'. \qquad (6.46)$$

This is the Chapman-Kolmogorov equation that relates the solution at a time $t' < t$ to the solution at a later time t. This expression forms the basis of path integral formulations of quantum mechanics and statistical mechanics (Feynman, 1965). Equation (6.46) holds because the underlying equation is of first order in time, as is the case for the diffusion equation and for the Schrödinger equation. We show in the next section that a similar relation applies to radiative transfer.

6.B The Chapman-Kolmogorov Equation for Radiative Transfer

Similar to equation (6.13) we define a Green's function for the specific intensity that satisfies equation (6.22). This Green's function is the response to energy injected in direction $\hat{\mathbf{n}}'$ at location $\mathbf{r}'$ at time $t = 0$. Because of causality this Green's function also satisfies equation (6.39). The derivation of the Green's function follows the same steps as for the diffusion equation that led to equation (6.41). The result is that the Green's function for radiative transfer satisfies

$$G^{RT}(\mathbf{r}, \hat{\mathbf{n}}, \mathbf{r}', \hat{\mathbf{n}}', t = 0^+) = \delta(\mathbf{r} - \mathbf{r}') \delta(\hat{\mathbf{n}} - \hat{\mathbf{n}}'). \qquad (6.47)$$

The additional terms that the radiative transfer equation (6.22) has compared to the diffusion equation (6.13) do not change the argument leading to equation (6.47), because in the derivation these additional terms are multiplied with 2ε, so that they vanish in the limit $\varepsilon \to 0$.

Next we consider, by analogy with equation (6.42), a solution

$$F(\mathbf{r}, \hat{\mathbf{n}}, t) = \int G^{RT}(\mathbf{r}, \hat{\mathbf{n}}, \mathbf{r}', \hat{\mathbf{n}}', t) I(\mathbf{r}', \hat{\mathbf{n}}', t = 0) dV' d^2 n'. \qquad (6.48)$$

Inserting this solution in the equation of radiative transfer and using equation (6.22) for the Green's function gives

$$\frac{\partial F(\mathbf{r}, \hat{\mathbf{n}}, t)}{\partial t} + v(\mathbf{r}) \hat{\mathbf{n}} \cdot \nabla F(\mathbf{r}, \hat{\mathbf{n}}, t) - \oint S(\mathbf{r}, \hat{\mathbf{n}}, \hat{\mathbf{n}}_0) F(\mathbf{r}, \hat{\mathbf{n}}_0, t) d^2 n' + q(\mathbf{r}) F(\mathbf{r}, \hat{\mathbf{n}}, t)$$
$$= \delta(t) I(\mathbf{r}, \hat{\mathbf{n}}, t = 0). \qquad (6.49)$$

For $t > 0$ the delta function in the right-hand side vanishes, so that $F(\mathbf{r}, \hat{\mathbf{n}}, t)$ satisfies the radiative transfer equation (6.19). Setting $t = 0$ in equation (6.48), and using equation (6.47), gives upon integration over $\mathbf{r}_0$ and $\hat{\mathbf{n}}_0$:

$$F(\mathbf{r}, \hat{\mathbf{n}}, t = 0) = I(\mathbf{r}, \hat{\mathbf{n}}, t = 0). \qquad (6.50)$$

This means that F and I both satisfy the equation of radiative transfer and the same initial condition; therefore these solutions are the same, and since the solutions of the radiative transfer equations are unique (Case, 1957), equation (6.48) implies that

$$I(\mathbf{r}, \hat{\mathbf{n}}, t) = \int G^{RT}(\mathbf{r}, \hat{\mathbf{n}}, \mathbf{r}', \hat{\mathbf{n}}', t) I(\mathbf{r}', \hat{\mathbf{n}}', t = 0) dV' d^2 n'. \qquad (6.51)$$

Because of invariance for translation in time, the time $t = 0$ can be replaced by an arbitrary time $t' < t$, which gives the Chapman-Kolmogorov equation for radiative transfer

$$I(\mathbf{r}, \hat{\mathbf{n}}, t) = \int G^{RT}(\mathbf{r}, \hat{\mathbf{n}}, \mathbf{r}', \hat{\mathbf{n}}', t - t') I(\mathbf{r}', \hat{\mathbf{n}}', t') dV' d^2 n'. \qquad (6.52)$$

7

Applications with Surface Waves Extracted from Ambient Seismic Noise

NIKOLAI M. SHAPIRO

Abstract

Signals of fundamental mode surface waves can be easily reconstructed from cross-correlations of the ambient seismic noise. This gives rise to a large class of methods that can be regrouped under a generic name of "ambient noise surface-wave tomography" and that are used to image the structure of the Earth's shallow layers (from top hundreds of meters to hundreds of kilometers). This chapter gives an overview of main approaches that have been recently developed for the seismic noise-based surface-wave imaging that continues to be a very dynamic field of research with some first-order issues on which we need to work.

7.1 Introduction

Seismic imaging is a class of inverse problems that exploits measurements made from seismic waves travelling through the Earth to infer the mechanical properties of the Earth's interior. The most traditional approach is to invert travel times of P and S body waves whose behavior in terms of arrival times and amplitudes can be described with ray tracing methods based on the eikonal equation. Other methods use surface waves, amplitudes, full waveforms, and even the part of the signal composed of scattered waves that can be inverted with a variety of theoretical approximations.

Traditional ways of solving seismic imaging inverse problems are based on signals from sources with well-known properties, i.e., their locations, mechanisms, and time histories. Artificial sources, mostly used in seismic exploration, are designed to have these fully controlled properties, and traditional passive seismology is mostly based on records from earthquakes whose contribution can be described by deterministic functions and well determined from observations.

Methods based on the deterministic description of seismic sources are very powerful. However, they are limited by the availability of such sources. Human-made sources are often too expensive, not energetic enough or difficult to implement, and earthquakes occur only in tectonically active areas and are relatively rare. Background vibrations of the Earth's surface often (and rather erroneously) called "seismic noise" represent a different kind of seismic signal that is excited by sources that cannot be described deterministically.

Theory of the seismic noise cross-correlations is described in Chapter 4 (Fichtner and Tsai, 2018). Its cornerstone is the "noise cross-correlation theorem" that states that in a linear system with a weak attenuation the time derivative of the cross-correlation of random wavefield recorded at two positions converges to the system Green's functions (e.g., Gouédard et al., 2008). It can be written as:

$$\lim_{T \to \infty} \frac{\partial}{\partial t} C_{AB}(t) = F(t) * [G_{AB}(t) + G_{BA}(-t)] \tag{7.1}$$

where $C_{AB}(t)$ is the cross-correlation of signals recorded in locations A and B computed in a window of length T, $G_{AB}(t)$ is the Green's function between these two locations, $G_{BA}(-t)$ is its reciprocal, and $F(t)$ is a time-domain function that accounts for the noise spectrum characteristics (e.g., Weaver and Lobkis, 2001; Snieder, 2004; Wapenaar, 2004; Roux et al., 2005; Campillo, 2006; Gouédard et al., 2008). In particular, $F(t)$ is a Dirac function for a pure white band noise. In other words, by computing noise cross-correlations between sufficiently long records at all available receivers, every receiver may be converted in a virtual source emitting a wavefield represented by noise correlation functions. During the last decade, application of this principle in seismology resulted in a new paradigm when the deterministic wavefields reconstructed from noise cross-correlations are used for imaging the Earth's interior instead of signals from earthquakes or from artificial seismic sources.

The idea to use random seismic noise for imaging the subsurface was first introduced to seismology by Aki (1957) and to seismic exploration by Claerbout (1968). However, a full exploitation of the potential of application of the noise cross-correlation theorem could only be achieved in the era of digital seismology when long digitally recorded seismic time series became available for massive analysis. Extraction of deterministic waves from correlations of passive seismic noise records was first demonstrated by Shapiro and Campillo (2004) and Sabra et al. (2005a) and immediately led to first applications of the noise-based seismic tomography (Shapiro et al., 2005; Sabra et al., 2005b).

The noise cross-correlation theorem is valid in the ideal case of a perfectly random and equipartitioned noise wavefield. However, the properties of the real seismic noise within the Earth are different from this ideal case. Seismic noise is mainly generated as a result of coupling of the solid Earth with the oceans and the atmosphere (e.g., Ardhuin et al., 2011, 2018). At high frequencies ($>$ 1 Hz) anthropogenic noise sources may become very important (Chapter 1 [McNamara and Boaz, 2018]). In both cases, the resulting noise wavefield is not stationary in space, in time, and in the frequency domain. Strong seismic sources of tectonic origin such as earthquakes and volcanoes contribute to the strong non-stationarity of the continuous seismic time series. As a consequence, the noise cross-correlation theorem cannot be directly applied to the seismological data.

The possibility to extract deterministic waves propagating between pairs of stations' noise cross-correlations requires presence of enough noise sources in the zones of constructive interference (e.g., Snieder, 2004; Roux et al., 2005). There is a strong difference between the configuration of these zones for surface waves and for body waves, with the latter mainly lying far from the surface at depth (Figure 7.1). This difference combined with the fact that most of the noise sources are located at the Earth's surface results in dominant reconstruction of surface waves (and mainly of their fundamental modes) from the noise correlation functions. Detailed conditions and applications for body-wave correlations are discussed in Chapter 8 (Nakata and Nishida, 2018). The inhomogeneous distribution of the noise at the Earth's surface leads to non-symmetric cross-correlations even for reconstructed surface waves (examples are shown in Chapter 5 [Ritzwoller and Feng, 2018]) and prevents correct measurements of their amplitudes (e.g., Stehly et al., 2006).

As described in Chapter 5 (Ritzwoller and Feng, 2018), the continuous seismic records must be pre-processed to reduce the effects of the non-stationarity of noise records and of inhomogeneous source distribution. However, even with the pre-processing, the real noise correlation functions are not exact equivalents of the Green's functions. Therefore, most of the existing imaging applications based on noise cross-correlations do not use exactly the noise correlation theorem.

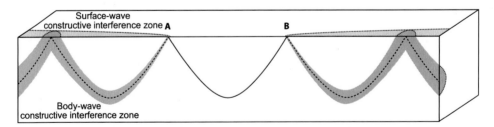

Figure 7.1. Schematic representation of zones of constructive interference for surface (light shading) and for body (dark shading) waves. Solid black line shows the ray for a refracted body wave connecting the two receiver locations A and B. Dashed lines show continuation of this ray in the regions behind the receivers.

Garnier and Papanicolaou (2009) have shown that the travel time of waves can be effectively estimated when the ray joining the two sensors continues into the noise source region, even when the full Green's function cannot be reconstructed.

7.2 Ambient Noise Travel Time Surface Wave Tomography

As explained in the previous section, the travel times of surface waves is the type of information that is most easily extracted from cross-correlation of ambient seismic noise. Therefore, one of the main applications based on noise cross-correlations is the ambient noise surface wave tomography (ANSWT) (e.g., Ritzwoller et al., 2011). This method is particularly advantageous when used with networks of seismographs. By computing cross-correlations of seismic noise from long enough time series between all pairs of stations, each of these stations can be converted into a virtual source emitting surface waves with travel times measured at all other stations. For a network composed of N stations, this provides us with travel times measured along $N(N - 1)/2$ paths connecting N collocated sources and receivers. This ensemble of surface-wave travel times measured from noise cross-correlations can be inverted with "standard" methods similar to measurements obtained from earthquakes.

Surface waves are called so because they propagate along the surface and their amplitude vanishes below a certain depth. The simplest expression for a surface wave at a single frequency propagating in direction X can be written as:

$$u_{sw}(x, z, t, \omega) \approx \xi(z, \omega)exp[i(\omega t - k(\omega)x)] = \xi(z, \omega)exp[i\omega(t - x/C(\omega))],$$
$$(7.2)$$

where t is time, ω is frequency, z is depth, u is the displacement vector, k is the wavenumber, and ξ is the eigenfunction describing the decay of the wave amplitude with depth (Figure 7.2). C is the phase velocity:

$$C(\omega) = \frac{\omega}{k(\omega)}.$$
$$(7.3)$$

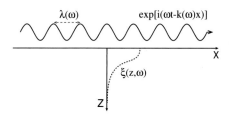

Figure 7.2. Schematic illustration of a surface wave as described in equation (7.2).

For such a wave, one can define a horizontal wavelength λ:

$$\lambda(\omega) = \frac{2\pi C(\omega)}{\omega}. \tag{7.4}$$

An important property of seismic surface waves in the Earth is their so-called "dispersion" with their wavenumbers k depending on frequency ω (e.g., Levshin et al., 1989). A simple physical meaning of this dispersion can be described as follows. The energy of a surface wave is concentrated in a near-surface layer where the value of its eigenfunction $\xi(z, \omega)$ is large (Figure 7.2). Typically, this thickness is close to a half wavelength. This implies that high-frequency (small-wavelength) surface waves are sensitive to very shallow parts of the structure, while at lower frequencies (longer wavelengths) they become sensitive to deeper layers. On average, the Earth's material becomes more rigid with depth and its wave speed increases. Therefore, surface wave phase velocity generally increases with the increasing wavelength or decreases with frequency. An important consequence of this is that the "dispersion" of surface waves reflects changes of elastic properties with depth. Therefore, measurements of surface wave phase and group velocities at different frequencies can be used to constrain the depth-dependent model of the underlying media.

In addition to the phase velocity $C(\omega)$ we can define the group velocity $U(\omega)$:

$$U(\omega) = \left(\frac{\partial k(\omega)}{\partial \omega}\right)^{-1}. \tag{7.5}$$

Phase velocity appears naturally in equation (7.2). At the same time, the energy or information is conveyed along a wave with the group velocity (e.g., Biot, 1957). In this sense, the former is more "physical." So far, if we want to estimate a time required for a surface wave to travel along a certain distance, we need to use the group velocity. The two velocities are simply interrelated:

$$U(\omega) = \frac{C(\omega)}{1 - \frac{\omega}{C(\omega)} \frac{\partial C(\omega)}{\partial \omega}}. \tag{7.6}$$

Note that this relationship includes partial derivatives over frequency. As a result, group velocities cannot be easily computed from phase velocity measurements at a discrete set of frequencies and must be measured independently. Moreover, the depth sensitivity of the phase and the group velocities is different (e.g., Ritzwoller et al., 2001). Therefore, the information obtained with these two types of measurements is complementary.

Traditionally, in the surface wave the "dispersion" is described as a function of period ($T = 1/f = 2\pi/\omega$) rather than frequency. Therefore, in many examples shown in this and other chapters, phase and group velocities and other surface wave properties are described as functions of period.

Finally, for a single source-receiver path, more than one travel time measurement can be made. Surface wave travel times (or velocities of propagation) measured at different periods form so-called dispersion curves. There are two main types of surface waves, Rayleigh and Love, and two types of propagation velocities, phase and group. Therefore, for a given source-receiver path and for a single mode of surface waves, it is possible to measure four types of dispersion curves.

Measuring the dispersion curves of a particular mode of surface waves requires isolating the signal corresponding to this mode from other waves present in the waveform. For this goal, the frequency time analysis (e.g., Levshin et al., 1989) is used. Details of this method are described in section 3.5 (Ritzwoller and Feng, 2018). Also, it can be directly used to measure group and phase velocities (e.g., Lin et al., 2008), as illustrated in Figure 5.11.

Rayleigh and Love waves are characterized by different polarizations, with the former observed in vertical (Z) and radial (R) components of the ground motions and the latter on the transverse (T) component. When computing cross-correlations between three-component records, the full correlation tensor contains 6 independent components (Figure 5.10). As described for example by Lin et al. (2008), to separate Rayleigh and Love waves, the cross-correlation tensor should be rotated to radial and transverse directions of motion based on azimuth and back azimuth measured when considering one of the stations as source and another as receiver. After such rotation, for components of the cross-correlation tensor, ZZ, RR, RZ, and ZR can be used to extract information about Rayleigh waves, and the TT component corresponds to Love waves. The mixed RT and TT cross-correlations are expected to vanish in isotropic media with reasonably homogeneous noise source distributions. Strong amplitude on these components can indicate either strongly inhomogeneous noise source distributions or a presence of anisotropy in the media (e.g., Durand et al., 2011).

Each individual cross-correlation contains signals on the positive (causal) and the negative (anti-causal) sides. In a case of perfect reconstruction, these two sides should be perfectly symmetric, as stated by the cross-correlation theorem. In reality, the symmetry in amplitudes is almost never achieved because of inhomogeneous distribution of the noise sources. When this source distribution inhomogeneity is not too strong, it results in differences in signal amplitudes on positive and negative sides, while travel measured from these two sides remain very similar. A significant difference of travel times measured from positive and negative sides indicates that the source distribution is really strongly heterogeneous and likely biases the travel time estimations. Therefore, a comparison of these travel time (dispersion) measurements obtained from positive and negative sides of cross-correlations can be used for quality control.

The most standard set of steps for the ANSWT can be described as follows:

1. Pre-processing continuous seismograms.
2. Computing cross-correlations between all available pairs of stations and components.
3. Measuring frequency-dependent phase and/or group travel times from negative and positive parts of ZZ, RR, RZ, and ZR correlations for Rayleigh waves and of TT correlations for Love waves ($t_{p,g}^{R,L}(\omega)$).
4. Quality control and final selection of measured phase and group travel times.
5. Regionalization of the measured dispersion curves with a 2D surface-wave tomography: construction of frequency-dependent phase and or group velocity maps for Rayleigh and Love waves ($C, U^{R,L}(\omega, x, y)$).
6. One-dimensional depth inversion of regionalized dispersion curves at every geographical location (x, y) to construct final 3D model of the media ($V_S(x, y, z)$).

A detailed description of the methods used for data pre-processing and for measuring surface-wave dispersion curves is given in Chapter 5 (Ritzwoller and Feng, 2018). In a case of "sparse" networks with the interstation distances being larger or comparable to wavelengths of used surface waves, regionalization of the dispersion curves can be formulated as a 2D tomographic problem, i.e., as a linear (or iterative linearized) inversion of measured phase and/or group travel times. In the case of "dense" networks with interstation distances smaller than the used wavelengths, the local phase velocities can be estimated directly from the spatial derivatives of the measured travel times.

The depth inversion of the surface wave dispersion curves for a 1D velocity profile is one of the most widely used inverse problems in seismology and can be solved with a multiplicity of methods. It can be formulated in a following way. The regionalization step produces at a particular geographical point the local dispersion curves that are used as input data:

$$d = \left[C^R(\omega), C^L(\omega), U^R(\omega), U^L(\omega) \right], \tag{7.7}$$

where C and U denote phase and group velocities, respectively, and indices R and L describe Rayleigh and Love waves. These dispersion curves are assumed to result from a locally one-dimensional model:

$$m = \left[c_{i,j,k,l}(z), \rho(z), Q(z) \right], \tag{7.8}$$

where z is depth, $c_{i,j,k,l}$ is the elastic tensor, ρ is density, and Q is the quality factor. The forward problem can then be written schematically as:

$$d = F(m) \tag{7.9}$$

and can be solved with a number of algorithms such as that of Herrmann (2013) for the flat-layer approximation and that of Woodhouse (1988) for the spherical geometry.

The information contained in the surface-wave dispersion curves is not sufficient to constrain the full set of elastic parameters present in equation (7.8). Surface-wave travel times are mostly sensitive to the shear-wave velocities in the media (e.g., Dahlen and Tromp, 1998). Therefore, in most surface-wave dispersion inversion algorithms the model is parameterized in terms of $V_S(z)$, and other parameters are commonly fixed or set to scale in some way with shear-wave velocities.

Inversion of equation (7.9) can be done with different approaches. One family of methods is based on computing linearized sensitivity kernels of surface wave phase and group velocities relative to elastic parameters of the media (e.g., Herrmann, 2013). The 1D depth inversion of the dispersion curves is a relatively low dimensional inverse problem and it therefore can be solved with Monte-Carlo type methods based on statistical exploration of the model parameters space (e.g., Shapiro and Ritzwoller, 2002; Yang et al., 2008; Moschetti et al., 2010). This approach consists in random generations of many models m (velocity profiles $V_S(z)$) followed by computation of dispersion curves based on equation 7.9 and their comparison with the data d to select an ensemble of models that fit well the observations. This family of methods became very popular in recent years with the increase of computing power that allows us to apply these methods based on solving many millions of forward problems. An example of Monte Carlo inversion of group velocity dispersion curves is provided in Figures 7.3c–d.

7.2.1 Ambient Noise Travel Time Surface Wave Tomography with "Sparse" Networks

The most standard 2D travel time tomography is based on ray approximation for surface waves (e.g., Woodhouse, 1974; Wang and Dahlen, 1995). With using this approximation, the forward problem for surface-wave tomography consists of predicting frequency-dependent travel times from a set of 2D phase or group velocity maps:

$$t(\omega) = \int_p \frac{ds}{v(\omega, x, y)}, \qquad (7.10)$$

where ω is the frequency, x and y are coordinates of a surface position, $t(\omega)$ is a phase or group travel time for Rayleigh or Love waves, $v(\omega, x, y)$ is the respective velocity map, p is the ray path, and s is the distance along the ray. For N interstation paths and a particular velocity type at one frequency, this gives a system of N equations that can be inverted to find $v(\omega, x, y)$ based on N travel time

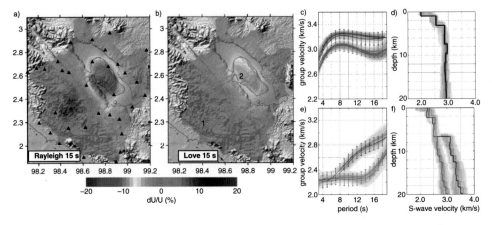

Figure 7.3. Ambient noise surface wave group velocity tomography revealing strong radial anisotropy beneath the Toba caldera (Jaxybulatov et al., 2014). (a) Group-velocity map of Rayleigh waves showing a strong negative anomaly beneath the caldera. Black triangles show positions of used seismic stations. (b) Group-velocity map of Love waves showing a much weaker anomaly beneath the caldera. (c and d) Inversion of the regionalized dispersion curves in location 1 (shown in (b)) for an isotropic 1D shear velocity model. (c) Observed and predicted dispersion curves with solid and dashed lines, respectively (dark grey (blue in the plate section) for Love waves, light grey (green in the plate section) for Rayleigh waves). Error bars of 200 m/s are estimated as average standard deviations of group velocity measurements. Gray lines, dispersion curves predicted from 1000 best-fitting 1D profiles (darker colors correspond to better misfits). (d) Predicted shear velocity profiles. The best-fitting isotropic 1D profile is shown with the thick solid line. (e and f) Inversion of the regionalized dispersion curves in location 2 (shown in (a)) for a radially anisotropic 1D shear velocity profile. Best-fit V_{SV} and V_{SH} velocity models are shown with light and dark grey solid lines (green and blue solid lines in the plate section), respectively, in (f). (A black-and-white version of this figure appears in some formats. For the color version, please refer to the plate section.)

measurements. As for most tomographic problems, this inversion involves additional regularization constraints to stabilize the solution (e.g., Barmin et al., 2001).

The ray approximation tends to break down in the presence of heterogeneities whose length scale is comparable to the wavelength of the wave and is, therefore, considered to be a high-frequency approximation. However, at the relatively short periods at which the ANSWT is mostly used, the finite-frequency effects in the 2D tomography are rather weak (e.g., Ritzwoller et al., 2002). Another approximation often used in the ANSWT and in the surface wave travel time tomography in general is that the nonlinearity related to the ray bending is often ignored. In most cases, the effects of the ray bending are indeed small. To take it into account, the tomographic problem must be solved iteratively by tracing 2D rays based on phase (and not group) velocity (e.g., Yoshizawa and Kennett, 2002; Rawlinson et al., 2008). When only phase velocities are used, the travel time tomography is purely 2D. Combining group-velocity measurements with accounting for the ray bending would require re-computing the 2D phase velocity maps from the 3D models at avery iteration. When the ray bending is ignored, the 2D tomography can be efficiently solved as a regularized linear inverse problem (e.g., Barmin et al., 2001).

The above-described approach to the ANSWT has been extensively used during the recent decade in many regions of the Earth to study the structure of the crust and the upper mantle at different scales. The ANSWT has been applied at a global scale (e.g., Nishida et al., 2009; Haned et al., 2016). Some of the most spectacular results of application of the ANSWT have been obtained with large-scale networks like in North America (e.g., Lin et al., 2008; Yang et al., 2008; Moschetti et al., 2010; Spica et al., 2016; Shen and Ritzwoller, 2016), in Europe (e.g., Yang et al., 2007; Stehly et al., 2009; Molinari et al., 2015), and in China (e.g., Yao et al., 2006, 2008; Liu et al., 2016; Shen et al., 2016). Also, this method turned out to be very efficient when using relatively short-period surface waves to study shallow subsurface and has been successfully applied to study many volcanic systems (e.g., Brenguier et al., 2007; Jaxybulatov et al., 2014; Mordret et al., 2015; Spica et al., 2017) and started to be applied with industrial data (e.g., Mordret et al., 2013a). Examples of group velocity maps for relatively short periods (15 s, wavelengths $\sim$ 50 km) Rayleigh and Love waves are shown in Figures 7.3a and b.

7.2.2 Noise-Based Surface Wave Applications with "Dense" Arrays

Many modern seismic networks are so dense that they sample the surface waves at the sub-wavelength scale. In this case, the travel time information can be used to measure the local properties of the wavefield, resulting in more accurate inferences about the wave propagation and, in consequence, about the underlying media. For example, a beamforming analysis (e.g., Rost and Thomas, 2002) with a sub-wavelength scale network can be used to measure simultaneously the local speed and the local direction of propagation of the surface waves. A large network can be then divided into many sub-arrays. The beamforming analysis of all possible noise cross-correlations results in a set of inter sub-array measurements of wave azimuths. Then a joint inversion of travel times and azimuths can be performed (Boué et al., 2014). A systematic application of double beamforming between different sub-arrays also makes it possible to better separate surface waves from the noise and body waves and to increase the amount and improve the accuracy of the travel time measurements (Roux et al., 2016; see also Chapter 8).

With a dense sub-wavelength sampling, an ensemble of waveforms extracted from interstation cross-correlations can be considered as a coherent wavefield (Figure 7.4b), and surface-wave travel time measurements across a large network can

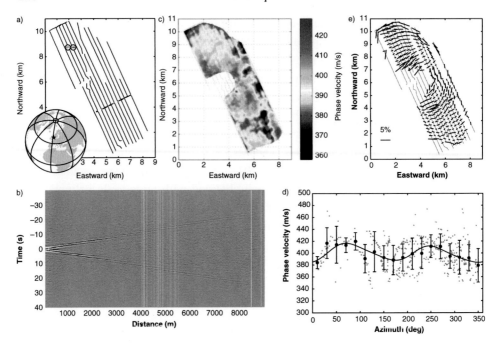

Figure 7.4. Ambient noise surface wave phase velocity tomography at Valhall oil field (Mordret et al., 2013c) revealing strong azimuthal anisotropy (Mordret et al., 2013b). (a) Map of the Valhall LoFS network. Each dot is a 4C sensor. The distance between the sensors is 50 m and is 300 m between the lines. In the inset, the location of the Valhall field is shown by the star. (b) Vertical-vertical component correlations between station 595 (position shown in (a)) and all other stations of the Valhall network filtered between 0.6 and 3 s and sorted with increasing inter-station distance. (c) Phase velocity map at 0.7 s obtained with the "Helmholtz tomography" based on equation (7.14). (d) Azimuthal distribution of the phase velocity at 0.7 s for the cell containing station 371 (position shown in (a)). The small gray dots are the phase velocity measurements for individual station pairs. The large black dots with error bars are the phase velocity averaged over 20°. The thick black curve is the fitted azimuthal variation for the averaged velocity measurements. (e) Fast directions and amplitudes of the azimuthal anisotropy at 0.7 s inferred from the "Helmholtz tomography" based on equation (7.14). (A black-and-white version of this figure appears in some formats. For the color version, please refer to the plate section.)

be used to estimate local travel time gradients. In a media with a smooth lateral heterogeneity, surface-wave phase travel times can be described with a ray-theory approximation (e.g., Woodhouse, 1974; Wang and Dahlen, 1995). In this case, an approximate 2D Eikonal equation can be used to estimate both the local phase velocity and the direction of wave propagation based on the local gradient of the phase travel times:

$$\frac{\bar{k}(r)}{C(\omega, r)} = \nabla\tau(\omega, r), \tag{7.11}$$

where r is the position, τ is the phase travel time, C is the phase velocity, ω is the frequency, and $\bar{k}$ is the unit vector describing the local direction of wave propagation. This gave rise to the Eikonal tomography (Lin et al., 2009) that has been used to infer the phase velocity distribution and the related azimuthal anisotropy at different scales and in many various settings (e.g., Lin et al., 2011a, 2013; Mordret et al., 2014). An example of regional-scale Eikonal tomography with the USArray data is shown in Figure 5.16.

Dense networks sample noise cross-correlations in time and space. This dense space sampling can be explored to measure local phase velocity based on the phase of the spectral correlation function (or spectral coherence function). This approach was first suggested by Keiiti Aki in 1957 with introduction of his well known SPAC method (Aki, 1957; see also Chapter 10). The idea of this method is that the spectral correlation function $CC(\omega)$ of a noise wavefield dominated by surface waves is related to the phase velocity C at frequency ω as:

$$CC(\omega) = J_0\left(\frac{\omega}{C(\omega)}r\right), \qquad (7.12)$$

where r is the interstation distance and J_0 is the zeroth-order Bessel function of the first kind. With a dense network, a range of distances (including those smaller than the wavelength) can be sampled and the phase velocity can be retrieved by equating the zero crossings of the real part of the spectrum of the noise cross-correlations with the zero crossings of the corresponding Bessel function. This approach has been successfully applied to the USArray data (Ekström et al., 2009) and to some smaller networks (e.g., Calkins et al., 2011).

A similar idea can be formulated slightly differently with so-called focal spot imaging (Hillers et al., 2016) that exploits the analogy between the noise cross-correlations and time reversal experiments (e.g., Derode et al., 2003). In this approach, amplitudes of the zero-lag time cross-correlations are interpreted as the focal spot of surface waves and its size is related to the local phase velocity. When applied to large and dense networks, this method can be used to estimate the phase-velocity maps.

7.2.3 Studies of Seismic Anisotropy Based on the Ambient Noise Surface Wave Tomography

While surface wave dispersion curves are primarily sensitive to study isotropic distributions of shear wave speed for crust and upper mantle applications, they can also be used to infer the information about the anisotropy. There are two main approaches to observing seismic anisotropy with surface waves. First, measurements of velocities of surface waves propagating in different directions can reveal the azimuthal anisotropy (Montagner and Nataf, 1986). This principle can be applied very efficiently in a case of travel times measured from interstation cross-correlations computed for large networks when the azimuthal distribution is well sampled. An example of such measurement at the Valhall oil field (Mordret et al., 2013b) is shown in Figure 7.4d. The local phase velocity clearly varies as a function of the azimuth θ in the form of an even-order sinusoid with $180°$ periodicity, as predicted by theoretical approximation for a slightly anisotropic media (e.g., Smith and Dahlen, 1973). This type of analysis can be applied at all stations of a large network to infer spatial variations of fast directions of propagation and of amplitudes of the azimuthal anisotropy (Figure 7.4e).

Another approach is to measure the "discrepancy" between the dispersion curves of Rayleigh and Love waves when they cannot be simultaneously explained with a simple isotropic model. This type of observation reflects the presence of the radial anisotropy that is the difference in speeds between the vertically and horizontally polarized elastic waves (e.g., Ekström and Dziewonski, 1998). In this case, the model velocity profile (equation (7.8)) is parameterized as:

$$m = \left[V_{SV}(z),\, V_{SH}(z)\right],\tag{7.13}$$

where V_{SV} and V_{SH} are velocities of vertically and horizontally polarized S-waves, respectively. An example of radial anisotropy inferred from the inversion of Rayleigh and Love wave group velocities beneath the Toba caldera in Indonesia (Jaxybulatov et al., 2014) is shown in Figures 7.3e–f.

The ANSWT has been widely used to study seismic anisotropy. The main difference of the noise-based studies compared to those based on earthquakes is that relatively short-period surface waves extracted from the ambient noise are sensitive to the crustal structure. Therefore, the ANSWT provided new information about the crustal anisotropy that has been difficult to study with earthquake-based methods. Examples include large-scale studies of the crustal anisotropy in the United States (e.g., Moschetti et al., 2010; Lin et al., 2011a) and China (e.g., Huang et al., 2010; Liu et al., 2016; Xie et al., 2017). At smaller scales, the ANSWT has been used to reveal the seismic anisotropy within volcanic systems (e.g., Jaxybulatov et al., 2014; Mordret et al., 2015) and within the overburden of oil and gas reservoirs (e.g., Mordret et al., 2013b; Tomar et al., 2017).

7.3 Using Surface-Wave Amplitudes and Waveforms Extracted from Seismic Noise

A direct interpretation of the noise correlation theorem when the noise cross-correlation is considered as a fully reconstructed Green's function would suggest that its amplitude could also be used to infer the properties of the Earth's crust and mantle. An important parameter, whose estimation requires measuring amplitudes, is the wave attenuation (quality factor). A few attempts to use noise correlations (or coherences) to estimate the distribution of the surface-wave attenuation have been made (e.g., Prieto et al., 2009; Lawrence and Prieto, 2011). At the same time, many studies demonstrated that the estimations of the attenuation from the noise cross-correlations could be very strongly biased by the unknown distribution of the noise sources (e.g., Cupillard and Capdeville, 2010; Lin et al., 2011b; Weaver, 2011; Stehly and Boué, 2017).

The direct interpretation of the noise correlation theorem also leads to the idea of directly inverting the extracted surface-wave waveforms based on their comparison with synthetic Green's functions computed in 3D Earth's models (e.g., Chen et al., 2014). This type of inversion is based, however, on an assumption that noise cross-correlation waveforms can be interpreted as signals emitted by well-localized force-like sources. This approximation can be acceptable when the corresponding misfit functions are based on waveform phases and when the amplitude information is ignored. A more accurate inversion of the noise correlation waveforms would require knowing the noise source distributions. In principle, a joint inverse problem for the media properties and the distribution of density of noise source can be formulated (e.g., Tromp et al., 2010; Fichtner et al., 2017; Sager et al., 2018). However, in many practical situations with most important sources of the noise contributing to cross-correlations lying outside the regions covered by the network, this kind of inverse problem remains poorly constrained.

Despite the difficulties with the methods based on the direct interpretation of the noise correlation theorem and/or with the inversion of the correlation waveforms and of their amplitudes, in some situations the amplitudes of the surface waves extracted from noise cross-correlations can be reasonably interpreted. One approach is to use the ellipticity of the Rayleigh waves. This ellipticity depends on frequency and is related to the local structure below the site where it is measured (e.g., Tanimoto and Rivera, 2008). A simple technique based on this idea has been introduced by Nakamura (1989) to study the site amplification of seismic waves when the horizontal to vertical spectra ratio (H/V) is measured directly from the noise (Chapter 10 [Hayashi, 2018]). However, with an arbitrary mixture of Rayleigh and Love waves in the noise, interpretation of such measurements becomes difficult (e.g., Bonnefoy-Claudet et al., 2006). This difficulty can be overcome by using noise cross-correlations of three-component records that allow us to separate Rayleigh and Love waves (e.g., Workman et al., 2017). Systematic application of this type of measurement to data of a dense network generates complementary information that can be added to travel times and helps to better constrain parameters such as density and Vp/Vs ratios (e.g., Lin et al., 2014) and the azimuthal anisotropy (Lin and Schmandt, 2014).

Although in the case of arbitrary source distribution the noise cross-correlations cannot be interpreted as media Green's functions (i.e., the response to a point force-like source), they still can be considered as representing a wavefield emitted by a virtual source with an unknown size and mechanism. An important property is that in a case of smoothly varying media this cross-correlation wavefield locally satis-fies the wave equation. When the Earth's noise is dominated by surface waves, the behavior of corresponding cross-correlations can be approximated with a 2D wave equation (e.g., Lin et al., 2013). With dense networks, the phase and amplitudes of this virtual wavefield can be interpreted simultaneously based on the full 2D Eikonal equation with the amplitude term (Lin and Ritzwoller, 2011):

$$\frac{1}{|C(\omega, r)|^2} = |\nabla \tau(\omega, r)|^2 - \frac{\nabla^2 A(\omega, r)}{A(\omega, r)\omega^2}, \tag{7.14}$$

where r is the position, τ is the phase travel time, C is the phase velocity, ω is the frequency, and A is the amplitude. On one side, this approach results in a more accurate interpretation of phase travel times with accounting for finite fre-quency effects. Examples of a phase velocity map and of a distribution of azimuthal anisotropy across the Valhall oil field inferred with the "Helmholtz" tomography based on equation (7.14) are shown in Figures 7.4c–e. Additionally, equation (7.14) provides us with a consistent framework for the interpretation of noise correlation amplitudes and makes it possible to infer information about their attenuation and local site amplification (Lin et al., 2012; Bowden et al., 2015).

7.4 Some Concluding Remarks

During the last decade, the Ambient Noise Surface-Wave Tomography (ANSWT) has been very rapidly developed and has become one of the most widely used approaches for seismic imaging at different scales. At the same time, the ANSWT is far from becoming a "standard" and "routine" method. As discussed in this chap-ter, the ANSWT regroups many possible approaches that can be applied depending on the properties of the seismic network used and on the distribution of noise sources. Overall, the ANSWT continues to be a very dynamic field of research with some first-order issues on which we need to work.

As in the case of all noise-based methods, the improvement of the ANSWT will require a better knowledge of the nature and of the distributions of noise sources. To advance in this direction, we need to work on further developing both the phys-ical models describing mechanisms of the noise generations and the methods to determine the effective source distributions from the analysis of seismic records. The next issue is the data pre-processing. We need to continue to improve the pre-processing methods and also to develop a better understanding of how they impact the final cross-correlations and the measured surface wave parameters. Most of the pre-processing methods used up to the current time were applied on records at individual stations and components and do not explore the internal coherences of different waves contained in the noise. These wavefield coherences can be explored with the array-based methods (e.g., Gallot et al., 2012; Carrière et al., 2014; Seydoux et al., 2017).

Finally, new schemes for the manipulation of the virtual wavefields and for the inversion of the surface waves reconstructed from noise cross-correlations will continue to emerge.

Acknowledgments

Supported by the Russian Ministry of Education and Science of the Russian Federation (grant N 14.W03.31.0033). I would like to thank Nori Nakata and two anonymous reviewers for their comments, which greatly improved this chapter.

References

Aki, K. 1957. Space and time spectra of stationary stochastic waves, with special reference to microtremors. *Bull. Earthq. Res. Inst. Tokyo Univ.*, **24**, 415–457.

Ardhuin, F., Gualtieri, L., and Stutzmann, E. 2018. Physics of ambient noise generation by ocean waves. Pages – of: Nakata, N., Gualtieri, L., and Fichtner, A. (eds.), *Seismic Ambient Noise*. Cambridge University Press, Cambridge, UK.

Ardhuin, F., Stutzmann, E., Schimmel, M., and Mangeney, A. 2011. Ocean wave sources of seismic noise. *J. Geophys. Res. Oceans*, **116**(C9), C09004.

Barmin, M. P., Ritzwoller, M. H., and Levshin, A. L. 2001. A fast and reliable method for surface wave tomography. *Pure Appl. Geophys.*, **158**(8), 1351–1375.

Biot, M. A. 1957. General theorems on the equivalence of group velocity and energy transport. *Phys. Rev.*, **105**(2), 1129–1137.

Bonnefoy-Claudet, S., Cornou, C., Bard, P.-Y., Cotton, F., Moczo, P., Kristek, J., and Donat, F. 2006. H/V ratio: a tool for site effects evaluation. Results from 1-D noise simulations. *Geophys. J. Int.*, **167**(2), 827–837.

Boué, P., Roux, P., Campillo, M., and Briand, X. 2014. Phase velocity tomography of surface waves using ambient noise cross correlation and array processing. *J. Geophys. Res. Solid Earth*, **119**(1), 519–529.

Bowden, D. C., Tsai, V. C., and Lin, F. C. 2015. Site amplification, attenuation, and scattering from noise correlation amplitudes across a dense array in Long Beach, CA. *Geophys. Res. Lett.*, **42**(5), 1360–1367.

Brenguier, F., Shapiro, N. M., Campillo, M., Nercessian, A., and Ferrazzini, V. 2007. 3-D surface wave tomography of the Piton de la Fournaise volcano using seismic noise correlations. *Geophys. Res. Lett.*, **34**(2), L02305.

Calkins, J. A., Abers, G. A., Ekström, G., Creager, K. C., and Rondenay, S. 2011. Shallow structure of the Cascadia subduction zone beneath western Washington from spectral ambient noise correlation. *J. of Geophys. Res.: Solid Earth*, **116**(B7).

Campillo, M. 2006. Phase and correlation in "random" seismic fields and the reconstruction of the Green function. *Pure Appl. Geophys.*, **163**(2-3), 475–502.

Carrière, O., Gerstoft, P., and Hodgkiss, W. S. 2014. Spatial filtering in ambient noise interferometry. *J. Acoust. Soc. Am.*, **135**(3), 1186–1196.

Chen, M., Huang, H., Yao, H., Hilst, R., and Niu, F. 2014. Low wave speed zones in the crust beneath SE Tibet revealed by ambient noise adjoint tomography. *Geophys. Res. Lett.*, **41**(2), 334–340.

Claerbout, J. F. 1968. Synthesis of a layered medium from its acoustic transmission response. *Geophysics*, **33**(2), 264–269.

Cupillard, P., and Capdeville, Y. 2010. On the amplitude of surface waves obtained by noise correlation and the capability to recover the attenuation: a numerical approach. *Geophys. J. Int.*, **181**(3), 1687–1700.

Dahlen, F. A., and Tromp, J. 1998. *Theoretical Global Seismology*. Princeton University Press, Princeton, NJ.

Derode, A., Larose, E., Campillo, M., and Fink, M. 2003. How to estimate the Green's function of a heterogeneous medium between two passive sensors? Application to acoustic waves. *Appl. Phys. Lett.*, **83**(15), 3054–3056.

Durand, S., Montagner, J. P., Roux, P., Brenguier, F., Nadeau, R. M., and Ricard, Y. 2011. Passive monitoring of anisotropy change associated with the Parkfield 2004 earthquake. *Geophys. Res. Lett.*, **38**(13), L13303.

Ekström, G., and Dziewonski, A. M. 1998. The unique anisotropy of the Pacific upper mantle. *Nature*, **394**(07), 168–172.

Ekström, G., Abers, G. A., and Webb, S. C. 2009. Determination of surfacewave phase velocities across USArray from noise and Aki's spectral formulation. *Geophys. Res. Lett.*, **36**(18), L18301.

Fichtner, A., and Tsai, V. 2018. Theoretical foundations of noise interferometry. Pages – of: Nakata, N., Gualtieri, L., and Fichtner, A. (eds.), *Seismic Ambient Noise*. Cambridge University Press, Cambridge, UK.

Fichtner, A., Stehly, L., Ermert, L., and Boehm, C. 2017. Generalized interferometry – I: theory for interstation correlations. *Geophys. J. Int.*, **208**(2), 603–638.

Gallot, T., Catheline, S., Roux, P., and Campillo, M. 2012. A passive inverse filter for Green's function retrieval. *J. Acoust. Soc. Am.*, **131**(1), EL21–EL27.

Garnier, J., and Papanicolaou, G. 2009. Passive sensor imaging using cross correlations of noisy signals in a scattering medium. *SIAM Journal on Imaging Sciences*, **2**(2), 396–437.

Gouédard, P., Stehly, L., Brenguier, F., Campillo, M., Colin de Verdière, Y., Larose, E., Margerin, L., Roux, P., Sánchez-Sesma, F. J., Shapiro, N. M., and Weaver, R. L. 2008. Cross-correlation of random fields: mathematical approach and applications. *Geophys. Prospect.*, **56**(3), 375–393.

Haned, A., Stutzmann, E., Schimmel, M., Kiselev, S., Davaille, A., and Yelles-Chaouche, A. 2016. Global tomography using seismic hum. *Geophys. J. Int.*, **204**(2), 1222–1236.

Hayashi, K. 2018. Near-surface engineering. Pages – of: Nakata, N., Gualtieri, L., and Fichtner, A. (eds.), *Seismic Ambient Noise*. Cambridge University Press, Cambridge, UK.

Herrmann, R. B. 2013. Computer programs in seismology: An evolving tool for instruction and research. *Seismol. Res. Lett.*, **84**(6), 1081–1088.

Hillers, G., Roux, P., Campillo, M., and BenZion, Y. 2016. Focal spot imaging based on zero lag crosscorrelation amplitude fields: Application to dense array data at the San Jacinto fault zone. *J. Geophys. Res. Solid Earth*, **121**(11), 8048–8067.

Huang, H., Yao, H., and van der Hilst, R. D. 2010. Radial anisotropy in the crust of SE Tibet and SW China from ambient noise interferometry. *Geophys. Res. Lett.*, **37**(21), L21310.

Jaxybulatov, K., Shapiro, N. M., Koulakov, I., Mordret, A., Landes, M., and Sens-Schonfelder, C. 2014. A large magmatic sill complex beneath the Toba caldera. *Science*, **346**(6209), 617–619.

Lawrence, J. F., and Prieto, G. A. 2011. Attenuation tomography of the western United States from ambient seismic noise. *J. Geophys. Res. Solid Earth*, **116**(B6), B06302.

Levshin, A. L., Yanovskaya, T. B., Lander, A. V., Bukchin, B. G., Barmin, M. P., Ratnikova, L. I., and Its, E. 1989. *Seismic Surface Waves in a Laterally Inhomogeneous Earth*. New York: Springer.

Liu, Z., Huang, J., and Yao, H. 2016. Anisotropic Rayleigh wave tomography of Northeast China using ambient seismic noise. *Phys. Earth Planet. Inter.*, **256**, 37–48.

Lin, F.-C., Li, D., Clayton, R. W., and Hollis, D. 2013. High-resolution 3D shallow crustal structure in Long Beach, California: Application of ambient noise tomography on a dense seismic array. *Geophysics*, **78**(4), Q45–Q56.

Lin, F., and Ritzwoller, M. H. 2011. Helmholtz surface wave tomography for isotropic and azimuthally anisotropic structure. *Geophys. J. Int.*, **186**(3), 1104–1120.

Lin, F., and Schmandt, B. 2014. Upper crustal azimuthal anisotropy across the contiguous U.S. determined by Rayleigh wave ellipticity. *Geophys. Res. Lett.*, **41**(23), 8301–8307.

Lin, F.-C., Moschetti, M. P., and Ritzwoller, M. H. 2008. Surface wave tomography of the western United States from ambient seismic noise: Rayleigh and Love wave phase velocity maps. *Geophys. J. Int.*, **173**(1), 281–298.

Lin, F., Ritzwoller, M. H., and Shen, W. 2011b. On the reliability of attenuation measurements from ambient noise crosscorrelations. *Geophys. Res. Lett.*, **38**(11), L11303.

Lin, F.-C., Ritzwoller, M. H., and Snieder, R. 2009. Eikonal tomography: surface wave tomography by phase front tracking across a regional broad-band seismic array. *Geophys. J. Int.*, **177**(3), 1091–1110.

Lin, F.-C., Ritzwoller, M. H., Yang, Y., Moschetti, M. P., and Fouch, M. J. 2011a. Complex and variable crustal and uppermost mantle seismic anisotropy in the western United States. *Nature Geoscience*, **4**(12), 55–61.

Lin, F., Tsai, V. C., and Ritzwoller, M. H. 2012. The local amplification of surface waves: A new observable to constrain elastic velocities, density, and anelastic attenuation. *J. Geophys. Res. Solid Earth*, **117**(B6), B06302.

Lin, F.-C., Tsai, V. C., and Schmandt, B. 2014. 3-D crustal structure of the western United States: application of Rayleigh-wave ellipticity extracted from noise cross-correlations. *Geophys. J. Int.*, **198**(2), 656–670.

McNamara, D., and Boaz, R. 2018. Visualization of the Seismic Ambient Noise Spectrum. Pages — of: Nakata, N., Gualtieri, L., and Fichtner, A. (eds.), *Seismic Ambient Noise*. Cambridge University Press, Cambridge, UK.

Molinari, I., Verbeke, J., Boschi, L., Kissling, E., and Morelli, A. 2015. Italian and Alpine threedimensional crustal structure imaged by ambientnoise surfacewave dispersion. *Geochem. Geophys. Geosyst.*, **16**(12), 4405–4421.

Montagner, J., and Nataf, H. 1986. A simple method for inverting the azimuthal anisotropy of surface waves. *J. Geophys. Res. Solid Earth*, **91**(B1), 511–520.

Mordret, A., Landès, M., Shapiro, N. M., Singh, S. C., and Roux, P. 2014. Ambient noise surface wave tomography to determine the shallow shear velocity structure at Valhall: depth inversion with a Neighbourhood Algorithm. *Geophys. J. Int.*, **198**(3), 1514–1525.

Mordret, A., Landès, M., Shapiro, N. M., Singh, S. C., Roux, P., and Barkved, O. I. 2013a. Near-surface study at the Valhall oil field from ambient noise surface wave tomography. *Geophys. J. Int.*, **193**(3), 1627–1643.

Mordret, A., Rivet, D., Landès, M., and Shapiro, N. M. 2015. Threedimensional shear velocity anisotropic model of Piton de la Fournaise Volcano (La Réunion Island) from ambient seismic noise. *J. Geophys. Res. Solid Earth*, **120**(1), 406–427.

Mordret, A., Shapiro, N. M., Singh, S. S., Roux, P., and Barkved, O. I. 2013c. Helmholtz tomography of ambient noise surface wave data to estimate Scholte wave phase velocity at Valhall Life of the FieldNoise tomography at Valhall. *Geophysics*, **78**(2), WA99–WA109.

Mordret, A., Shapiro, N. M., Singh, S., Roux, P., Montagner, J., and Barkved, O. I. 2013b. Azimuthal anisotropy at Valhall: The Helmholtz equation approach. *Geophys. Res. Lett.*, **40**(11), 2636–2641.

Moschetti, M. P., Ritzwoller, M. H., Lin, F., and Yang, Y. 2010. Seismic evidence for widespread western-US deep-crustal deformation caused by extension. *Nature*, **464**(04), 885–889.

Nakamura, Y. 1989. A method for dynamic characteristics estimation of subsurface using microtremor on the ground surface. *Q. Rep. Railw. Tech. Res. Inst.*, **30**, 25–30.

Nakata, N., and Nishida, K. 2018. Body wave exploration. Pages – of: Nakata, N., Gualtieri, L., and Fichtner, A. (eds.), *Seismic Ambient Noise*. Cambridge University Press, Cambridge, UK.

Nishida, K., Montagner, J.-P., and Kawakatsu, H. 2009. Global surface wave tomography using seismic hum. *Science*, **326**(5949), 112–112.

Prieto, G. A., Lawrence, J. F., and Beroza, G. C. 2009. Anelastic Earth structure from the coherency of the ambient seismic field. *J. Geophys. Res. Solid Earth*, **114**(B7), B07303.

Rawlinson, N., Hauser, J., and Sambridge, M. 2008. Seismic ray tracing and wavefront tracking in laterally heterogeneous media. *Advances in Geophysics*, **49**, 203–273.

Ritzwoller, M. H., and Feng, L. 2018. Overview of pre- and post-processing of ambient noise correlations. Pages – of: Nakata, N., Gualtieri, L., and Fichtner, A. (eds.), *Seismic Ambient Noise*. Cambridge University Press, Cambridge, UK.

Ritzwoller, M. H., Lin, F.-C., and Shen, W. 2011. Ambient noise tomography with a large seismic array. *Comptes Rendus Geoscience*, **343**(8), 558–570. Nouveaux développements de l'imagerie et du suivi temporel à partir du bruit sismique.

Ritzwoller, M. H., Shapiro, N. M., Barmin, M. P., and Levshin, A. L. 2002. Global surface wave diffraction tomography. *J. Geophys. Res. Solid Earth*, **107**(B12), ESE 4–1–ESE 4–13.

Ritzwoller, M. H., Shapiro, N. M., Levshin, A. L., and Leahy, G. M. 2001. Crustal and upper mantle structure beneath Antarctica and surrounding oceans. *J. Geophys. Res. Solid Earth*, **106**(B12), 30645–30670.

Rost, S., and Thomas, C. 2002. Array seismology: Methods and applications. *Rev. Geophys.*, **40**(3), 1008.

Roux, P., Moreau, L., Lecointre, A., Hillers, G., Campillo, M., Ben-Zion, Y., Zigone, D., and Vernon, F. 2016. A methodological approach towards high-resolution surface wave imaging of the San Jacinto Fault Zone using ambient-noise recordings at a spatially dense array. *Geophys. J. Int.*, **206**(2), 980–992.

Roux, P., Sabra, K., Kuperman, W., and Roux, A. 2005. Ambient noise cross-correlation in free space: theoretical approach. *J. Acoust. Soc. Am.*, **117**, 79–84.

Sabra, K. G., Gerstoft, P., Roux, P., Kuperman, W. A., and Fehler, M. C. 2005a. Extracting timedomain Green's function estimates from ambient seismic noise. *Geophys. Res. Lett.*, **32**(3), L03310.

———. 2005b. Surface wave tomography from microseisms in Southern California. *Geophys. Res. Lett.*, **32**(14), L14311.

Sager, K., Ermert, L., Boehm, C., and Fichtner, A. 2018. Towards full waveform ambient noise inversion. *Geophys. J. Int.*, **212**(1), 566–590.

Seydoux, L., de Rosny, J., and Shapiro, N. M. 2017. Pre-processing ambient noise cross-correlations with equalizing the covariance matrix eigenspectrum. *Geophys. J. Int.*, **210**(3), 1432–1449.

Shapiro, N., and Campillo, M. 2004. Emergence of broadband Rayleigh waves from correlations of the ambient seismic noise. *Geophys. Res. Lett.*, **31**(4), 7614.

Shapiro, N. M., and Ritzwoller, M. H. 2002. MonteCarlo inversion for a global shear velocity model of the crust and upper mantle. *Geophys. J. Int.*, **151**(1), 88–105.

Shapiro, N. M., Campillo, M., Stehly, L., and Ritzwoller, M. H. 2005. High-resolution surface-wave tomography from ambient seismic noise. *Science*, **307**(5715), 1615–1618.

Shen, W., and Ritzwoller, M. H. 2016. Crustal and uppermost mantle structure beneath the United States. *J. Geophys. Res. Solid Earth*, **121**(6), 4306–4342.

Shen, W., Ritzwoller, M. H., Kang, D., Kim, Y., Lin, F.-C., Ning, J., Wang, W., Zheng, Y., and Zhou, L. 2016. A seismic reference model for the crust and uppermost mantle beneath China from surface wave dispersion. *Geophys. J. Int.*, **206**(2), 954–979.

Smith, M. L., and Dahlen, F. A. 1973. The azimuthal dependence of Love and Rayleigh wave propagation in a slightly anisotropic medium. *J. Geophys. Res.*, **78**(17), 3321–3333.

Snieder, R. 2004. Extracting the Green's function from the correlation of coda waves: A derivation based on stationary phase. *Phys. Rev. E*, **69**(4), 046610.

Spica, Z., Perton, M., and Legrand, D. 2017. Anatomy of the Colima volcano magmatic system, Mexico. *Earth Planet. Sci. Lett.*, **459**, 1–13.

Spica, Z., Perton, M., Calò, M., Legrand, D., Córdoba-Montiel, F., and Iglesias, A. 2016. 3-D shear wave velocity model of Mexico and South US: bridging seismic networks with ambient noise cross-correlations (C1) and correlation of coda of correlations (C3). *Geophys. J. Int.*, **206**(3), 1795–1813.

Stehly, L., Campillo, M., and Shapiro, N. M. 2006. A study of the seismic noise from its long-range correlation properties. *J. Geophys. Res. Solid Earth*, **111**(10), 1–12.

Stehly, L., Fry, B., Campillo, M., Shapiro, N. M., Guilbert, J., Boschi, L., and Giardini, D. 2009. Tomography of the Alpine region from observations of seismic ambient noise. *Geophys. J. Int.*, **178**(1), 338–350.

Stehly, L., and Boué, P. 2017. On the interpretation of the amplitude decay of noise correlations computed along a line of receivers. *Geophys. J. Int.*, **209**(1), 358–372.

Tanimoto, T., and Rivera, L. 2008. The ZH ratio method for longperiod seismic data: sensitivity kernels and observational techniques. *Geophys. J. Int.*, **172**(1), 187–198.

Tomar, G., Shapiro, N. M., Mordret, A., Singh, S. C., and Montagner, J.-P. 2017. Radial anisotropy in Valhall: ambient noise-based studies of Scholte and Love waves. *Geophys. J. Int.*, **208**(3), 1524–1539.

Tromp, J., Luo, Y., Hanasoge, S., and Peter, D. 2010. Noise crosscorrelation sensitivity kernels. *Geophys. J. Int.*, **183**(2), 791–819.

Wang, Z., and Dahlen, F. A. 1995. Validity of surface-wave ray theory on a laterally heterogeneous earth. *Geophys. J. Int.*, **123**(3), 757–773.

Wapenaar, K. 2004. Retrieving the elastodynamic Green's function of an arbitrary inhomogeneous medium by cross correlation. *Phys. Rev. Lett.*, **93 25**, 254301.

Weaver, R. L. 2011. On the amplitudes of correlations and the inference of attenuations, specific intensities and site factors from ambient noise. *Comptes Rendus Geosci.*, **343**(8), 615–622. Nouveaux développements de l'imagerie et du suivi temporel à partir du bruit sismique.

Weaver, R. L., and Lobkis, O. I. 2001. Ultrasonics without a source: Thermal fluctuation correlations at MHz frequencies. *Phys. Rev. Lett.*, **87**(13), 134301 1–4.

Woodhouse, J. H. 1974. Surface waves in a laterally varying layered structure. *Geophys. J. R. Astron. Soc.*, **37**(3), 461–490.

———. 1988. The calculation of eigenfrequencies and eigenfunctions of the free oscillations of the Earth and the sun. Pages 321–370 of: Doornbos, D. J. (ed), *Seismological. Algorithms, Computational Methods and Computer Programs*. Academic, London, UK.

Workman, E., Lin, F.-C., and Koper, K. D. 2017. Determination of Rayleigh wave ellipticity across the Earthscope Transportable Array using single-station and array-based processing of ambient seismic noise. *Geophys. J. Int.*, **208**(1), 234–245.

Xie, J., Ritzwoller, M. H., Shen, W., and Wang, W. 2017. Crustal anisotropy across eastern Tibet and surroundings modeled as a depth-dependent tilted hexagonally symmetric medium. *Geophys. J. Int.*, **209**(1), 466–491.

Yang, Y., Ritzwoller, M. H., Levshin, A. L., and Shapiro, N. M. 2007. Ambient noise Rayleigh wave tomography across Europe. *Geophys. J. Int.*, **168**(1), 259–274.

Yang, Y., Ritzwoller, M. H., Lin, F., Moschetti, M. P., and Shapiro, N. M. 2008. Structure of the crust and uppermost mantle beneath the western United States revealed by ambient noise and earthquake tomography. *J. Geophys. Res. Solid Earth*, **113**(B12), B12310.

Yao, H., Beghein, C., and Hilst, R. D. V. D. 2008. Surface wave array tomography in SE Tibet from ambient seismic noise and twostation analysis – II. Crustal and uppermantle structure. *Geophys. J. Int.*, **173**(1), 205–219.

Yao, H., Hilst, R. D. V. D., and Hoop, M. V. D. 2006. Surfacewave array tomography in SE Tibet from ambient seismic noise and twostation analysis – I. Phase velocity maps. *Geophys. J. Int.*, **166**(2), 732–744.

Yoshizawa, K., and Kennett, B. L. N. 2002. Determination of the influence zone for surface wave paths. *Geophys. J. Int.*, **149**(2), 440–453.

8

Body Wave Exploration

NORI NAKATA AND KIWAMU NISHIDA

Abstract

Surface and body waves are two different types of waves. The former propagate along the surface and the latter inside media. In elastic media, propagation velocities of the surface and body waves are strongly and weakly dispersive in frequencies, respectively. However, they are not completely independent, or they are indeed strongly related. We can describe surface waves based on multiply reflected body waves (i.e., reverberations). Summations of normal modes can be used for representing both surface and body waves. In this chapter, we discuss how we can find body waves in ambient noise, how to use them, what we can do with them, and what is different compared to surface waves. Because of the amount of energy, coherency of waves, and excitation mechanisms of ambient noise, most applications of seismic interferometry use surface waves, especially fundamental-mode Rayleigh waves, and fewer studies successfully extract body-wave exploration. We must, moreover, consider carefully the adequacy of the basic assumptions for seismic interferometry (e.g., equipartition of the wavefield), because the body waves are more transparent (less scattered) than the surface waves. Nevertheless, we consider that body-wave retrieval from ambient-noise correlations is an important step to improve the spatial resolution in deeper parts and extend the depth sensitivity of seismic response of the medium. One of keys for body-wave extraction is how we can smartly use the spatial coherency of body waves. Due to the increase in the number of seismometers, computational resources, and our understanding of ambient-noise correlation functions, we can reach the stage of successfully reconstructing body waves that are useful for understanding Earth structure.

8.1 Introduction

Seismic interferometry is a powerful tool to extract Earth response from complicated wavefields such as background ambient noise at various scales (e.g., Curtis et al., 2006). Using this technique, we can theoretically extract Green's functions between receivers (Lobkis and Weaver, 2001; Snieder, 2004; Wapenaar and Fokkema, 2006), and can reveal seismic velocity and attenuation structures in the subsurface (e.g., Shapiro et al., 2005; Lin et al., 2009; Lawrence and Prieto, 2011) as well as monitor time-lapse variations of them (e.g., Brenguier et al., 2008; Mehta et al., 2008; Mainsant et al., 2012). Although the theory of seismic interferometry is not limited to either body or surface waves (Aki, 1957; Claerbout, 1968; Wapenaar, 2004) and we can potentially reconstruct full Green's functions from ambient seismic fields, surface waves are generally easier to extract with seismic interferometry because their energy is dominant in the ambient-noise fields.

Despite successes of surface-wave extraction, only a few studies have retrieved body waves from continuous ground motion records and used them for imaging and monitoring subsurface structures; however, body waves provide information that is different from what we can learn from surface waves. The sensitivities of surface and body waves to the structure are different. Surface waves propagate along the surface, and the depth sensitivity follows the exponential decay of their amplitudes as an evanescent wave (Zhou et al., 2004). On the other hand, body waves propagate in 3D, and these waves hold the information of the Earth's structure along the paths, which penetrate into the deep Earth (Dahlen and Baig, 2002). Therefore, body waves can be more sensitive than surface waves to finer structure in the deeper part. With this fact, for example for tomography, we can invert surface and body waves separately (Boschi and Dziewonski, 1999) or jointly (Liu and Zhao, 2016) to obtain more accurate structural models.

In this chapter, we first discuss characteristics of random wavefields. We then discuss body-wave exploration in ambient noise with a comparison to surface-wave extraction, which includes noise sources, discovery of them with beamforming analyses, and their wave phenomena in subsurface structure. Next, we focus on the body-wave retrieval between receivers using cross-correlation and suggest additional processing steps to enhance the signal-to-noise ratio (SNR) of weak body waves in ambient noise. Then we introduce several key applicational studies of body-wave exploration in different scales and discuss possible future research opportunities.

8.2 Keys for Characterizing the Random Wavefield

A reconstruction of the full Green's function including body and surface waves requires random wavefields excited by seismic sources on a closed surface surrounding the receivers (e.g., Wapenaar et al., 2006). For the reconstruction of the direct surface wave, it is sufficient to have a source anywhere in their *stationary zone* (or stationary phase region; see Chapter 4 for details) on the Earth's surface along the major arc between the pair of receivers (Figure 7.1 in Shapiro, 2018). For the reconstruction of the body waves, however, the source must be at the appropriate stationary zones, most of which are buried in the Earth's interior (Figure 7.1 in Shapiro, 2018). It is difficult to meet this condition from a practical point of view (Forghani and Snieder, 2010) due to the fact that the dominant excitation sources of ambient noise are on the Earth's surface (e.g., ocean swell). In general, body-wave amplitudes are underestimated by seismic interferometry using ambient noise, because fewer sources are located within their stationary zone at depths.

For understanding body-wave retrievals from the random wavefield using seismic interferometry, the physics of the excitation (see also Chapter 3) and the propagation is important. This section shows a brief summary of key factors for the characterizations: (1) the equivalent body force system of the source, (2) locations of source, and (3) the scattering properties of the medium.

First, let us consider the equivalent force system of the excitation sources of ambient noise. On Earth, the typical frequency of the ambient noise ranges from 1 mHz (= 10^{-3} Hz) to 100 Hz. Below 1 Hz, the dominant sources of this wavefield are oceanic gravity waves (e.g., Nishida, 2017). These signals are stationary stochastic within approximate time scales of several hours. Based on the typical frequencies of these wavefields, they are categorized into seismic hum from 1 to 20 mHz, primary microseisms (PM) from 0.02 to 0.1 Hz, and secondary microseisms (SM) from 0.1 to 1 Hz. Above 1 Hz, ambient seismic wavefields are linked more to human activities (e.g., Bonnefoy-Claudet et al., 2006). Below 0.1 Hz, the force system characterized by random shear traction sources on the seafloor is dominant (Nishida, 2014). They can be explained by topographic coupling between ocean swell and seismic waves at shallow depths (Ardhuin et al., 2015; Hasselmann, 1963). On the other hand, above 0.1 Hz, the excitation sources for SM can be characterized by the vertical single force on the sea surface known as the Longuet-Higgins mechanism (Longuet-Higgins, 1950; Kedar et al., 2008; Gualtieri et al., 2013) (see Chapter 3 for details).

Because the P-wave radiation from the horizontal single force in PM is stronger horizontally than vertically due to the radiation pattern, P-wave microseisms in local and regional scales are dominant. On the other hand, the vertical single force in SM radiates stronger P-waves into the vertical direction, which propagates in teleseismic distance, so teleseismic P-wave microseisms are dominant in this frequency.

Next, we consider source localization of microseisms. Strong excitation areas of SM tend to be localized where the standing part of ocean swell activities is high because the excitation amplitudes of SM are proportional to the square of the primary ocean wave height according to the Longuet-Higgins mechanism. Because excitation amplitudes of PM are roughly proportional to the primary ocean wave height, the SM amplitudes are proportional to the square of PM amplitudes.

Observed amplitude ratios between PM and SM at coastal stations also support the model of excitation (Nishida, 2017). The excitations of SM are amplified by the source-side site effect due to acoustic resonance of the water layer (Gualtieri et al., 2014), which can be also interpreted as water reverberations. The strong excitation area favors the resonant water depth as shown by Nishida (2017).

In spite of the source-side site effect of SM, the sources exist in both pelagic and coastal regions. On the other hand, the excitation sources of PM are localized along coastal regions, because the most probable excitation mechanism of PM is linear topographic coupling between seismic waves and ocean swell at shallow depth along coastal lines.

A strong localized source outside the stationary zones can cause spurious arrivals resulting from imperfect cancellations of nonphysical amplitudes for seismic interferometry (Snieder et al., 2008). We note that the near-surface sources even in the stationary zones can cause spurious arrivals before direct P-waves. They are originated from cross-terms between surface waves and body waves due to the cross-correlations (Forghani and Snieder, 2010). The cross-term cannot be canceled out when the angular coverage of the sources is limited on the surface.

Last, let us consider the effects of scattering during seismic-wave propagations. If the medium is less scattered (transparent), randomness of the wavefield is mostly controlled by the source characteristics. The transparency of the medium for body waves strongly depends on their frequency ranges. With increasing frequency, the scattering tends to randomize the wavefield.

For example, the scattering mean free path of teleseismic body waves in the mantle is on the order of 1000 km (Sato et al., 2012). When the distance between a station and the source area of microseisms is longer than 3000 km, P-wave microseisms are generally dominant above 0.2 Hz, because the corresponding surface waves attenuate due to the strong scattering in the crust with mean free path on the order of 100 km (Hillers et al., 2013). In this case, the teleseismic P-wave holds the information of the localized source (Nishida, 2017), because the teleseismic P-wave is less scattered in the mantle. On the other hand, a regional P-wave of the PM is mostly trapped in the crust due to the radiation pattern. The azimuthal distribution of body-wave propagations tends to be homogenized due to the scattering properties in the crust.

Earthquake coda wave, which is generated by multiple scattering due to the heterogeneities of the medium during the propagations, is also another source for random seismic wavefield. The coda waves can be studied with similar techniques although they are not the focus of this book. The strong heterogeneities in the crust and the upper mantle cause the multiple scattering. The coda wave of teleseismic body waves is feasible for retrieving coherent body waves using seismic interferometry caused by the scattering points, which can be recognized as random sources outside a closed surface surrounding the receivers (Tonegawa et al., 2009). The angular coverage of the sources on the closed surface is important for the better reconstruction of body-wave propagations. Careful processing is necessary below 0.04 Hz for body-wave retrieval, because late coda ($\sim$ 10 hours from the earthquake) affects the body-wave reconstruction from ambient noise (Boué et al., 2014).

8.3 Characteristics of Body Wave Microseisms

In this section we discuss characteristics of body wave microseisms. For example, beamforming results (see Chapter 2 for details) above 0.05 Hz in Japan (Figure 8.1) show dominance of the fundamental-mode surface waves. The spectra of the radial component show a corresponding concentric circle in the slowness of crustal P-waves (Pg), whereas the spectra of vertical components show a weak first over-tone branch of Rayleigh waves. The spectra of the transverse component show the first and second overtones of Love waves above 0.1 Hz. The energy of these higher modes is trapped in the crust, which can be also categorized as "trapped body wave." Because they are trapped in the shallow layer, they can be also interpreted as superposition of "normal mode." The beamforming results in RR and ZZ components above 0.1 Hz also show bright spots, which are teleseismic body waves categorized as "diving wave." Here, we summarize seismic observations of body wave microseisms in past studies.

8.3.1 Body Waves Trapped in the Crust and/or Sediment

The Pg wave, which was trapped in the crust, shows less heterogeneous azimuthal coverage in the beamforming results in RR components above 0.1 Hz. This is because the wavefield is nearly in 2D (Figure 8.1). Scattering due to the strong lateral heterogeneities in the crust tends to homogenize the energy partitions during the propagation. Although the azimuthal distribution shows slight two- and four-lobed patterns, they are still suitable for seismic interferometry.

Trapped body waves and overtones of surface waves were also observed in other regions. Beamforming analyses of SM at inland arrays in the United States (Toksöz and Lacoss, 1968; Haubrich and McCamy, 1969) showed the dominance of over-tones above 0.2 Hz. The first overtone of Rayleigh waves was also revealed from 0.14 to 0.25 Hz in New Zealand (Brooks et al., 2009). A beamforming analysis of vertical seismograms recorded at an inland array, which spans 25 km in Australia (Gal et al., 2015), shows dominance of fundamental Rayleigh waves from 0.3 to 0.5 Hz, and dominance of overtones from 0.5 to 0.7 Hz. Cross-correlation analyses of SM revealed Moho-reflected P-waves (0.5–1 Hz) at Yellowknife in north Canada

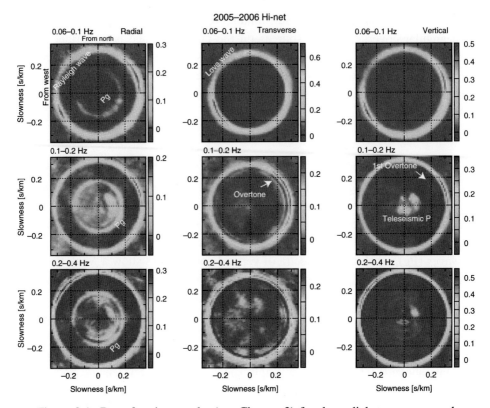

Figure 8.1. Beamforming results (see Chapter 2) for the radial, transverse, and vertical components in three frequency ranges (0.06–0.1 Hz, 0.1–0.2 Hz, and 0.2–0.4 Hz) using Hi-net 1 Hz velocitymeters in Japan. The beamforming results show observed dominance of Love and Rayleigh waves. First, daily beamforming results were calculated. From 0.06 to 0.1 Hz, we used about 700 stations, whereas in the other frequency ranges, we used about 200 stations in the western part of Japan. After computing daily beamforming results, we normalize the daily spectrum by the maximum in the slowness–slowness domain, and stack them over two years from 2005 to 2006. (A black-and-white version of this figure appears in some formats. For the color version, please refer to the plate section.)

and at Kimberley in south Africa (Zhan et al., 2010), and both Moho-reflected P- and S-waves in the northern Fennoscandian region (0.5–2 Hz) (Poli et al., 2012b). An autocorrelation analysis revealed Moho-reflected P-waves (0.5–1 Hz) at the Sierra Nevada in the United States (Tibuleac and von Seggern, 2012).

Shear-coupled PL (S-PL) waves (Oliver, 1961) revealed by the cross-correlation analysis of PM (Nishida, 2013) is another type of trapped body wave, which are leaking P-waves in the crustal wave coupled with S-waves in the uppermost mantle. The excitation can be explained by the shear traction source.

8.3.2 Diving Wave: Teleseismic Body Waves

Observation of the teleseismic P-waves of SM is ubiquitous in every region (Nishida, 2017). For example, an inland array in Kazakhstan (Vinnik, 1973) shows dominance of teleseismic body waves between 0.15 and 0.3 Hz, when the source-receiver distance is longer than 4000 km; especially when the distance between a station and the source area of microseisms is longer than 3000 km, P-wave microseisms are generally dominant above 0.2 Hz. The dominance of teleseismic body waves is attributed to differences in the propagation characteristics between body and surface waves. The teleseismic body waves are less scattered because they travel into deeper mantle structures, which are less heterogeneous. For example, the scattering mean free path of teleseismic body waves in the mantle is on the order of 1000 km (Sato et al., 2012). In contrast, surface waves in these frequencies are attenuated by scattering during the propagation because of strong lateral heterogeneities in the crust, with a mean free path on the order of 100 km (Hillers et al., 2013). Even at coastal stations, teleseismic body waves become significant as shown in Figure 8.1.

The detection of teleseismic P-wave microseisms with frequencies below 0.1 Hz is more challenging because of weaker amplitudes, particularly in the frequency ranges of seismic hum and primary microseisms (Nishida, 2013, 2014; Boué et al., 2013). Among the observed body waves, shear-coupled PL waves are dominant, and weaker phases such as P, PKP, PcP, and SH can also be detected in the spatio-temporal domain below 0.1 Hz (Nishida, 2013). These observations can be attributed to the force system, which is characterized by random shear traction on the seafloor.

8.4 One More Step for Body-Wave Extraction After Cross-Correlation

Because the body waves in ambient noise are weaker, azimuthally heterogeneous, or limited in some frequency range, extraction of them between stations by using ambient-noise seismic interferometry is challenging. Here we list several additional processing steps to enhance SNR of extracted body waves. One of the common ideas of the processing techniques discussed below is to use the spatial coherency of body waves among receivers. Because the signal of body waves at each correlation function is weak, averaging over long time periods might not be enough for the body-wave extraction, and spatial averaging further strengthens the signals and suppresses unwanted noise or off-target waves. Some of these processing steps can be used for surface-wave extraction as well. Also, we can apply them before cross-correlation if the filters are linear, but it is computationally more expensive.

8.4.1 Binned Stack

If we assume that the Earth structure can be represented as a 1D structure, which is often a good assumption for the global scale (Shearer, 1991), we can use binned stack as spatial averaging. When $u(\mathbf{x}_r, \mathbf{x}_s, t)$ indicates the cross-correlation functions between receivers at $\mathbf{x}_r$ and $\mathbf{x}_s$, the binned correlation function $\hat{u}$ can be expressed as

$$\hat{u}(D, t) = \sum_{D-d_0/2 < d \le D+d_0/2} u(\mathbf{x}_r, \mathbf{x}_s, t), \qquad (8.1)$$

where D is the discrete distance after binning;

$$D \equiv \text{floor}\left(\frac{d(\mathbf{x}_r, \mathbf{x}_s) + d_0/2}{d_0}\right),$$

d is the offset between $\mathbf{x}_r$ and $\mathbf{x}_s$, d_0 is the size of the bin, and floor is the floor function. With the binned stack, we average over all correlation functions that have the receiver offset in between $D - d_0/2$ and $D + d_0/2$. Often one needs to normalize the amplitude by the number of traces in each bin after the binned stack because the number of traces varies at each bin. When we use 2D receivers on the ground surface, the correlation functions u are in 5D such as 2D in $\mathbf{x}_r$, 2D in $\mathbf{x}_s$, and 1D in time. Using the binned stack, 5D cross-correlation data becomes 2D data, and if the assumption works well, we increase the stacking number and SNR. Also, the binned stack makes visualization easier. To smooth the wavefields between bins, we can also use a linear interpolation, which splits the data into nearby two bins with proper weights (Claerbout, 2014). In equation (8.1), we ignore the lateral variations of correlation functions for averaging and assume the 1D structure as discussed, so that the subsurface structure in the horizontal directions is averaged and homogenized after the binned stack. Therefore, this stack is not useful for 3D structural imaging. We can keep the lateral variations by applying this stack into subgroups of the array, or use the double beamforming discussed in the next section.

Binned stack is very powerful when many stations are available; for example, geophone arrays on local scales (e.g., industrial arrays), regional high-quality arrays (e.g., USArray and Hi-net), and concatenation of multiple global broadband networks. Using a 2500 vertical-component geophone array, we find clear P-wave signals propagating with the velocity of 2.0 km/s (Figure 8.2). These signals are weak and not visible in each correlation function (Figure 8.2b). Because this area is at the Los Angeles basin with sedimentary layers, the assumption of the 1D subsurface structure is a good approximation.

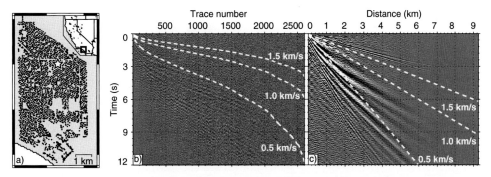

Figure 8.2. Example to improve the SNR after binned stack. (a) Location of geophones used at Long Beach, California. The black dots show the location of 2500 vertical-component geophones, and the white star indicates the reference receiver for cross-correlation used in panel (b). (b) Cross-correlation gather. The trace numbers are sorted by the interreceiver distance from the reference receiver. (c) Bin-stacked gather using all correlation pairs (i.e., six million pairs). The size of bin is 50 m. The white dashed lines indicate travel times with constant velocities. Modified after Nakata et al. (2015).

8.4.2 Double Beamforming

One of the assumptions in the binned stack is the lateral homogeneity of the structure. When we are interested in the lateral complexity (e.g., scattering of higher frequency waves) or if the structure is laterally complicated (e.g., volcanoes, subduction zone), this assumption does not hold well. In the complex structure, identifying the type of wavefields and extracting on-target waves in these structures are not trivial because the coherency of wavefields decreases.

Assuming that we have a lot of receivers as an array and make clusters of receivers as a sub-array (Figure 8.3), double beamforming (DBF) techniques are powerful for enhancing SNR of target waves between two sub-arrays. With cross-correlation techniques, one sub-array can be considered as a source array, and waves propagate from the source array to the other sub-array. As reviewed in Chapter 2, beamforming is useful to identify different wave types and their incoming azimuths. We can apply this technique to either receiver or source arrays, and in DBF, we compute beamforming simultaneously at both receiver and source arrays (Weber and Wicks, 1996; Rost and Thomas, 2002). DBF has been used for identifying global reflections (Krüger et al., 1993; Poli et al., 2015), understanding local- and laboratory-scale wave phenomena (De Cacqueray et al., 2011; Nakata et al., 2016), multi-path shallow-water tomography (Roux et al., 2008, 2011), and surface-wave tomography (Boué et al., 2014; Roux et al., 2016).

Similar to the binned stack, we start with the correlation functions (u) obtained from continuous ambient-noise records. Assuming that $\mathbf{x}_r$ and $\mathbf{x}_s$ represent receiver

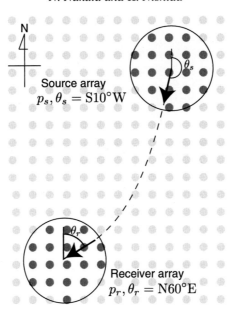

Figure 8.3. Coordinate for DBF. The azimuths are defined from the North clockwise. The light gray dots show the location of receivers, and the dark gray the receivers used for DBF analyses. The black arrows are the angle on source and receiver arrays, and the dashed line indicates the expected ray path (e.g., shortest path). The black circle represents the array size. The size of the arrays or the number of receivers does not have to be the same on source and receiver arrays.

locations in two sub-arrays, we can apply array signal processing for both subarrays. Usually, the receivers are distributed on the ground in 1D or 2D, and sometimes 3D, for example at mines; the dimensions of **x** correspond to the dimensions of receivers. Here we consider that the receivers are distributed in 2D as the most common case. We can use various types of beamforming for DBF, such as those based on plane-wave projection (2D *tau-p*), Capon, and MUSICAL (Le Touzé et al., 2012). To transform u to the double-beam domain, we scan the slowness and azimuth domain (or slowness–slowness domain) at the sub-arrays by computing

$$B(p_r, \theta_r, p_s, \theta_s, t) = \frac{1}{N_r N_s} \sum_{\mathbf{x}_r} \sum_{\mathbf{x}_s} u(\mathbf{x}_r, \mathbf{x}_s, t + \tau_r(\mathbf{x}_r, p_r, \theta_r) - \tau_s(\mathbf{x}_s, p_s, \theta_s)),$$

(8.2)

where B is wavefields in the double-beam domain, τ is the time lag that is a function of the location of receivers in each sub-array of r and s, slowness (p), and azimuth (θ), and N is the number of receivers in each array. Here, the azimuths at the source and receiver arrays are defined by the outgoing and incoming waves relative to the north direction, respectively (see Figure 8.3). The time lag τ is a relative time delay from a reference point for each plane wave. We set the reference at the center of each array ($\mathbf{x}^c$), and thus the time lag is defined as

$$\tau_{r/s} = p_{r/s}(\mathbf{x}_{r/s} - \mathbf{x}_{r/s}^c) \cdot \begin{pmatrix} \sin\theta_{r/s} \\ \cos\theta_{r/s} \end{pmatrix}. \tag{8.3}$$

Due to the summation over all combinations of receivers (equation (8.2)), we improve the SNR by at most $\sqrt{N_r N_s}$ compared to the receiver-by-receiver correlation (u) when the signal consists of plane waves and the noise is white. Therefore, a large number of receivers is helpful to enhance target waves.

Because we assume the receivers are distributed in 2D (on the ground surface), u and B are 5D functions. When the receivers are in 1D or 3D, the dimensions of beams should be 3D or 7D, respectively. The slowness measured with 2D arrays on the surface is the inverse of the phase velocity for surface waves and of the local apparent velocity for body waves.

Dense arrays have an important role for DBF to avoid aliasing artifacts (Roux et al., 2008), as similar to the single beamforming (Chapter 2) or Fourier transform. The Nyquist wavenumber k_n is defined as

$$k_n = 2\pi p_n f_n = \pi/\Delta x, \tag{8.4}$$

where f is the frequency and Δx is the minimum station spacing. For example, when the receiver spacing is $\Delta x = 2$ km and our target frequency is up to 0.5 Hz, the spatial aliasing occurs when $pf > 0.5$ km^{-1}. The slowness resolution Δp is given by

$$\Delta pf = 1/(2X), \tag{8.5}$$

where X is the array size (e.g., maximum offset of receivers).

By specifying the azimuth and/or slowness of extracted wavefields by applying DBF to arrays of receivers, we can improve the SNR of the waves. Figure 8.4 shows a field-data example of DBF to retrieve P-waves propagating through a volcanic structure. Compared to the one-receiver-pair correlation, the binned stack uses much more receiver pairs (e.g., 49×49 functions are used in Figure 8.4c), and hence the extracted waves have much higher SNR. For example, we can extract well the body wave at 1.7 s from array C to array B. For the wave between C and A, however, the process of the binned stack is not enough to extract coherent waves, and DBF shows prominent coherency of hourly waves. Binned stack can be considered as a special case of DBF, where we assume vertically incident waves (i.e., the apparent velocity is infinite). Therefore, when we specify appropriate azimuth and velocity, DBF works better than binned stack, although we use the same number of receiver pairs for these process methods. We still have strong spurious waves

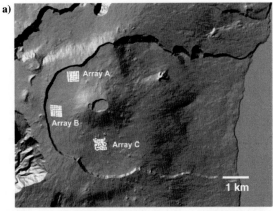

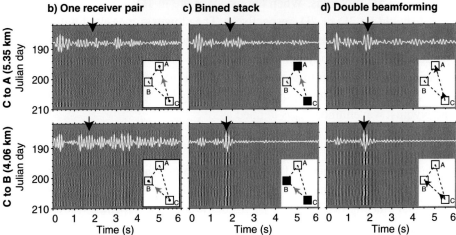

Figure 8.4. Example to improve the SNR after DBF. (a) Geometry of receiver arrays at the Piton de la Fournaise volcano on La Réunion island. Each array contains 49 receivers and the receiver spacing is about 80 m. (b) Hourly cross-correlation functions (gray-scale image) between two receivers shown by the black dot in the inset, and the averaged function over 30 days (white line). (c,d) Same as the panel (b), but using all receivers in the arrays and computing (c) binned stack and (d) DBF. For DBF, the azimuth of the direct path and velocity greater than 1.5 km/s are used. The top and bottom rows show the correlation functions from arrays C to A and arrays C to B, respectively. The black arrows at the top of each panel highlight the direct body waves between arrays. Modified after Nakata et al. (2016).

around 0.5 s, which are caused by the noise sources located on the east side of the island, and which are too strong to suppress by the DBF analysis (as we discussed above, ideal improvement of SNR is $\sqrt{N_r N_s}$). Since the fluctuation of location and power of noise sources causes the variation of amplitudes of hourly correlations, weak signals are observed on, for example, days 196–198.

Instead of computing DBF, or single beamforming, in the time-slowness domain, we can apply a filter in the FK domain. Note that the wavenumber is the product of slowness and angular frequency. FK filtering is feasible for enhancing SNR of the body-wave propagations (Yilmaz, 2001). When we consider only stochastic stationary surface-wave propagations, the cross-spectra (frequency-domain representations of cross-correlations) at a frequency against the receiver offsets can be represented by a Bessel function in Cartesian coordinates or a Legendre function in spherical coordinates. This method is known as Aki's spatial autocorrelation method (SPAC) (Aki, 1957; Hayashi, 2018). As a natural extension of SPAC, the body-wave propagations can be represented by a superposition of Bessel functions, which gives us the FK spectra (Nishida, 2013). The cross-correlations can be interpreted as the representation of the FK spectrum in the spatial–time domain. Filtering in the FK domain could be feasible for discriminating a specific phase (e.g., suppression of the FK spectra with phase velocity lower than those of fundamental toroidal or spheroidal modes).

8.4.3 Migration

DBF discussed in the previous section is a projection of wavefields based on plane waves, and we search the azimuth and slowness to interpret the reconstructed wavefields (often we do a grid-search) or steer the wavefields to improve SNR of target waves based on their azimuth and slowness. This means that we assume a simple structure beneath each array for DBF. When we have more complicated subsurface velocity models than 1D models, we can use wavefield migration or back-projection of correlation functions. The migration technique has been mainly used to image the subsurface reflectivity with scattered waves in exploration seismology (Claerbout, 1985; Biondi, 2006). Because the cross-correlation functions, in theory, contain all wavefields including reflections, we can use these waves for migration to map the boundary of subsurface layers, or we can first apply moveout correction and then post-stack migration (Hohl and Mateeva, 2006; Draganov et al., 2009). This technique requires surface-related multiples, and when ambient noise sources are located at the surface, the waves at least need to be reflected twice (or three times including reflection at the surface) (Forghani and Snieder, 2010). Thus, the signal becomes weak and difficult to extract. The pre-stack migration often computes cross-correlation after wavefield extrapolation (Claerbout, 1971), and hence we can directly migrate the continuous ambient-noise records and perform correlation at each image point (Artman, 2006; Vidal and Wapenaar, 2014). The small energy of reflected waves in ambient noise and the capability of computers to solve wave equations for large amounts of data make this method difficult to apply.

Autocorrelation of ambient noise is widely used to map the reflectors and measure the time-lapse changes (Sens-Schönfelder and Brenguier, 2018). Note that this is different from the Spatial Autocorrelation method (SPAC, Hayashi, Chapter 10). Based on Claerbout (1968), autocorrelation of continuous records provides zero-offset response of the structure, which contains reflected and multiply scattered waves. Importantly, after autocorrelation, all direct-wave signals (both surface and body waves) focus at the zero-lag time, and hence it is easier to retrieve scattered waves. With migration of autocorrelated wavefields, one can image strong reflectors such as subduction slab or reservoir cap/bed rocks (Ito and Shiomi, 2012; Boullenger et al., 2015), although the SNR is usually very low and we often need a priori information of the location of reflectors and careful interpretation of the migrated images.

8.5 Body-Wave Extraction at Different Scales and Their Applications

Compared to surface-wave applications, using body waves in ambient noise is not common. Here we discuss several key studies of body-wave applications at different scales. Although target waves are different according to the scales, the signal-processing techniques are similar to each other and mostly discussed in the previous section. Body-wave extraction and applications can be classified into two categories: exploration of direct or reflected waves. One of the keys is the separation of body and surface waves. With subsurface receivers, where no strong surface waves exist, body-wave imaging is more successful than using surface receivers (Vasconcelos and Snieder, 2008; Olivier et al., 2015), but these receivers are not common except in a few places in the world (e.g., at petroleum reservoirs, mines, and active faults). Therefore, we focus on body-wave extraction with surface receivers in this section.

8.5.1 Global Scale

Global seismic phases provide images that help us to better understand the interior of the Earth such as mantle convection, plate tectonics, dynamics of the inner core, and the history of the Earth's structure. Caused by the depth sensitivity of surface waves, body waves are primary resources to map the 3D heterogeneities of the deep structure. Since the distribution of large earthquake locations is mostly limited around plate boundaries, the resolution of the structural images obtained from earthquake waveforms varies. Ambient noise sources, and earthquake coda, are strong enough to generate global body-wave reflections, refractions, and diffractions; therefore, we can potentially increase the seismic illumination for imaging of the deep Earth. Note that due to the nature of the body-wave propagation and source mechanisms, P-waves of SM are less scattered, hard to use for seismic inter-ferometry, and longer-period body waves have weak energy. The noise sources to generate strong body waves contain rich information to study the coupling of atmosphere, ocean, and solid earth.

So far, the applications of ambient-noise body waves at the global scale mostly focus on the discoveries of global seismic phases from ambient noise (Nishida, 2013; Lin et al., 2013a; Lin and Tsai, 2013; Boué et al., 2013; Poli et al., 2015), mechanisms of the noise sources for such waves (Nishida, 2013; Boué et al., 2014; Nishida and Takagi, 2016), and inner-core imaging (Wang et al., 2015; Huang et al., 2015). Wave focusing caused by caustics enhances the amplitude of seismic waves propagating through or near the inner cores (Snieder and Sens-Schönfelder, 2015). For this scale, although the processing scheme is similar to the regional or local scale, we need a longer time interval ($\sim$ years) of data to obtain stationary corre-lation functions. Alternatively, with an assumption of 1D structure, we can apply spatial stacking such as binned stack.

Nishida (2013) retrieves global body-wave propagation from nine years of con-tinuous broadband records in three components (transverse, radial, and vertical; Figure 8.5), and discusses the mechanisms and distribution of noise sources by comparing the observed correlation functions and synthetic waveforms. He uses the frequency range at strong seismic hum (5 to 40 mHz), and one can extract body-wave reflections with much higher frequencies; for example, Boué et al. (2014) retrieve waves around 0.1 to 0.2 Hz.

Earthquake codas are another type of randomly scattered waves, and seismic interferometry is powerful to extract coherent parts of these waves (Huang et al., 2015; Wang et al., 2015). As Boué et al. (2014) discussed, we can extract global phases from only a few days of continuous records from earthquake codas; how-ever, we need to validate that the retrieved wavefields are not biased by the location of earthquakes, since earthquake coda in longer periods (< 0.1 Hz) may not be scat-tered strongly (Sens-Schönfelder et al., 2015). At the global scale, the separation of ambient noise from earthquake codas is not trivial. The effects of the bias can be visible in the correlation waveforms such as the existence of *ScS* in the verti-cal components (Lin et al., 2013a; Boué et al., 2014; Poli et al., 2017). When we carefully select the continuous data, we can get rid of most of such pseudo *ScS* arrivals (Nishida, 2013). The amplitude balance of each phase in the correlation functions is, however, still different from the synthetic data or typical earthquake records. Nishida (2013) interprets the strong *PL* wave (Figure 8.5c) as being caused by shear-traction sources in the frequency range of the seismic hum. Because the attenuation in the deeper Earth is smaller, core phases with large apparent veloc-ities (i.e., nearly vertical incidence) would be enhanced by cross-correlation in addition to the caustics effect (Snieder and Sens-Schönfelder, 2015). As discussed in Chapter 4, we need to be careful to use the amplitudes in the ambient-noise

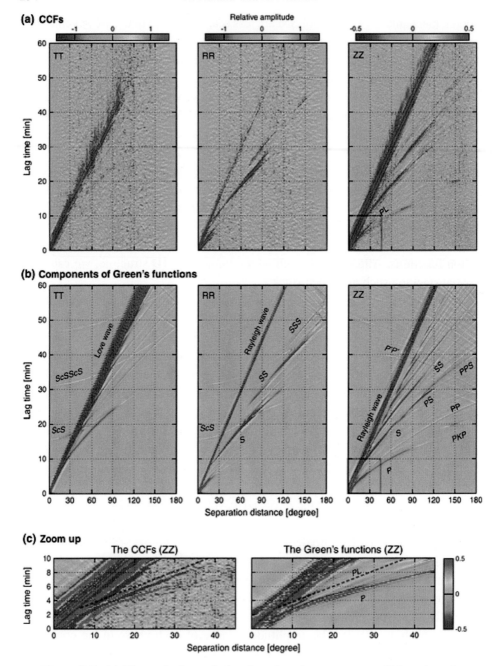

Figure 8.5. (a) Bin-stacked correlation functions in transverse, radial, and vertical components (TT, RR, and ZZ, respectively). (b) TT, RR, and ZZ components of the synthetic Green's functions obtained with the spherical Earth model (Dziewonski and Anderson, 1981). (c) Zoom up in the near-offset wavefields to show P and SPL waves. Modified after Nishida (2013). (A black-and-white version of this figure appears in some formats. For the color version, please refer to the plate section.)

correlation functions, because the amplitudes are easily biased by noise sources, subsurface structure, and signal processing.

8.5.2 Regional Scale

The imaging targets of the regional scale can be the Earth's crust, the upper mantle, and major discontinuities such as the Moho and the mantle transition zone (e.g., 410-km and 660-km depth). Applications of ambient-noise correlation at this scale are very much dominated by surface waves (see Chapter 7). We speculate that this is because (1) surface waves are stronger and easier to extract, similar to other scales, (2) earthquakes and their coda are useful for imaging (e.g., Bostock, 1997; Campillo and Paul, 2003; Tonegawa et al., 2009), and (3) if the targets are the crust, surface waves have high spatial resolution according to their frequency range of waves (e.g., Lin et al., 2008; Young et al., 2011). Despite these trends, there are efforts to extract regional waves from ambient noise. With an array in Africa, Zhan et al. (2010) retrieve *SmS* and its multiples from ambient noise. Similarly, Poli et al. (2012b) find Moho reflections (*PmP*, *SmS*, and *SmS2*) using cross-correlation of ambient noise observed in Finland. They sort all correlation functions in a distance-time plot, instead of computing binned stack, to find these waves; therefore, their results and discussion focus mainly on the general trends for the whole survey area. The arrival times of reflected waves from ambient noise agree well with earthquake synthetic data, and they also find that this region has low S velocities and Q values, which are comparable results from previous studies (Pedersen and Campillo, 1991). Poli et al. (2012a) use the same dataset in Finland and find reflected waves from mantle discontinuities (*P410P* and *P660P*; Figure 8.6). They use the slowness-time diagram and binned stack to enhance SNR of the reflected

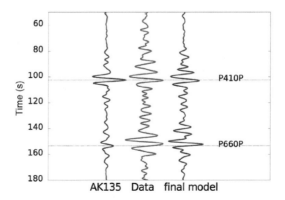

Figure 8.6. Extracted reflected waves from 410-km and 660-km discontinuities from ambient noise (middle), synthetic with AK135 model (left), and final model for this region (right). Modified after Poli et al. (2012a).

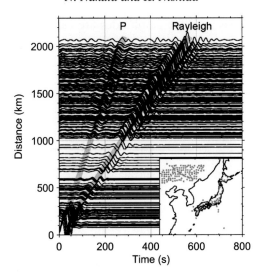

Figure 8.7. Extracted P and Rayleigh waves propagating from the reference receiver highlighted by the black star to other stations in the gray dots in the inset. We use DBF to enhance SNR of direct-path waves by clustering a small array around each receiver. The period band used is 20–60 s.

waves. The shape and impedance of reflected waves are significantly different from the synthetic waves computed based on the AK135 1D global Earth model, and they propose another 1D model for this region (Figure 8.6). Slowness beamforming is useful to extract reflections from these discontinuities (Feng et al., 2017).

Because the difference of arrival times of P and Rayleigh waves is usually large enough on this scale, we can clearly identify both waves in the time domain and potentially use them for imaging of structure. Figure 8.7 shows P and Rayleigh waves retrieved from ambient noise observed at F-net in Japan and NECESSArray in China (Ni, 2009), which contain a total of about 230 broadband stations. We repeatedly apply DBF for each receiver pair by making small sub-arrays for improving the SNR of waves propagating in the direct path. Based on the slowness of the DBF, the P-waves are mostly refracted waves. The body-wave sources can be considered as a combination of waves trapped in the crust as well as waves propagating through the upper mantle based on the station distance, frequency range of the waves, and possible source locations. Since we obtain P-wave propagation between each individual receiver pair, we can use them for imaging subsurface with wavefield tomography.

8.5.3 Local Scale

On the local scale ($\sim$ 10 km), high-frequency (1–50 Hz) body waves have been extracted from ambient noise. Compared to larger scales, where body-wave noise is mainly generated by the coupling between atmosphere, ocean, and solid Earth as discussed above, anthropogenic noise sources are important on this scale such as traffic, construction, and other cultural noise (Nakata et al., 2015). Near-coast breaking ocean waves, rivers, lakes, and winds also generate high-frequency noise (Zhang et al., 2009; Gimbert et al., 2014; Gimbert and Tsai, 2015; Poppeliers and Mallinson, 2015; Xu et al., 2017). Due to the attenuation of high-frequency waves, the most successful applications are close to coast lines or civilized areas.

For reflected-wave extraction, Draganov et al. (2007, 2009, 2013) applied cross-correlation techniques to 11 hours of continuous ambient noise records in Libya and imaged subsurface structures using retrieved P-wave reflected waves after suppressing surface waves with frequency-wavenumber filters (Figure 8.8). The

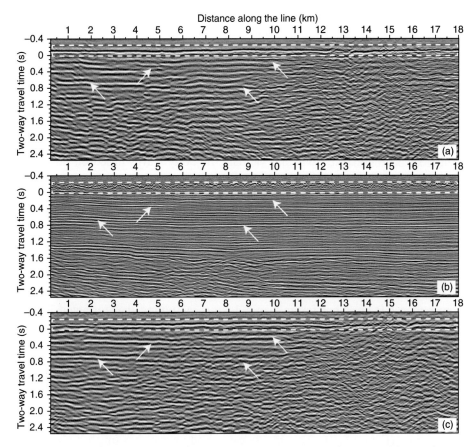

Figure 8.8. Time-migrated reflection images using (a) continuous ambient-noise records, (b) active sources, and (c) filtered ambient-noise records for vertically incident waves. The signals highlighted by the dashed box around 0 second, and right- and left-going white arrows are discussed in the main text. Modified after Draganov et al. (2013).

Earth's surface is highlighted by the dashed white line, and the ambient-noise image shows much higher coherency than the active-source image, which implies that the ambient noise has a potential to image very near surface and static correction for active sources. More importantly, several key subsurface layers are correctly imaged from ambient noise (the white arrows in Figure 8.8). The improvement due to the selection of ambient-noise windows for vertically incident waves (Figure 8.8c) is limited, and only few signals become clearer (e.g., right-going white arrow), which perhaps emphasizes the difficulties of the selection of noise panels for body-wave reflection imaging. Even though the subsurface images obtained from ambient noise have a clear layered structure up to 0.8 s two-way travel time, the ambient-noise images are less coherent and have much lower frequencies than the image obtained with active sources. This difference can be explained by the frequency range of ambient noise, which is usually neither as high nor as broad as active sources (Nakata et al., 2011). Draganov et al. (2013) use the velocity model obtained by active sources for migrating reflected waves constructed from ambient noise. When such a velocity model is not available, we can use surface waves to obtain an accurate near-surface velocity model (see Chapter 10).

Direct/refracted body waves have been also extracted from ambient-noise correlation. Roux et al. (2005) show one of the first studies to extract P-wave refractions, which propagate within a 11-km distance from Parkfield, California. They apply binned stack to enhance SNR and validate the body waves based on their polarities with different components. With an assumption of isotropic media, Takagi et al. (2014) separate P and Rayleigh waves based on the fact that the polarity of these waves is opposite in different components of the extracted Green's tensor. Compared to reflected waves, arrival times of direct (diving/refraction) waves are predictable (e.g., distance divided by 6 km/s $\pm$ 3 km/s). Therefore, we can make a filter to isolate such direct body waves after we compute cross-correlations. Nakata et al. (2015) use a time window to isolate waves that propagate slower than 6.0 km/s but faster than 1.1 km/s, assuming that refracted P-waves are contained in this window (Figure 8.9a). Then they apply a matched filter to find receiver pairs that contain strong body waves and a noise-suppression filter to increase SNR of the P-waves (Figure 8.9b). With the matched filter, they retain only 50% of pairs to isolate body-wave energy; therefore, the number of traces decreases in Figure 8.9b. After these filters, we can extract clear diving P waves, which are useful for subsurface imaging, for example with travel time tomography (Figure 8.10). This image using body waves has much higher spatial resolution than using surface waves (Lin et al., 2013b) and successfully captures geologic features in this region (Nakata et al., 2015). Because only 10 days of ambient noise data are used to create the subsurface image in Figure 8.10, we can

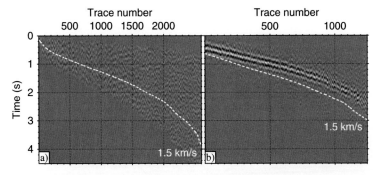

Figure 8.9. The same virtual shot gather as in Figure 8.2b after applying (a) a band-pass filter and (b) the P-wave isolation and noise suppression filters. The frequency range for both panels is from 3.0 to 15.0 Hz. The white dashed line shows the arrival time of a traveling wave with the apparent velocity of 1.5 km/s. The number of traces reduces after the trace selection. Modified after Nakata et al. (2015).

potentially image time-lapse subsurface structures using ambient-noise body-wave tomography.

At a local scale, clear S-waves are not successfully retrieved yet. In addition to the weak S-waves in ambient noise, the difficulty is caused by the availability of components of receivers; for example, the Long Beach array has only the vertical component, and hence we often do not observe horizontal motions. This situation is changing due to the recent development of sensors, and three-component local dense array data will be available in the near future.

8.6 Concluding Remarks

The way to use body waves in ambient noise has not matured yet, which also means that we still have a lot of opportunities to study this area, although we have made significant progress on various scales in the last five years. Based on the up-to-date studies, body waves exist in ambient noise in a wide frequency range, and we can explore/extract them by using arrays of seismometers. Compared to the surface waves, the body waves are often hard to find due to the discrete noise sources in time and space, weak energy, and/or violation for the stationary-phase approximation. The differences of the structural heterogeneities and mechanisms of noise sources limit the frequency ranges of extractable teleseismic body waves. To overcome these limitations, we can use spatial averaging of seismic data to enhance SNR besides averaging over long time periods. In addition to studying the mechanisms to generate such body waves, we can use these waves for structural imaging such as estimation of subsurface velocities, locating impedance contrasts, and mapping geological anomalies.

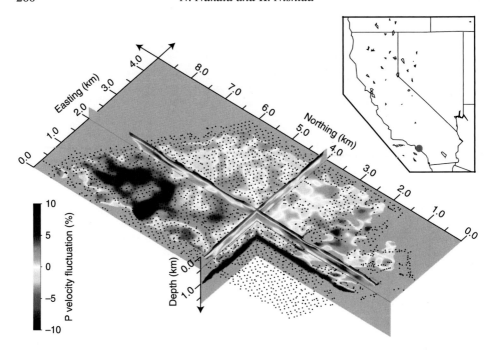

Figure 8.10. P-wave velocity model in 3D view obtained from the extract P-waves shown in Figure 8.9b. The color illustrates the fractional fluctuation of P-wave velocities, and the cold color indicates faster velocities than the laterally averaged velocity. The gray area shows the poorly resolved area according to the ray coverage of travel-time tomography (Nakata et al., 2015). The black dots are the location of the stations projected at the depth of the horizontal slice (the stations are deployed at the ground surface). The red dot (in the color version; grey dot in the greyscale version) in the inset shows the location of the survey. Modified after Nakata and Beroza (2015). (A black-and-white version of this figure appears in some formats. For the color version, please refer to the plate section.)

At the global scale, since the distribution of large earthquakes is mostly concentrated around plate boundaries, we can potentially increase the illumination of seismic waves for structural imaging by using body waves in ambient noise. To detect such body waves, we can use binned stack and find averaged body waves according to offsets of receivers. For imaging, however, we need to find the body waves more locally in space, which means we can only use fewer stations compared to the binned stack. This limitation requires us to apply advanced array signal processing techniques as well as noise-suppression filters. At regional and local scales, we have a better receiver coverage and can use spatial coherency of wavefields without aliasing compared to the global scale, where the receiver distribution is also heterogeneous in space. Hence we could retrieve body waves from individual receiver pairs or local sub-arrays, and these waves could be used for imaging. However, understanding the noise source distribution and/or mechanisms is still limited.

Theoretical approaches for deeply understanding ambient-noise correlation, particularly for body waves, are also necessary. Body waves have different stationary zones than surface waves, and the zones are usually smaller, if noise sources are mainly distributed in the near surface. Therefore, the bias of travel time and/or amplitudes of extracted body waves in correlation functions caused by non-stationary phases can be more significant than surface waves. For example, Weaver et al. (2009) discuss the bias caused by non-equipartitioned random waves for correlation functions and explain why the surface-wave (i.e., 2D) applications with travel times work well. This type of approach would be useful for body-wave extraction, in which we have much more limited source distribution and limited area of stationary-phase zones. In summary, body-wave exploration in ambient noise is a challenging and interesting topic, and we have opportunities to learn more about ambient noise source mechanisms (e.g., seismic hum, microseisms, and higher-frequency noise) and the Earth's interior at various scales.

Acknowledgments

Deyan Draganov and Piero Poli kindly provide their figures. We are grateful to the seismological data management centers for making the data used in this chapter available and people involved in the deployment of seismometers. We thank Florent Brenguier, Lucia Gualtieri, and an anonymous reviewer for their fruitful comments and discussion to improve the quality of this chapter.

References

Aki, K. 1957. Space and time spectra of stationary stochastic waves, with special reference to microtremors. *Bull. Earthq. Res. Inst.*, **35**, 415–456.

Ardhuin, F., Gualtieri, L., and Stutzmann, E. 2015. How ocean waves rock the Earth: Two mechanisms explain microseisms with periods 3 to 300 s. *Geophys. Res. Lett.*, **42**(3), 765–772.

Artman, B. 2006. Imaging passive seismic data. *Geophysics*, **71**(4), S1177–S1187.

Biondi, B. 2006. *3D Seismic Imaging.* Society of Exploration Geophysics.

Bonnefoy-Claudet, S., Cotton, F., and Bard, P.-Y. 2006. The nature of noise wavefield and its applications for site effects studies: A literature review. *Earth-Science Reviews*, **79**(3), 205–227.

Boschi, L., and Dziewonski, A. M. 1999. High- and low-resolution images of the Earth's mantle: Implications of different approaches to tomographic modeling. *J. Geophys. Res.*, **104**(B11), 25567–25594.

Bostock, M. G. 1997. Anisotropic upper-mantle stratigraphy and architecture of the Slave craton. *Nature*, **390**, 392–395.

Boué, P., Roux, P., Campillo, M., and Briand, X. 2014. Phase velocity tomography of surface waves using ambient noise cross correlation and array processing. *J. Geophys. Res.*, **119**, 519–529.

Boué, P., Poli, P., Campillo, M., and Roux, P. 2014. Reverberations, coda waves and ambient noise: Correlations at the global scale and retrieval of the deep phases. *Earth Planet. Sci. Lett.*, **391**(apr), 137–145.

Boué, P., Poli, P., Pedersen, M. C. C., Briand, X., and Roux, P. 2013. Teleseismic correlations of ambient seismic noise for deep global imaging of the Earth. *Geophys. J. Int.*, **194**, 844–848.

Boullenger, B., Verdel, A., Paap, B., Thorbecke, J., and Draganov, D. 2015. Studying CO_2 storage with ambient-noise seismic interferometry: A combined numerical feasibility study and field-data example for Ketzin, Germany. *Geophysics*, **80**(1), Q1–Q13.

Brenguier, F., Campillo, M., Hadziioannou, C., Shapiro, N. M., Nadeau, R. M., and Larose, E. 2008. Postseismic relaxation along the San Andreas Fault at Parkfield from continuous seismological observations. *Science*, **321**, 1478–1481.

Brooks, L. A., Townend, J., Gerstoft, P., Bannister, S., and Carter, L. 2009. Fundamental and higher-mode Rayleigh wave characteristics of ambient seismic noise in New Zealand. *Geophys. Res. Lett.*, **36**(23), 2–6.

Campillo, M., and Paul, A. 2003. Long-range correlations in the diffuse seismic coda. *Science*, **299**, 547–549.

Claerbout, J. 2014. *Geophysical image estimation by example*. lulu.com.

———. 1968. Synthesis of a layered medium from its acoustic transmission response. *Geophysics*, **33**(2), 264–269.

———. 1971. Toward a unified theory of reflector mapping. *Geophysics*, **36**(3), 467–481.

———. 1985. *Imaging the Earth's Interior*. Blackwell Science Inc.

Curtis, A., Gerstoft, P., Sato, H., Snieder, R., and Wapenaar, K. 2006. Seismic interferometry — turning noise into signal. *The Leading Edge*, **25**(9), 1082–1092.

Dahlen, F. A., and Baig, A. M. 2002. Fréchet kernels for body-wave amplitudes. *Geophys. J. Int.*, **150**, 440–466.

De Cacqueray, B., Roux, P., Campillo, M., Catheline, S., and Boué, P. 2011. Elastic-wave identification and extraction through array processing: An experimental investigation at the laboratory scale. *J. Appl. Geophys.*, **74**, 81–88.

Draganov, D., Campman, X., Thorbecke, J., Verdel, A., and Wapenaar, K. 2009. Reflection images from ambient seismic noise. *Geophysics*, **74**(5), A63–A67.

———. 2013. Seismic exploration-scale velocities and structure from ambient seismic noise (> 1 Hz). *J. Geophys. Res.*, **118**, 4345–4360.

Draganov, D., Wapenaar, K., Mulder, W., Singer, J., and Verdel, A. 2007. Retrieval of reflections from seismic background-noise measurements. *Geophys. Res. Lett.*, **34**, L04305.

Dziewonski, A. M., and Anderson, D. L. 1981. Preliminary reference Earth model. *Phys. Earth Planet. Inter.*, **25**, 297–356.

Feng, J., Yao, H., Poli, P., Fang, L., Wu, Y., and Zhang, P. 2017. Depth variations of 410-km and 660-km discontinuities in eastern North China Craton revealed by ambient noise interferometry. *Geophys. Res. Lett.* (in press).

Forghani, F., and Snieder, R. 2010. Underestimation of body waves and feasibility of surface-wave reconstruction by seismic interferometry. *The Leading Edge*, **29**(2), 790–794.

Gal, M., Reading, A. M., Ellingsen, S. P., Gualtieri, L., Koper, K. D., Burlacu, R., Tkalvci'c, H., and Hemer, M. A. 2015. The frequency dependence and locations of short-period microseisms generated in the Southern Ocean and West Pacific. *J. Geophys. Res.*, **120**(8), 5764–5781.

Gimbert, F., and Tsai, V. C. 2015. Predicting short-period, wind-wave generated seismic noise in coastal regions. *Earth Planet. Sci. Lett.*, **426**, 280–292.

Gimbert, F., Tsai, V. C., and Lamb, M. P. 2014. A physical model for seismic noise generation by turbulent flow in rivers. *J. Geophys. Res.*, **119**, 2209–2238.

Gualtieri, L., Stutzmann, E., Capdeville, Y., Ardhuin, F., Schimmel, M., Mangeney, A., and Morelli, A. 2013. Modeling secondary microseismic noise by normal mode summation. *Geophys. J. Int.*, **193**(3), 1732–1745.

Gualtieri, L., Stutzmann, E., Farra, V., Capdeville, Y., Schimmel, M., Ardhuin, F., and Morelli, A. 2014. Modeling the ocean site effect on seismic noise body waves. *Geophys. J. Int.*, **197**, 1096–1106.

Hasselmann, K. 1963. A statistical analysis of the generation of microseisms. *Rev. Geophys.*, **1**(2), 177–210.

Haubrich, R. A., and McCamy, K. 1969. Microseisms: Coastal and pelagic sources. *Rev. Geophys.*, **7**(3), 539–571.

Hayashi, K. 2018. Near-surface engineering. Pages – of: Nakata, N., Gualtieri, L., and Fichtner, A. (eds.), *Seismic Ambient Noise*. Cambridge University Press, Cambridge, UK.

Hillers, G., Campillo, M., Ben-Zion, Y., and Landès, M. 2013. Interaction of microseisms with crustal heterogeneity: A case study from the San Jacinto fault zone area. *Geochemistry, Geophys. Geosystems*, **14**(7), 2182–2197.

Hohl, D., and Mateeva, A. 2006. Passive seismic reflectivity imaging with ocean-bottom cable data. *SEG Expanded Abstracts*, **25**, 1560–1564.

Huang, H.-H., Lin, F.-C., Tsai, V. C., and Koper, K. D. 2015. High-resolution probing of inner core structure with seismic interferometry. *Geophys. Res. Lett.*, **42**, 10622–10630.

Ito, Y., and Shiomi, K. 2012. Seismic scatterers within subducting slab revealed from ambient noise autocorrelation. *Geophys. Res. Lett.*, **39**, L19303.

Kedar, S., Longuet-Higgins, M., Webb, F., Graham, N., Clayton, R., and Jones, C. 2008. The origin of deep ocean microseisms in the North Atlantic Ocean. *Proc. R. Soc. A Math. Phys. Eng. Sci.*, **464**(2091), 777–793.

Krüger, F., Weber, M., Scherbaum, F., and Schlittenhardt, J. 1993. Double beam analysis of anomalies in the core-mantle boundary region. *Geophys. Res. Lett.*, **20**(14), 1475–1478.

Lawrence, J. F., and Prieto, G. A. 2011. Attenuation tomography of the western United States from ambient seismic noise. *J. Geophys. Res.*, **116**, B06302.

Le Touzé, G., Nicolas, B., Mars, J. I., Roux, P., and Oudompheng, B. 2012. Double-Capon and double-MUSICAL for arrival separation and observable estimation in an acoustic waveguide. *EURASIP Journal on Advances in Signal Processing*, **2012**, 187.

Lin, F.-C., Li, D., Clayton, R. W., and Hollis, D. 2013b. High-resolution 3D shallow crustal structure in Long Beach, California: Application of ambient noise tomography on a dense seismic array. *Geophysics*, **78**, Q45–Q56.

Lin, F.-C., Moschetti, M. P., and Ritzwoller, M. H. 2008. Surface wave tomography of the western United States from ambient seismic noise: Rayleigh and Love wave phase velocity maps. *Geophys. J. Int.*, **173**, 281–298.

Lin, F.-C., Ritzwoller, M. H., and Snieder, R. 2009. Eikonal tomography: surface wave tomography by phase front tracking across a regional broad-band seismic array. *Geophys. J. Int.*, **177**, 1091–1110.

Lin, F.-C., and Tsai, V. C. 2013. Seismic interferometry with antipodal station pairs. *Geophys. Res. Lett.*, **40**, 4609–4613.

Lin, F.-C., Tsai, V. C., Schmandt, B., Duputel, Z., and Zhan, Z. 2013a. Extracting seismic core phases with array interferometry. *Geophys. Res. Lett.*, **40**(6), 1049–1053.

Liu, X., and Zhao, D. 2016. P and S wave tomography of Japan subduction zone from joint inversions of local and teleseismic travel times and surface-wave data. *Phys. Earth Planet. Inter.*, **252**, 1–22.

Lobkis, O. I., and Weaver, R. L. 2001. On the emergence of the Green's function in the correlations of a diffuse field. *J. Acoust. Soc. Am.*, **110**(6), 3011–3017.

Longuet-Higgins, M. S. 1950. A theory of the origin of microseisms. *Philos. Trans. R. Soc. Lond. Ser. A*, **243**, 1–35.

Mainsant, G., Larose, E., Brönnimann, C., Jongmans, D., Michoud, C., and Jaboyedoff, M. 2012. Ambient seismic noise monitoring of a clay landslide: Toward failure prediction. *J. Geophys. Res.*, **117**, F01030.

Mehta, K., Sheiman, J. L., Snieder, R., and Calvert, R. 2008. Strengthening the virtual-source method for time-lapse monitoring. *Geophysics*, **73**(3), S73–S80.

Nakata, N., and Beroza, G. C. 2015. Stochastic characterization of mesoscale seismic velocity heterogeneity in Long Beach, California. *Geophys. J. Int.*, **203**, 2049–2054.

Nakata, N., Boué, P., Brenguier, F., Roux, P., Ferrazzini, V., and Campillo, M. 2016. Body and surface wave reconstruction from seismic-noise correlations between arrays at Piton de la Fournaise volcano. *Geophys. Res. Lett.*, **43**, 1047–1054.

Nakata, N., Chang, J. P., Lawrence, J. F., and Boué, P. 2015. Body-wave extraction and tomography at Long Beach, California, with ambient-noise interferometry. *J. Geophys. Res.*, **120**, 1159–1173.

Nakata, N., Tsuji, T., and Matsuoka, T. 2011. Acceleration of computation speed for elastic wave simulation using a Graphic Processing Unit. *Explor. Geophys.*, **42**, 98–104.

Ni, J. 2009. Collaborative Research: Northeast China Extended Seismic Array: Deep subduction, mantle dynamics and lithospheric evolution beneath Northeast China. *International Federation of Digital Seismograph Networks*.

Nishida, K. 2013. Global propagation of body waves revealed by cross-correlation analysis of seismic hum. *Geophys. Res. Lett.*, **40**, 1691–1696.

———. 2014. Source spectra of seismic hum. *Geophys. J. Int.*, **199**, 416–429.

———. 2017. Ambient seismic wave field. *Proc. Jpn. Acad., Ser. B*, **93**(7), 423–448.

Nishida, K., and Takagi, R. 2016. Teleseismic S wave microseisms. *Science*, **353**, 919–921.

Olivier, G., Brenguier, F., Campillo, M., Lynch, R., and Roux, P. 2015. Body-wave reconstruction from ambient seismic noise correlations in an underground mine. *Geophysics*, **80**(3), 1–15.

Oliver, J. 1961. On the long period character of shear waves. *Bull. Seismol. Soc. Am.*, **51**(1), 1–12.

Pedersen, H., and Campillo, M. 1991. Depth dependence of Q beneath the Baltic shield inferred from modeling of short period seismograms. *Geophys. Res. Lett.*, **18**(9), 1755–1758.

Poli, P., Campillo, M., and de Hoop, M. 2017. Analysis of intermediate period correlations of coda from deep earthquakes. *Earth Planet. Sci. Lett.*, **477**, 147–155.

Poli, P., Campillo, M., Pedersen, H., and Group, L. W. 2012a. Body-wave imaging of Earth's mantle discontinuities from ambient seismic noise. *Science*, **338**, 1063–1065.

Poli, P., Pedersen, H. A., Campillo, M., and the POLENET/LAPNET working group. 2012b. Emergence of body waves from cross-correlation of short period seismic noise. *Geophys. J. Int.*, **188**, 549–558.

Poli, P., Thomas, C., Campillo, M., and Pedersen, H. A. 2015. Imaging the D" reflector with noise correlations. *Geophys. Res. Lett.*, **42**, 60–65.

Poppeliers, C., and Mallinson, D. 2015. High-frequency seismic noise generated from breaking shallow water ocean waves and the link to time-variable sea states. *Geophys. Res. Lett.*, **42**, 8563–8569.

Rost, S., and Thomas, C. 2002. Array seismology: Methods and applications. *Rev. Geophys.*, **40**, 1008.

Roux, P., Cornuelle, B. D., Kuperman, W. A., and Hodgkiss, W. S. 2008. The structure of raylike arrivals in a shallow-water waveguide. *J. Acoust. Soc. Am.*, **124**, 3430–3439.

Roux, P., Iturbe, I., Nicolas, B., Virieux, J., and Mars, J. I. 2011. Travel-time tomography in shallow water: Experimental demonstration at an ultrasonic scale. *J. Acoust. Soc. Am.*, **130**, 1232–1241.

Roux, P., Moreau, L., Lecointre, A., Hillers, G., Campillo, M., Ben-Zion, Y., Zigone, D., and Vernon, F. 2016. A methodological approach towards high-resolution surface wave imaging of the San Jachinto Fault Zone using ambient-noise recordings at a spatially dense array. *Geophys. J. Int.*, **206**, 980–992.

Roux, P., Sabra, K., Gerstoft, P., and Kuperman, W. 2005. P-waves from cross correlation of seismic noise. *Geophys. Res. Lett.*, **32**, L19303.

Sato, H., Fehler, M. C., and Maeda, T. 2012. *Seismic wave propagation and scattering in the heterogeneous earth*. 2nd edn. Springer.

Sens-Schönfelder, C., and Brenguier, F. 2018. Noise-based monitoring. Pages – of: Nakata, N., Gualtieri, L., and Fichtner, A. (eds.), *Seismic Ambient Noise*. Cambridge University Press, Cambridge, UK.

Sens-Schönfelder, C., Snieder, R., and Stähler, S. C. 2015. The lack of equipartitioning in global body wave coda. *Geophys. Res. Lett.*, **42**(18), 7483–7489.

Shapiro, N. 2018. Applications with surface waves extracted from ambient seismic noise. Pages — of: Nakata, N., Gualtieri, L., and Fichtner, A. (eds.), *Seismic Ambient Noise*. Cambridge University Press, Cambridge, UK.

Shapiro, N. M., Campillo, M., Stehly, L., and Ritzwoller, M. H. 2005. High-resolution surface-wave tomography from ambient seismic noise. *Science*, **307**, 1615–1618.

Shearer, P. M. 1991. Imaging global body wave phases by stacking long-period seismograms. *J. Geophys. Res.*, **96**(B12), 20353–20364.

Snieder, R. 2004. Extracting the Green's function from the correlation of coda waves: A derivation based on stationary phase. *Phys. Rev. E*, **69**, 046610.

Snieder, R., and Sens-Schönfelder, C. 2015. Seismic interferometry and stationary phase at caustics. *J. Geophys. Res.*, **120**, 4333–4343.

Snieder, R., Van Wijk, K., Haney, M., and Calvert, R. 2008. Cancellation of spurious arrivals in Green's function extraction and the generalized optical theorem. *Phys. Rev. E - Stat. Nonlinear, Soft Matter Phys.*, **78**(3), 1–8.

Takagi, R., Nakahara, H., Kono, T., and Okada, T. 2014. Separating body and Rayleigh waves with cross terms of the cross-correlation tensor of ambient noise. *J. Geophys. Res.*, **119**, 2005–2018.

Tibuleac, I. M., and von Seggern, D. 2012. Crust-mantle boundary reflectors in Nevada from ambient seismic noise autocorrelations. *Geophys. J. Int.*, **189**(1), 493–500.

Toksöz, M. N., and Lacoss, R. T. 1968. Microseisms: mode structure and sources. *Science*, **159**(3817), 872–873.

Tonegawa, T., Nishida, K., Watanabe, T., and Shiomi, K. 2009. Seismic interferometry of teleseismic S-wave coda retrieval of body waves: an application to the Philippine Sea slab underneath the Japanese Islands. *Geophys. J. Int.*, **178**, 1574–1586.

Vasconcelos, I., and Snieder, R. 2008. Interferometry by deconvolution: Part 2 - Theory for elastic waves and application to drill-bit seismic imaging. *Geophysics*, **73**(3), S129–S141.

Vidal, C. A., and Wapenaar, K. 2014. Passive seismic interferometry by multi-dimensional deconvolution-decorrelation. *SEG Expanded abstract*, 2224–2228.

Vinnik, L. 1973. Sources of microseismic P waves. *Pure Appl. Geophys.*, **103**, 282–289.

Wang, T., Song, X., and Xia, H. H. 2015. Equatorial anisotropy in the inner part of Earth's inner core from autocorrelation of earthquake coda. *Nature Geosci.*, **8**, 224–227.

Wapenaar, K. 2004. Retrieving the elastodynamic Green's function of an arbitrary inhomogeneous medium by cross correlation. *Phys. Rev. Lett.*, **93**, 254301.

Wapenaar, K., and Fokkema, J. 2006. Green's function representations for seismic interferometry. *Geophysics*, **71**(4), SI33–SI46.

Wapenaar, K., Draganov, D., and Robertsson, J. 2006. Introduction to the supplement on seismic interferometry. *Geophysics*, **71**(4), SI1–SI4.

Weaver, R., Froment, B., and Campillo, M. 2009. On the correlation of non-isotropically distributed ballistic scalar diffuse waves. *J. Acoust. Soc. Am.*, **126**(4), 1817–1826.

Weber, M., and Wicks, C. W. 1996. Reflections from a distant subduction zone. *Geophys. Res. Lett.*, **23**(12), 1453–1456.

Xu, Y., Koper, K. D., and Brulacu, R. 2017. Lakes as a source of short-period (0.5–2 s) microseisms. *J. Geophys. Res.*, **122**, 8241–8256.

Yilmaz, O. 2001. *Seismic Data Analysis*. 2nd edn. Investigations in Geophysics, vol. 10. Tulsa, USA: Society of Exploration Geophysicists.

Young, M. K., Rawlinson, N., Arroucau, P., Reading, A. M., and Tkalčić, H. 2011. High-frequency ambient noise tomography of southeast Australia: New constraints on Tasmania's tectonic past. *Geophys. Res. Lett.*, **38**, L13313.

Zhan, Z., Ni, S., Helmberger, D., and Clayton, R. W. 2010. Retrieval of Moho-reflected shear wave arrivals from ambient seismic noise. *Geophys. J. Int.*, 408–420.

Zhang, J., Gerstoft, P., and Shearer, P. 2009. High-frequency P-wave seismic noise driven by ocean winds. *Geophys. Res. Lett.*, **36**, L09302.

Zhou, Y., Dahlen, F. A., and Nolet, G. 2004. Three-dimensional sensitivity kernels for surface wave observables. *Geophys. J. Int.*, **158**, 142–168.

9

Noise-Based Monitoring

CHRISTOPH SENS-SCHÖNFELDER AND FLORENT
BRENGUIER

Abstract

A principal advantage of using ambient vibration for seismological investigations
is that it is persistent in time and ubiquitous in space. In contrast to earthquakes or
active sources that either occur erratically or require logistical efforts, the ambient
seismic field allows for continuous observations with existing data and instruments.
This possibility provides seismology with access to the fourth dimension: time. It
permits us to study the effects of dynamic processes on the elastic properties of the
subsurface. Such effects are known from laboratory experiments with ultrasound
and can be observed in the Earth due to the continuous excitation of ambient noise.
The present chapter reviews observations of subsurface processes made in a continuous mode with seismic noise. As for seismological investigations of structure, the
most important material parameter to be investigated is the propagation velocity
of seismic waves. Spectacular observations from volcanoes have revealed velocity changes with a systematic relation to volcanic activity indicating a predictive
potential for eruptions. Interpreted as caused by the opening and closing of cracks,
those changes can be related to deformation detected at the surface with geodetic
methods. Similar mechanisms involving static stress changes have been invoked for
deep velocity changes observed in fault zones where the most dramatic changes are
expected during large earthquakes and within underground mines where drilling
produces static stress perturbations. Naturally, the large-scale stress redistribution
during earthquakes is connected to strong ground motion and dynamic strain, especially in fault areas where changes of the static stress are expected. Dynamic strain
in heterogeneous Earth materials lowers the elastic moduli, leading to a widespread
decrease of seismic velocity after large earthquakes. This damaging effect is known
in materials science and has been studied in detail partly because of the transient
recovery process that re-increases the velocity. The damage process induced by
seismic waves has been invoked to explain coseismic velocity changes after the
Tohoku-Oki earthquake located around active volcanoes that are very sensitive to

damage because of pressurized fluids. Even though the relation between seismic velocity and mechanical strength of a material has not been proven in detail, the near-surface damage observed from the propagation of seismic noise is related to the increased occurrence rate of landslides after earthquakes. We conclude the discussion of seismic observations with a review of models mostly originating from material sciences for the internal processes in Earth materials that are responsible for the various observations of temporal changes.

9.1 Introduction

While the interaction of seismic waves with the propagation medium has been used to investigate the structure of the Earth from the first moments of seismology, attempts using wave propagations to study changes of material properties due to dynamic processes have been rare. Systematic investigations of temporal changes have been limited to 4D seismic surveys in hydrocarbon exploration. One reason for this imbalance is the difference in the spatial and temporal variations of the elastic material properties. On a global scale the velocity of shear waves changes from 1 km/s or less near the surface to more than 6 km/s in the mantle. Locally, the spatial variations of the seismic velocity inferred by tomography amount to several percent (Chapter 7, Shapiro, 2018). Compared to this variability, the temporal variations are small enough to treat the elastic properties of Earth's materials as material *constants*. However, as we know now thanks to precise noise-based observations, temporal variations of seismic velocity are ubiquitous.

Noise-based monitoring that developed in the last decade is a very effective tool not just because it is cost effective as it uses standard seismic data, but also because the possibility of monitoring continuously for long time spans allows us to identify and isolate the effects of different processes. This in turn provides the basis for studying the response of a material to some perturbing process – leading to a new class of material properties that describe the susceptibility of the material to perturbations.

In this chapter we show how the use of the seismic ambient field revolutionized the study of dynamic properties of Earth's materials. We review important observations of variable material properties that were obtained with impulsive sources (section 9.1.1). We then describe the technical basics of noise-based monitoring in section 9.2. Applications of the ambient-field-based monitoring are discussed in section 9.4 for different targets. In section 9.5 we link the field observations to laboratory experiments and discuss current models used to explain the different observations of material changes to understand the link between seismological observations and their physical origin. A short summary follows in section 9.6.

9.1.1 Observations of Dynamic Earth's Material Properties

In 1984 Poupinet et al. (1984) demonstrated for the first time that the seismic velocity in the Earth undergoes measurable temporal changes. They analyzed local earthquakes that occurred before and after an $M_L = 5.9$ earthquake on the Calaveras fault in California. Among the multitude of events in this active area there were earthquake pairs that had very similar waveforms on different stations indicating almost identical repeating sources referred to as doublets. Poupinet and coauthors recognized that the waveforms of these doublets were not only very similar at the arrival of the P- and S-waves but also much later in the scattered coda wavefield. The complex waveform in the coda was perfectly reproduced by the second event of the doublet even though it appeared impossible to explain its details (Figure 9.1). Poupinet's remarkable achievement was to recognize that at the time of the $M_L = 5.9$ event, minute changes in the coda waveforms appeared that were impossible to observe in individual waveforms but could be measured when two waveforms were compared to each other. The waveforms of doublets that occurred before and after the earthquake differed by a tiny phase shift that increases with increasing travel time, i.e., lapse time in the seismogram. As we will discuss in more detail in section 9.2.1, such changes can be explained by changes in the wave velocity.

A time-domain version of the concept of using scattered waves for monitoring was proposed by Snieder et al. (2002). Alluding to the scattered wavefield in the late part of a seismogram called "coda" in analogy to the terminal part of a music piece, scientists called this measurement comparing waveform phases "coda wave interferometry" (CWI).

The method of Poupinet et al. (1984) and Snieder et al. (2002) was later applied by a number of authors to earthquake doublets and records of repeatable active sources. Seismic velocity changes following the $M_L = 6.9$ Loma Prieta earthquake were investigated by Schaff and Beroza (2004) and Rubinstein and Beroza (2004) together with the Morgan Hill event. Rubinstein and Beroza (2005) included borehole sensors to constrain the depth range of the material that experienced the velocity reduction. Velocity changes related to the Izmit and Dücze earthquakes on the North-Anatolian fault in Turkey were investigated by Peng and Ben-Zion (2006). At the Merapi and Iwate volcanoes velocity changes were observed by Ratdomopurbo and Poupinet (1995) and Yamawaki et al. (2004), respectively. To circumvent the dependence on the occurrence of repeating earthquakes, repeatable sources were also used at these volcanoes. Nishimura et al. (2000) and Nishimura et al. (2005) deployed explosives to monitor velocity changes associated with volcanic processes in Mt. Iwate and nearby earthquakes in Japan. At Merapi, air gun shots in an artificial pool were used to acquire repeatable signals (Wegler et al., 2006). The healing process of the Landers and the Hector Mine rupture zone were

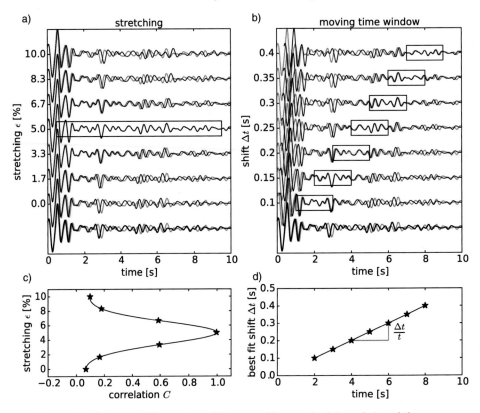

Figure 9.1. Synthetic illustration of the stretching method (a and c) and the moving time window method (b and d) to measure the apparent velocity change. Top panels show the waveforms of any measure of the oscillation like, displacement, acceleration or tilt on any component. In panels a and b the trace u_k of the reference state in gray is identical throughout. Black shows the modified traces. In panel a the black trace u_l is stretched by different amounts of which one best mimics the observed waveform change (5%, panel c). In panel b the black trace is shifted by different amounts. In this case it matches the reference trace at different lapse times shown by the boxes. The required shift increases proportional to lapse time maintaining a constant relative time shift $\Delta t / t$ (panel d).

monitored using explosives by Li and Vidale (2001), Li et al. (2003), and Vidale and Li (2003). Velocity changes associated with nearby earthquakes were observed using a downhole acoustic source at the SAFOD borehole (Niu et al., 2008).

All the investigations require exactly repeatable sources to resolve the tiny amplitude of temporal variations of material properties. This leads to either a very irregular temporal sampling when repeating earthquakes are used or short and sparse time series resulting from the high costs of repeatable active sources. Few authors have succeeded in observing temporal changes with repeated tomography not requiring repeatable sources (Patanè et al., 2006; Koulakov et al., 2013).

An elegant way to obtain dense and regular temporal sampling without the high technical and logistic effort required for active sources is to monitor changes in the propagation properties of the ambient seismic wave field. This concept was introduced by Sens-Schönfelder and Wegler (2006) and will be discussed in the following sections. We will first discuss the fundamental steps in noise-based monitoring. The specific requirements of using the ambient field for monitoring will be discussed in section 9.2.2.

9.2 Methodological Steps in Noise-Based Monitoring

The key to continuous high-precision observations of seismic velocity changes is the combination of two seismological methods that utilize the phase information in complex waveforms. In a first step, a repeatable seismic signal is generated from the ambient seismic noise by means of seismic interferometry (Chapter 4, Fichtner and Tsai, 2018). The second step then evaluates variations in these retrieved waveforms to infer underlying medium changes. To achieve the required accuracy, it is usually the late part of these signals containing the scattered coda waves that is analyzed with coda wave interferometry. Referring to the phase-comparison of passively retrieved signals (images) Sens-Schönfelder and Wegler (2006) introduced this approach under the name *passive image interferometry*.

In the following sections we will discuss the two steps in reverse order because the way in which changes in the waveforms are evaluated influences the requirements on the waveform retrieval. We will first discuss coda wave interferometry and then signal retrieval from the ambient field.

9.2.1 Coda Wave Interferometry

Very precise physical measurements can be obtained from the direct comparison of waveforms. This follows from the fast variations of the wave phase that provide high contrasts even for tiny changes. In optics, where the phase of the electromagnetic waves is usually not known, the comparison between two waves is done in a dedicated apparatus – the interferometer. In seismology we can directly measure the phase of seismic waves, which allows us to emulate the interferometric measurement between recorded waveforms in a computer at any later time. Two types of changes between waveforms can be observed: a phase shift resulting from later arrivals of waves in one waveform compared to the other and a random change in waveform amplitude that leads to a loss of similarity.

As the amplitude of a waveform perturbation increases with increasing propagation distance of the wave in the perturbed medium, one might want to use records at long distances to obtain high precision. However, as shown by Poupinet et al. (1984) and Snieder et al. (2002) one can make use of the heterogeneity of the subsurface and investigate waves that travel on a crumpled path and arrive after a long propagation close to their source to form the coda portion of a seismogram.

Now let us assume the propagation velocity in the subsurface has slightly changed homogeneously in space from v to $v + \Delta v$ between the recording of two waveforms in an identical source-receiver setup. This small change will only marginally affect the propagation path, but it will either speed up or slow down the waves. The travel time of waves will change from t to $t + \Delta t$. Δt will increase with increasing t, resulting in a constant relative travel time change that equals the negative relative velocity change:

$$\frac{\Delta t}{t} = -\frac{\Delta v}{v} = \epsilon. \tag{9.1}$$

As we have assumed here that the velocity change is homogeneous in space – which will rarely be the case – we will use the term *apparent velocity change*, indicating that we mean a homogeneous perturbation that would generate the observed velocity change. We will see later (and in Chapter 6 [Snieder et al., 2018]) more detailed information than the apparently homogeneous change can be obtained by using measurements in different frequencies, lapse time windows, or station configurations.

Under the assumption of a homogeneous velocity change, the *relative* travel time change ϵ is constant for all lapse times. Effectively the time axis of the waveforms resulting from such a change will be stretched or compressed versions. The *stretching method* introduced by Sens-Schönfelder and Wegler (2006) directly simulates this stretching (Figures 9.1a and c) to measure the relative travel time change and finally to estimate the causative apparent change of seismic velocity. In practice the change underlying the difference between two waveforms is estimated by calculating the correlation coefficient $C(\epsilon)$ between one waveform u_k that sensed the medium in state k and a number of modified versions of the other waveform u_l that represents the medium in state l

$$C_{k,l}(\epsilon) = \sum_{t=t_{min}}^{t_{max}} \frac{u_k(t)u_l(t(1+\epsilon))}{\sqrt{u_k^2(t)u_l^2(t(1+\epsilon))}}. \tag{9.2}$$

The modification simulating the stretching is simply an interpolation of the waveform with a slightly modified sampling interval $dt(1 + \epsilon)$. This is illustrated in Figure 9.1a. The value of ϵ_{max} for which $C_{k,l}(\epsilon)$ assumes its maximum indicates the stretching that best simulates the apparent velocity change $\frac{\Delta v}{v}$ (Figure 9.1c). $1 - C_{k,l}(\epsilon_{max})$ is the so-called decorrelation that indicates the remaining waveform change that cannot be explained with the homogeneous velocity change simulated with the stretching. As discussed by Dai et al. (2013), the stretching can also be performed in the frequency domain.

Alternatively to using a long time window and stretching the waveforms, one can use a number of short time windows as illustrated in Figures 9.1b and d. In these time windows of duration t_w, centered at lapse times t_i, the distortion due to the velocity change is small and one can shift the traces in these windows to quantify the similarity between the reference waveform u_k and the current waveform u_l locally at t_i

$$C_{k,l}^i(\Delta t) = \sum_{t=t_i-t_w/2}^{t_i+t_w/2} \frac{u_k(t)u_l(t+\Delta t)}{\sqrt{u_k^2(t)u_l^2(t+\Delta t)}} . \tag{9.3}$$

Finally a linear function is fitted to the values of the time shifts $\Delta t_{max}(t_i)$ that maximize $C_{k,l}^i(\Delta t)$ of the different time windows at t_i to obtain an averaged value that represents a constant relative travel time change $\Delta t/t$ corresponding to the apparent velocity change (Figure 9.1d). Also values of the time window shifts $\Delta t_{max}(t_i)$ can be obtained from phase differences in the frequency domain as suggested by Poupinet et al. (1984). This led to the name "moving window cross-spectral analysis" or its abbreviation "MWCS" (Ratdomopurbo and Poupinet, 1995) that is sometimes used to refer to the moving window method in the literature.

A detailed comparison of the short time window and the stretching methods was published by Hadziioannou et al. (2009). It indicates that under noisy conditions the stretching method has superior stability. On the other hand, Zhan et al. (2013) showed that the stretching method can be biased by changes in the frequency content of the signal if it is applied to an isolated arrival such as the ballistic wave. The precision of the stretching method was evaluated by Weaver et al. (2011) as a function of the signal-to-noise ratio, frequency content, and time window used for the analysis.

We have seen how a change between two similar waveforms representing the propagation medium in two different states can be quantified in terms of an apparent velocity change and a remaining decorrelation. To obtain a monitoring time series that describes the state of the medium continuously over time, a seismic signal is required that repeatedly samples the medium. At this point the ambient seismic wavefield comes into play.

9.2.2 Reconstruction of Repeatable Signals

Coda wave interferometry as described above can be applied to any repeatedly measured signal. But continuous monitoring with CWI requires continuously repeating signals with identical source-receiver configurations. This is a major handicap in seismology, where the wave sources are usually earthquakes with uncontrolled and strongly clustered occurrences in space and time. The insight that the ambient seismic field, which is generated mostly by atmospheric and oceanic sources at the Earth's surface (Chapter 3 [Ardhuin et al., 2018]) can help to overcome this limitation, paved the road for time-dependent seismology.

The ambient seismic field "as is" is of little use either for investigations of structure or for monitoring. It is dominated by fluctuations resulting from the superposition of unknown simultaneously acting noise sources. Yet, the correlation properties of the ambient field allow us to separate randomness of the excitations from deterministic propagation effects. The discovery that Rayleigh waves can be reconstructed from the ambient seismic wavefield (Shapiro and Campillo, 2004) bears an enormous potential for surface wave tomography (Chapters 4 and 7 [Fichtner and Tsai, 2018; Shapiro, 2018]). Ambient noise tomography has become a standard tool in seismology that increases the resolution of tomographic studies and makes the investigation independent of the occurrence of earthquakes. For the observation of temporal material changes, noise interferometry has a further – very significant – advantage. The ambient field is omnipresent, which allows for continuous observation.

In Chapters 4 and 5 (Fichtner and Tsai, 2018; Ritzwoller and Feng, 2018) the Green's function extraction from ambient seismic noise field is discussed in detail. Due to the confinement of noise sources to the surface, the Green's function extraction is often focused on the surface wave part used in tomography. Of crucial importance for successful Green's function reconstruction is the distribution of noise sources. Many studies (e.g., Froment et al., 2010; Colombi et al., 2014) have investigated the traveltime bias of surface waves or the appearance of spurious arrivals (e.g., Snieder et al., 2008) when the theoretical requirements of the source distribution for Green's function extraction are not perfectly met.

Yet, the applicability of the noise correlation technique for monitoring has two additional requirements: (1) The ambient seismic field should allow us to retrieve a deterministic signal from "short" noise sequences. (2) The full Green's function including scattered waves must be retrieved from the ambient field. The first requirement determines the sampling rate of the monitoring time series. If we need to correlate a year of seismic noise to obtain a stable approximation of the Green's function, the sampling rate of the monitoring would be one year. Stable and independent Green's functions must be retrievable at a much higher rate to observe interesting processes. The second requirement corresponds to the experience of investigations using impulsive sources: changes in the propagation properties can best be observed when coda waves are analyzed that travel through the medium for long times. Fortunately, the ambient field allows us to meet these

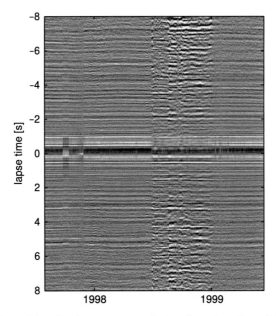

Figure 9.2. Ensemble of noise cross-correlation functions from Merapi volcano for stations at 100 m distance. The direct wave at 0.3 s is followed by coda waves that are coherent over the study period at much larger lapse times. Temporal variations of the velocity are visible as slight bending of the horizontal phases with travel time maxima in the first half of 1998. (Reprinted from Sens-Schönfelder, 2008, with permission from Oxford University Press.)

requirements. Figure 9.2 shows an example of the repeatedly calculated noise correlation functions (NCFs) assembled side by side in a matrix where one axis is time of the year and the other axis is correlation lag corresponding to the travel time of the waves. The figure shows the direct surface waves and arrivals at late lapse times in the seismic coda that are coherent over the whole time of the measurements.

When using coda waves for monitoring variations of the velocity, the requirements on the distribution of sources are relaxed compared to measurements of absolute velocities from ballistic waves. In fact the requirements can drastically be changed if the velocity change is homogeneous in space. In this case it is sufficient that noise sources are stationary without any constraint on their distribution (Hadziioannou et al., 2009). But since we know that source regions, especially in the microseism frequency band, move seasonally (see Chapter 3 [Ardhuin et al., 2018]), the spatial stationarity is far from reality and it is useful to understand the effects of source location changes on the different parts of the noise correlation signal.

Let us assume we can separate the signal that arrives at the stations into a ballistic arrival B that travels from the source directly to the receiver without being scattered and the remaining scattered part S. The ballistic part arrives after a well-defined travel time, whereas the scattered field excites the receiver for a long time after the ballistic arrival. Correlating these parts results in correlations between two ballistic arrivals (B, B), between a ballistic and a scattered part (B, S), and the correlation between two scattered fields (S, S). The apparent travel time change measured in these correlation functions due to a displacement of the source x is the difference between the travel time changes of the two correlated signals $\frac{d\tau}{dx} = \frac{dt_1}{dx} - \frac{dt_2}{dx}$. For the ballistic part this is simply $\frac{1}{c}\frac{dr}{dx}$ where r is the source-receiver distance. This leads to the strong sensitivity of tomography studies to source changes (Chapters 4 and 7 [Fichtner and Tsai, 2018; Shapiro, 2018]).

The propagation of a scattered wave is independent of the source location after it encounters the first scatterer, which means that any perturbation of traveltimes in S stems from the path R to the first scatterer: $\frac{dt}{dx} = \frac{1}{c}\frac{dR}{dx}$. The practical difference between the imprint of x on B and S is that a scattered arrival never occurs isolated. There are numerous scatterers that cause a superposition of different contributions – with different $\frac{dt}{dx}$ that are averaged in the correlation process. We therefore have

$$\frac{d\tau}{dx}(B, B) > \frac{d\tau}{dx}(B, S) > \frac{d\tau}{dx}(S, S). \tag{9.4}$$

Since the first (B, B) part with the strongest bias from source locations does not contribute to the coda part of a noise correlation function, we can expect coda wave measurements to be less sensitive to changes in source locations than direct waves.

In numerical experiments and analyses of field data, Froment et al. (2010) could therefore show that the estimation of travel time changes is far more stable in the coda than in ballistic waves. This was confirmed by Colombi et al. (2014).

9.2.3 Monitoring Changes

We have seen in the previous section that noise correlation functions can be obtained at a rate that allows for continuous monitoring and that coda wave interferometry allows us to estimate apparent velocity changes with high precision as a relative measurement between two of these signals. However, for long-term monitoring there will be hundreds or thousands of traces and different possibilities exist to combine those into a single curve representing the evolution of the propagation medium. The simplest approach is to create a single reference trace per station pair against which all other signals are compared. Often a long-term stack of the NCFs is used as a reference. Using a long-term stack is motivated by the need for a high-quality, noise-free reference trace.

In some cases, a long-term stack of NCFs might, however, deteriorate the reference function if travel time changes resulting, for example, from velocity changes become large compared to the dominant period of the signal. For $tdv/v > \pi/2\omega$ the phase change increases over $\pi/2$ and the individual NCFs do not stack constructively anymore. This occurs especially for late lapse time t or measurements at high-frequency ω. For this case, Sens-Schönfelder (2008) and Richter et al. (2014) suggested constructing the reference in an iterative way. In a first step, a rough reference is constructed from a long-term stack over all available times and apparent velocity changes are measured with respect to this preliminary reference. Then the individual NCFs are corrected for this preliminary velocity change by stretching. The corrected NCFs are then stacked again to obtain a final reference correlation function that does not suffer from phase shifts anymore. This reference trace is then used for the final measurements of the apparent velocity variations.

Another effect that influences the accuracy of the interferometric monitoring is related to permanent structural changes of the propagation medium that cannot be compensated by a simple stretching of the NCFs. As shown by Snieder (2006) a random displacement of scatterers results in a drop of the correlation between the coda signals recorded before and after the occurrence of the change. Spatially heterogeneous velocity changes, the addition of new scatterers, or permanent changes of the noise source distribution will have a similar effect on the NCFs. As a consequence of such perturbations the waves practically sample a different medium after the perturbing event. Volcanic eruptions can induce such changes by the emplacement of dykes and dome or caldera collapse. To compensate for these changes of the medium, Sens-Schönfelder et al. (2014) suggested using multiple reference traces constructed from different periods between the perturbing events. The different measurements corresponding to different references are then combined into a single curve after the apparent velocity changes between the different references have been corrected.

Brenguier et al. (2014) proposed using every single NCF as a reference for measurements of apparent velocity changes. For a set of n independent NCFs this process results in $n(n-1)/2$ measurements that are inverted in a least squares sense to obtain the $n-1$ independent observations of relative velocity changes. During this inversion a regularization can be introduced that applies a weighting to the measurements according to the temporal distance between the two traces that are compared. This allows us to focus adaptively either on fast varying signals or on long-term variations.

9.3 Spatial Sensitivity

An important topic for the coda-wave based monitoring with ambient seismic noise is the spatial sensitivity of the observations. The spatial sensitivity or sensitivity kernel describes the locations within the investigated volume at which a certain measurement is able to detect a perturbation of the material properties. Knowing this sensitivity for a set of measurements ultimately allows us to locate a perturbation spatially.

There are two main reasons why the sensitivity of noise-based coda measurements is different from the tomographic imaging, as discussed in Chapters 4 and 7 (Fichtner and Tsai, 2018; Shapiro, 2018). On the one hand, the noise correlation functions are (depending on the frequency range) mostly dominated by surface waves, and on the other hand the coda waves are scattered and sample the medium on complicated paths that are generally unknown. Moreover, scattering that generates the seismic coda converts the seismic energy back and forth between surface waves and body waves. Together with the frequency dependence of the scattering strength this causes a rather complicated dependency of the spatial sensitivity on the lapse time and frequency at which a measurement is performed. In general the sensitivity of coda wave measurements is spatially distributed within the single scattering ellipse and peaked at the locations of the two seismic stations. The details, however, depend on the wavefield characteristics, the frequency and lapse time of the measurement, and finally also the medium properties. In Chapter 6 (Snieder et al., 2018), a mathematical framework is presented that allows us to calculate the spatial sensitivity under the assumption that the propagation of the scattered coda waves can be described statistically.

9.4 Applications of Noise-Based Monitoring

Since the first investigations with the noise-based monitoring technique it has been applied in a large number of different environments. In this chapter we will discuss some key applications that demonstrate the technical capabilities of the methodology and the progress in understanding natural processes. We first discuss changes that are related to environmental influences like temperature and hydrology, because these effects often superimpose other signals of potentially greater interest. In section 9.4.2 we will introduce observations of volcanic processes using noise-based monitoring. Earthquake-related observations will be discussed in section 9.4.3.

9.4.1 Observations of Environmental Changes

Most seismic sensors are installed at the Earth's surface or in shallow boreholes. Since the sources of the ambient seismic field are also located at the surface, the ambient noise monitoring is very sensitive to changes that occur in the shallow subsurface, where numerous meteorological influences occur. Understanding such effects helps in interpreting other signals; for example, those signals originating from tectonic processes that are superimposed by environmental signals.

This effect was observed in the first application of noise-based monitoring (Sens-Schönfelder and Wegler, 2006). At Merapi volcano, Indonesia, investigations using earthquake multiplets (Ratdomopurbo and Poupinet, 1995) and active sources (Wegler et al., 2006) observed variations of the seismic velocity. Using ambient noise Sens-Schönfelder and Wegler (2006) could construct a continuous time series of velocity variations with a 1-day sampling over nearly two years (Figure 9.3). This amount of detail allowed us to identify changes in the groundwater level (Figure 9.3b) as the cause of the velocity changes. The hydrological changes in the volcanic edifice alter the propagation velocities at frequencies above 0.5 Hz

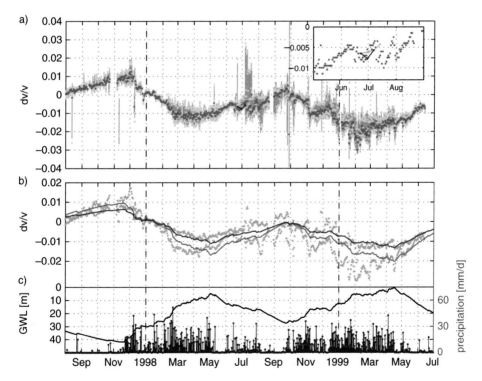

Figure 9.3. Seismic velocity changes observed at Merapi volcano, Indonesia. (a) Velocity changes observed for different combinations of array elements and their mean. Inset shows zoom into a comparison with active measurements by Wegler et al. (2006) shown as solid lines. (b) Measurements in 2–4 s (light grey; red in the plate section) and 6–8 s (dark grey; blue in the plate section) lapse time windows as dots. Continuous curves indicate model predictions based on precipitation and modeled ground water level (c). (Reprinted from Sens-Schönfelder and Wegler, 2006, with permission from Wiley.) (A black-and-white version of this figure appears in some formats. For the color version, please refer to the plate section.)

annually by about 1 percent. The different amplitudes observed at different lapse times (Figure 9.3b) clearly indicate a contribution of body wave propagation in the strongly scattering volcanic environment (Obermann et al., 2013a). This lapse time dependency was used to assess the depth range of the change.

The concept of deriving a model for the ground water level from precipitation data was also applied by Hillers et al. (2014) to explain velocity changes observed in a 1100 m deep bore hole in Taiwan between 1 Hz and 4 Hz. In contrast the observed lapse time dependence in the borehole with larger amplitudes for later times was opposite to the observations from the surface measurements at Merapi, confirming that the material changes occur in the shallow subsurface.

Very often, observations of velocity changes are interpreted as being related to environmental conditions solely based on their periodicity of a year or a day. Clear annual changes were observed by Hobiger et al. (2012, 2016) in several Japanese fault areas and by Hillers et al. (2015a) at the San Jacinto fault in California. In the Sichuan basin the influence of the monsoon was documented by Obermann et al. (2014). The influence of thermo-elastic stress on velocity changes in the Los Angeles basin and surrounding areas was discussed by Meier et al. (2010), who conclude that stresses induced by spatially variable surface temperatures can have a significant effect on velocities deeper in the crust. Thermoelastic stress was also invoked as the most likely explanation for annual and daily velocity changes observed in the dry Atacama desert in Northern Chile by Richter et al. (2014) because of the lack of precipitation and the almost perfect resemblance of the daily temperature to the velocity (except for a small phase shift). In the sedimentary basin in Northern Germany, regular annual velocity changes that follow local measurements of the ground water level are superimposed by sudden perturbations of large amplitude when the daily maximum air temperature drops below 0°C due to freezing of the ground. At first the seismic waves speed up because of the stiffening of the frozen soil, and then they significantly drop when the soil thaws because of the infiltration of meltwater (Gassenmeier et al., 2015). Analyzing seasonal velocity changes in all parts of Japan, Wang et al. (2017) discovered significant regional differences in the causative processes. Precipitation and resulting pore pressure changes dominate in the south, and a complex superposition of effects from precipitation, snow load, and sea level changes determine the seasonal seismic velocities changes in northern Japan.

Mordret et al. (2016) investigated velocity changes in Greenland and attributed the seasonal velocity changes to variations in the loading from the snowpack and pore pressure perturbations induced in the subsurface below the glacial cover. Since the pore pressure counteracts the confining pressure, Mordret et al. (2016) observed a delayed decrease of velocity following the accumulation of snow and the resulting pore pressure maximum.

The observation of periodic velocity changes in almost all environments investigated (including the lunar surface [Sens-Schönfelder and Larose, 2008]) demonstrates the sensitivity of the near-surface materials to environmental influences. Still, it seems not very exciting at first sight to monitor velocity changes in response to excitations that can easily be observed directly such as ground water level or temperature variations. An interesting aspect of using seismic velocity variations to study environmental processes such as hydrological changes concerns the spatial sensitivity of the measurements. In contrast to direct point measurements, the coda measurements represent a spatial average over a larger area, which might be a very useful complementary representation. A further reason to study superficial changes is that they always superimpose changes of deeper origin that cannot be observed directly.

9.4.2 Observations of Volcanic Processes

Volcano monitoring is probably the most successful application of noise-based monitoring. As sources of very significant natural hazards, volcanoes are investigated with various geodetic, geophysical, and geochemical methods that help to develop models for their internal structure and processes (Lees, 2007; Koulakov et al., 2013; Jaxybulatov et al., 2014). For the effective mitigation of hazards, the volcano investigations need to be augmented by monitoring techniques that provide information about the state of the volcanic system and changes thereof on time scales from years to hours. An important indicator for the activity of the volcano is the pressure buildup in the magma system and eventually the migration of magma that needs to be monitored closely. Geodetic techniques which can easily provide this temporal resolution only measure surface effects of changes within the volcanic edifice. This results in ambiguity and reduction of sensitivity. Seismic waves that travel through the interior of the volcano are sensitive to perturbations of the stress state that alter their velocity and to the fluid content via the ratio of P- and S-velocities. However, the temporal resolution of seismic tomography is of the order of years (Patanè et al., 2006), and the precision of the order of 1% is far too low to resolve changes of 0.1% or less that are induced by stress changes of the order of kPa. Here the seismic noise–based monitoring with a temporal resolution of usually 1 day and a precision below 0.1% finds its ideal application.

Brenguier et al. (2008b) were the first to observe changes of the seismic velocity induced by volcanic processes using ambient seismic noise–based monitoring. Using data from the permanent seismic monitoring network at Piton de la Fournaise during 1999 and 2000, Brenguier et al. (2008b) observed long-term velocity variations interpreted as environmental effects superimposed by short-term fluctuations that occurred a few days before eruptions. Since then, Piton de la Fournaise was subject to numerous studies that confirmed the systematic decrease of seismic velocity prior to eruptions. The 2007 eruption and dome collapse were studied by Clarke et al. (2013), and Rivet et al. (2014) investigated the long-term evolution of the seismic velocities at Piton de la Fournaise. Sens-Schönfelder et al. (2014) analyzed data of the UnderVolc experiment (Brenguier et al., 2012) from 2010 to 2011 and correlated the spatially resolved velocity changes with GPS displacement and volcanic activity. They found a clear correspondence between periods of inflation and velocity decreases. This occurs mostly prior to eruptions but also during a magma intrusion that did not reach the surface. Eruptive periods are characterized by spatially heterogeneous velocity increase, and periods of relaxation without volcanic and seismic activity are marked by a large-scale increase of velocity (Figure 9.4). These observations confirm the interpretation that velocity changes are caused by the dilation or compression of the surface of the volcano in response to pressure changes at depth. The decorrelation of the NCFs was investigated by Obermann et al. (2013b) to localize the two eruptions that occurred in 2010.

The strong interest in Piton de la Fournaise is due to the extraordinary activity with more than one eruption per year and the good instrumentation for a long period of time. It appears that the volcanic system at Piton de la Fournaise contains a shallow magma reservoir close to sea level 2800 m below the summit (Battaglia et al., 2005; Massin et al., 2011), which could be the common source for the eruptions (Peltier et al., 2009). The situation at Piton de la Fournaise might therefore be more favorable for the observation of eruption-related velocity changes than at other volcanoes.

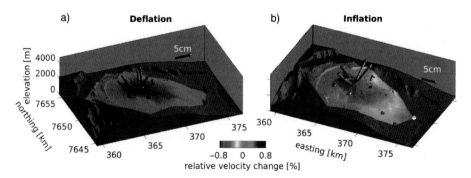

Figure 9.4. Seismic velocity changes observed at Piton de la Fournaise volcano during a phase of deflation in the first half of 2010 (a) and a phase of inflation in October 2010. (b) Arrows indicate displacement of GPS stations in corresponding time intervals. (Reprinted from Sens-Schönfelder et al., 2014, with permission from Elsevier.) (A black-and-white version of this figure appears in some formats. For the color version, please refer to the plate section.)

At Kilauea in Hawaii, short-term seismic velocity changes were found to correlate with deformation documented by tilt measurements (Donaldson et al., 2017). Interestingly, periods of inflation correspond to increase of the velocity and decreasing velocities indicate deflation, which is exactly the opposite to what is observed at Piton de la Fournaise. This apparent contradiction can be resolved by considering the spatial stress distribution resulting from pressure sources at different depths. The shallow reservoir at Kilauea with a connection to a lava lake leads to widespread compression during inflation. On the other hand, the reservoir 2.8 km below the steep topography of Piton de la Fournaise causes extension of the edifice upon inflation.

At the andesitic Merapi volcano, Budi-Santoso and Lesage (2016) analyzed velocity changes associated with the large 2010 eruption using families of repeating earthquakes at different depths as well as NCF. Observations are significantly more complex compared to the basaltic Piton de la Fournaise and Kilauea. Shallow multiplets showed a velocity decrease about 10 days prior to the eruption, whereas noise-based measurements showed positive as well as negative excursions in this period. After the first eruption the event-based time series stopped because the seismic activity changes and NCF observations showed consistent decrease of velocity in this period that lasted for 10 days until the largest eruption. Budi-Santoso and Lesage (2016) conclude that the variable changes prior to the first eruption are due to the spatial heterogeneity of the stress changes. They suggest to interpret any velocity change irrespective of its sign as an indicator of volcanic activity.

Mordret et al. (2010) investigated changes at Mt. Ruapehu in New Zealand associated with two eruptions in 2006 and 2007. While a localized 0.8% velocity decrease was observed starting 2 days prior to the 2006 eruption, no change in velocity was observed during the 2007 eruption.

While the observation of velocity changes during or before volcanic activity can help to understand the volcanic processes, the response of the volcanic system to other perturbations can help to probe the state of the system and provide additional characterization. Lesage et al. (2014) found that the seismic velocity of Volcán de Colima in Mexico was more sensitive to the shaking from regional earthquakes than to its own volcanic activity. The exceptional sensitivity of volcanic systems to dynamic strain was also observed by Brenguier et al. (2014) along the volcanic arc in Japan after the Tohoku earthquake in 2011. These observations will be discussed in the following section.

9.4.3 Earthquake-Related Observations

Changes of material properties during the strong shaking caused by earthquakes have been known for a long time. In the most extreme cases, the complete loss of strength can be observed as liquefaction during earthquakes or the seismic triggering of distant volcanic eruptions. A less dramatic expression of material changes induced by dynamic strain is the change of the site response measured during a strong mainshock and smaller aftershocks as observed by Field et al. (1998) during the Northridge earthquake. Yet, it is only the continuous high-precision, noise-based measurements that have allowed for routine observations of earthquake-induced subsurface changes and the subsequent processes that lead to a recovery of the initial state.

The coseismic decrease of the elastic moduli was first observed using the ambient seismic field at a single seismic station in Japan during the Mw 6.6 Mid-Niigata earthquake (Wegler and Sens-Schönfelder, 2007). Using a network of stations, Brenguier et al. (2008a) first observed the postseismic relaxation following the Mw 6.0 Parkfield earthquake with a similar time dependence as the postseismic GPS deformation.

After this initial observation, the two effects of an instantaneous decrease of the seismic velocity during a large earthquake and a postseismic partial or complete recovery of variable duration have been observed in many studies worldwide. A large number of observations exists in Japan due to the numerous earthquakes in a continuously well-instrumented area (Wegler et al., 2009; Sawazaki et al., 2009; Nakata and Snieder, 2011; Hobiger et al., 2012; Takagi et al., 2012; Hobiger et al., 2013; Sawazaki and Snieder, 2013). Some of these investigations estimate the velocity change between a downhole sensor and a surface sensor installed at the KiK-net stations using deconvolution of signals from distant earthquakes.

Comparing the responses of six large crustal earthquakes in Japan, Hobiger et al. (2016) showed that all events with magnitudes between $M_w = 6.2$ and 6.9 induced coseismic velocity drops of variable amplitude up to 0.5% in the vicinity of the fault. Recovery times were estimated between 0.1 and 1 year with large variations between different stations and frequency bands. For the Iwate-Miyagi earthquake, Hobiger et al. (2012) investigated the depth distribution of the material that experienced the velocity change. They observed a relatively clear correlation between measurements in different frequency bands and concluded, from the increasing amplitude of the coseismic change with increasing frequency, that the velocity changes occur mostly close to the surface. But Hobiger et al. (2012) also found stations where the frequency dependence requires changes that affect layers down to 2500 m. A similar observation associated with velocity changes during the Mw 7.9 2008 Wenchuan earthquake (Chen et al., 2010) was reported by Froment et al. (2013), who showed that the changes observed at periods above 10 s required the involvement of material at significant depth.

The interpretation of earthquake-related seismic velocity changes usually relies on either of two hypotheses that are often invoked and depends on the depth of their occurrence (Sens-Schönfelder and Wegler, 2011). Shallow changes measured at high frequencies are mostly interpreted as a consequence of near-surface damage during strong coseismic shaking and postseismic healing. In contrast, changes located deeper than a few hundred meters have been related to changes of the ambient stress level, with a coseismic drop and a postseismic re-increase. Physical details of these processes will be discussed in section 9.5.3. Key to gaining more insight into the physical origin of velocity changes are investigations that compare the responses of the subsurface to the excitation in a quantitative way. This is possible either by comparing the response of a specific region to various excitations or by studying the response of different materials to comparable excitations.

A unique possibility for the latter approach arose with the Tohoku earthquake in 2011. This devastating event offshore western Japan shook large parts of the best instrumented region of the world. Brenguier et al. (2014) mapped the coseismic decrease of the seismic velocity throughout western Japan and compared it to maps of the static strain and ground shaking during the main shock (Figure 9.5). Despite

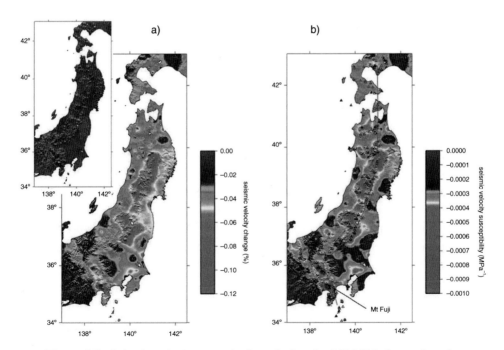

Figure 9.5. Seismic velocity perturbations during the 2011 Tohoku earthquake. (a) Seismic velocity changes during the main shock. (b) Seismic velocity susceptibility to dynamic strain. Inset shows velocity variations 5 days before the earthquake. (Reprinted from Brenguier et al., 2014, with permission from AAAS.) (A black-and-white version of this figure appears in some formats. For the color version, please refer to the plate section.)

the differences between the spatial distribution of the near-field strain perturbation and the broader spread shaking induced by the large-amplitude surface waves, neither phenomenon can explain the spatial distribution of the velocity perturbation, because the variable susceptibility of the subsurface materials is missing in such a direct comparison. The strong velocity changes distant from the fault zone indicate that static strain cannot play a significant role in the decrease of velocities, as it is orders of magnitude below the dynamic strain. Therefore, Brenguier et al. (2014) used the dynamic strain $\Delta\xi$ derived from peak ground velocity measurements of strong motion sensors as excitation and calculated the velocity susceptibility $\Delta v/\Delta\xi$ that describes how sensitive the material is to dynamic strain. This velocity change – scaled by the dynamic excitation – perfectly highlights the volcanic front and the Mt. Fuji region where the volcanic activity is located. This finding indicates that coseismic velocity reduction is caused dominantly by dynamic strain and that the effects of a perturbation depend on the location. In Japan the presence of over-pressurized fluids in the active volcanic systems was put forward as the relevant ingredient in the raising of the velocity susceptibility.

The role of pressurized fluids as a modulator for the velocity susceptibility is supported by the distribution of seismic velocity changes at depths greater than 5 km in the subduction zone off Costa Rica after an Mw 7.6 earthquake. Chaves and Schwartz (2016) observed strong changes only in an area where pressurized fluids were indicated by independent magnetotelluric and teleseismic receiver function studies.

Other evidence for the dynamic strain of passing elastic waves as the cause of the commonly observed seismic velocity decrease is provided by long-term studies that observe various events at the same location to rule out differences in susceptibility. Richter et al. (2014) observed systematic reductions of the seismic velocity in northern Chile after shaking from various regional earthquakes that correlated in amplitude very well with the local peak ground acceleration at the respective station. On the other hand, Richter et al. (2014) showed that there is no correlation between peak ground acceleration and amplitude of the velocity decrease if different stations are compared – again indicating the role of the spatially different velocity susceptibilities that lead to different amplitudes of velocity changes at locations that experienced the same level of excitation. As mentioned above, Lesage et al. (2014) investigated velocity variations during 15 years at Volcán de Colima in Mexico. In contrast to volcanic activity that had no pronounced effect on velocities they observed systematic velocity decreases in the volcanic edifice during shaking from regional tectonic earthquakes. The amplitude of the decrease was proportional to the logarithm of the peak amplitude of the seismic waves at the volcano. Outside the volcanic edifice the susceptibility was significantly lower and did not permit measurement of coseismic changes. The superior sensitivity of hydrothermal systems to dynamic perturbations and pore pressure changes was confirmed by Taira et al. (2017) at the Salton Sea geothermal field in California.

Convincing support for the dynamic strain from passing waves as the excitation mechanism of coseismic velocity changes comes from the modeling of Gassenmeier et al. (2016). These authors used a continuous measure of the local shaking to predict changes of the seismic velocity in northern Chile over a period of 8 years that includes numerous earthquakes (Figure 9.6). To characterize the effect of shaking in the measurement interval, they integrated the absolute value of the local ground acceleration. This integral measure of shaking explained all small

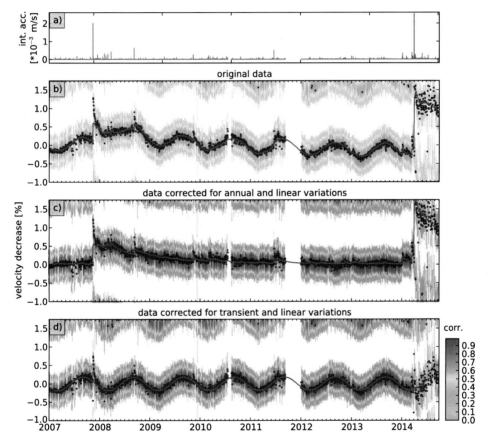

Figure 9.6. Seismic velocity variations at station Patache in the Atacama desert in northern Chile. (a) Integrated daily acceleration as measure of the shaking. (b) Observed seismic velocity change (dots) and model prediction based on the shaking. (c) Shaking induced and (d) seasonal component of the signal. (Reprinted from Gassenmeier et al., 2016, with permission from Oxford University Press.) (A black-and-white version of this figure appears in some formats. For the color version, please refer to the plate section.)

excursions of velocity variations that remained after the correction of a sinusoidal seasonal signal. Most of the larger perturbations could be related to shaking from regional earthquakes. The use of the integral value was motivated by a large aftershock of the Iquique earthquake that occurred hours after the main shock and which was located closer to the investigated station. This aftershock caused similar peak ground motion to the main shock, and clearly both events needed to be considered to account for the velocity signal.

9.4.4 Geotechnical Observations

Monitoring the propagation velocity of seismic waves can complement geotechnical monitoring of man-made structures. Applications range from the small scale where active ultrasonic sources can be used, to reservoir monitoring where the ambient vibrations are exploited. For a review of monitoring processes in concrete with controlled sources the reader is referred to Planès and Larose (2013). With active sources the stress changes were for example observed in an underground mine (Grêt et al., 2006), in a concrete bridge (Stähler et al., 2011), and in large test specimens (Zhang et al., 2016) where detailed observations of progressive damage were obtained. Traffic noise was used by Salvermoser et al. (2015) to demonstrate the softening of the structure with increasing temperature.

Traffic noise combined with waves emitted by a wind turbine was used by Planès et al. (2017) to monitor the behavior of a sea levee during a tidal cycle. They found a delayed anti-correlated response of the shear wave velocity to the sea level height that was interpreted as reflecting the pore pressure change in the levee. Maximum variations were located in two sections of the 300 m profiles that coincided with sand boils, suggesting the potential to monitor the internal erosion (Planès et al., 2016). A similar relation between areas of elevated seepage and pronounced velocity decreases was reported by Olivier et al. (2017) from a tailings dam in Australia built to contain the waste slurry remaining after processing ore. Olivier et al. (2017) used the ambient noise of unspecified sources in a selective way that neglects noise segments with unfavorable properties.

Of particular interest is the possibility of using the noise-based method for reservoir monitoring where the alternatives for observing processes in the reservoir are limited. Obermann et al. (2015) analyzed a seismic network installed around the St. Gallen geothermal site in Switzerland where gas was unexpectedly released from the formation after a fluid injection. Well control operations to stop the gas inflow required to continue operations despite increasing seismic activity led to an M_L 3.5 earthquake. Obermann et al. (2015) show that the pressure increase at injection depth could be monitored as changes in the ambient noise correlation functions already at the beginning of the fluid injection. This loss of correlation was restricted to the stations in the vicinity of the injection site. At the Basel (Switzerland) deep geothermal energy project, Hillers et al. (2015b) observed a velocity increase and a loss of waveform correlation that peaked about 15 days after a fluid injection at 5 km depth. This observation was interpreted as indicating deformation in the formation and the overlying material.

In situ observations of structural changes in the subsurface are possible using the ambient vibration in mines. Olivier et al. (2015b) showed that the Green's function retrieval at high frequencies in mining environments can be significantly improved by a selective stacking strategy that neglects unfavorable noise segments contaminated for example by monochromatic sources. Using this stacking method Olivier et al. (2015a) were able to monitor the superimposed effects of a widespread velocity decrease during a mining blast and the localized velocity increase in an area that experienced stress increase due to the newly created void. Dales et al. (2017) used the wavefield generated by three different ore crushers that constitute persistent and dominant sources to monitor medium changes between these sources and seismic receivers.

In general, applications of noise-based monitoring in the geotechnical context are of considerable interest as they can be of immediate use for engineers. The ambient noise in such environments, however, requires special care to be taken during the signal extraction.

9.5 Processes and Models for Changes of Seismic Material Properties

In the previous sections we have discussed observations that provide clear evidence for the importance of dynamic strain during coseismic velocity drop. The influence of static stress is seen at the Piton de la Fournaise volcano during volcanic activity. Postseismic recovery as well as the velocity susceptibility are clearly documented features of seismic velocities in Earth's material. In this section, we will discuss possible physical processes responsible for these observations.

9.5.1 Classical Nonlinearity

Variable seismic velocities that are influenced by the ambient stress level are not compatible with linear elasticity. Mathematically an extension of the stress-strain relation from the linear Hooke's law to the second or higher orders leads to the stress-strain relation

$$\sigma = M(\varepsilon + \beta\varepsilon^2 + \delta\varepsilon^3 + \ldots) \tag{9.5}$$

where M is the modulus and β and δ are the nonlinearity coefficients that are related to material properties describing the nonlinearity. For the velocity of sound it follows that

$$c = \sqrt{\rho^{-1}d\sigma/d\varepsilon} \approx c_0(1 + \beta\varepsilon + (3\delta - \beta^2/2\varepsilon^2 + \ldots) \tag{9.6}$$

(Ostrovsky and Johnson, 2001) where we see that the velocity depends on the applied strain.

Such an extension is physically reasonable because the assumption of a harmonic potential that allows for the linear Hooke's law only holds for small strain. The potential of the atomic forces in homogeneous materials such as crystals or fluids is asymmetric between the strong short-range repulsive forces and the attractive forces that hold the medium together. This is the so-called classical or atomic nonlinearity that can be observed in most homogeneous materials with typical values of nonlinear moduli of the order of 100–1000 GPa. The small strain ($\varepsilon \ll 1$) usually involved in dynamic experiments justifies the use of linear elasticity as a good approximation for dynamic observations in homogeneous materials.

9.5.2 Mesoscopic Nonlinearity

Earth materials are heterogeneous assemblages of different, yet possibly more homogeneous compounds like grains or crystals. In between these compounds exists a network of cracks, fissures, and contacts that has a significant influence on the elastic properties of the bulk material which are usually very different from the elastic properties of the individual constituents. Guyer and Johnson (1999) refer to these structures as the bond system of the material. A detailed review of the resulting behavior of geomaterials can be found in Ostrovsky and Johnson (2001).

Several of the observations that manifest these internal interfaces are discussed in Guyer and Johnson (2009) and Ostrovsky and Johnson (2001). In general, the sound velocity in such micro-heterogeneous materials is significantly lower than the velocity in any of its constituents. For sandstone and quartz these velocities differ by a factor of 2 to 4, which means an order of magnitude in the corresponding elastic modulus. But more important for the interpretation of velocity changes is the fact that micro-heterogeneity strongly increases the sensitivity of the sound velocity to stress changes. Expressed as the fractional change of velocity per fractional change of pressure, the sensitivity of rocks can be orders of magnitude larger than that of its minerals. In addition to these amplitude effects, the micro-heterogeneity induces a hysteresis in the stress-strain relation and a time dependence that involves various time scales and leads to memory inside the hysteresis loops and dynamics that follow the logarithm of time.

An effect of primary interest in the ultrasonic community is dynamic softening. When a micro-heterogeneous material is excited dynamically, i.e., with positive and negative strain that alternates faster than the medium can adapt to the new condition, it softens (Guyer et al., 1999; TenCate et al., 2004; Johnson and Sutin, 2005;

Pasqualini et al., 2007). This can also be observed in granular materials (Jia et al., 2011) and is very similar to many field observations of coseismic velocity changes.

The physics of these types of materials has been extensively studied in laboratory experiments with sandstone and concrete (Ostrovsky and Johnson, 2001). It is well accepted that the mentioned nonlinear behavior originates at internal inhomogeneities such as cracks and contacts. A clear indication for this is the increase of the nonlinear effects with increasing damage in the material (Van Den Abeele et al., 2000; Tremblay et al., 2010). Even homogeneous materials such as aluminum can show pronounced nonlinear behavior when they contain cracks (Rivière et al., 2014). This combination of soft and strong material phases causes large strain in the soft part of the material, which might occupy only a small fraction of the volume. The nonlinear behavior of the material depends on the weak contacts, while the linear elastic response is governed by the strong contacts (Tournat et al., 2004). Consequently the mesoscopic nonlinearity disappears when cracks are closed and contacts are compressed by the application of an external load (Rivière et al., 2016).

9.5.3 Models for the Elastic Behavior of Micro-Heterogeneous Materials

Rocks with abundant micro-heterogeneity and granular materials such as unconsolidated sediments belong to the class of materials that exhibit mesoscopic structural nonlinear elasticity. Models mostly developed to understand and explain related laboratory experiments can thus help us to understand the field observations of velocity changes. We will restrict ourselves here to brief discussions of only a few models to convey some common elements that are required to reproduce essential observations.

The precise physical origin of mesoscopic nonlinearity is still unclear, and potentially a variety of mechanisms is involved, which makes the search for *the* physical process elusive. Many models used to explain the behavior of geomaterials are therefore phenomenological. At first approaches focused on modeling hysteresis and the associated endpoint memory in quasi-static measurements of stress-strain relationships. The concept, termed Preisach-Mayergosz (PM) model, was developed by Guyer et al. (1995, 1997) among others. According to this idea, the macroscopic behavior of the material is governed by a large ensemble of hysteretic elastic units (HEU), of which each unit can be in an open (soft) or closed (strong) state. One can think of these units as micro-contacts with two possible configurations. Hysteresis occurs because the units switch between the two configurations at different pressures depending on the sign of the pressure change. To close a HEU, the pressure has to increase above a certain pressure P_c. To bring the unit into the open state again, the pressure has to decrease below a lower pressure threshold P_o ($P_o < P_c$). Using an ensemble of HEUs with various values of P_c and P_o allows one to explain hysteresis and memory in a micro-heterogeneous material. However, the PM model does not tell us anything about the nature of the HEU that creates the hysteresis. An essential element of this model is the presence of a variety of units on the size scale of micro-contacts that can be switched between two states with different properties. The model offers a possible explanation for the strong stress sensitivity of the seismic velocity in rocks, but it neither includes the time dependence that is required to explain postseismic relaxation nor does it explain the dynamic softening.

Transitions between the two states of the Preisach-Mayergosz model by thermo-dynamic fluctuations were included by Gusev and Tournat (2005). Zaitsev et al. (2014) presented a model for slow relaxation in granular materials that relies on thermodynamic transitions of the nanoscale contacts between two states. The two states in their model result from weakly loaded adhesive contacts (Barthel, 2008). Such contacts can be in an open state in which a weak external force keeps the two sides of the contact at a distance where the attractive force is weak. But the contacts can also be in a closed state where the adhesive force pulls the contacts together until the attractive force is balanced not only by the external force but additionally by the repulsive force of the contacting sides. Transitions between the states are possible but require the overcoming of a certain energy barrier that determines the probability of the transition. Tensile shocks will open contacts, and the equilibrium is reestablished by the thermal fluctuations that will preferentially close contacts after the perturbation. By using a large number of contacts and a variety of barrier heights Zaitsev et al. (2014) succeeded in qualitatively modeling damage during perturbations and the recovery process based on the interaction of contacts of a few nanometers in size.

A simple model for dynamic softening is a bilinear stress-strain relation (Lyakhovsky et al., 2009) in which the material has a smaller modulus during a tensional phase than during compression. This is clearly motivated by the contri-bution of cracks to the bulk modulus that is different for tension and compression. Increasing damage further lowers the tensional modulus as it increases the num-ber of cracks. Such a model can reproduce the amplitude-dependent softening of a material and hysteresis, but it does not allow for relaxation.

A model that includes relaxation together with dynamic softening and hystere-sis was presented by Vakhnenko et al. (2006). Their soft ratchet model consists of defects (ruptured cohesive bonds) that are created and annihilated at certain rates depending on the external stress. Vakhnenko et al. (2006) could quantita-tively reproduce a variety of observations with their cohesive bond model that involves the smallest contact scales that are subject to thermodynamic fluctuations with different barrier heights.

A common feature of the models for mesoscopic nonlinear effects is the spatial scale on which they operate. They all involve the very small structural scales on which thermodynamic fluctuations can introduce time dependence. An obvious question arising here is: What is the relevance of such processes for seismological observations? Up to now there is no clear answer to this question, but we put forward the following thoughts.

The relaxation process of the post-seismic velocity re-increase was shown to be linear on a logarithmic time scale. This has been clearly shown in laboratory and field experiments. Snieder et al. (2017) showed that this behavior can be explained with a superposition of exponential processes with a range of characteristic times. The limitation of the range of characteristic times guarantees the finiteness of the $\log(t)$-recovery for $t \to 0$ and $t \to \infty$. The spectrum of timescales can result from a variety of processes with different characteristic times or from similar processes occurring in cascades in which one level of the cascade is activated before the next one becomes active. But can observations on time scales of seconds done in the lab really be related to field observations that show changes lasting for years? At first there is an overlap of the longest laboratory time scales and the shortest time scales in the field. This is indicated by techniques that can observe changes faster than noise correlation techniques, which require time for convergence. Measurements of site responses (Wu et al., 2009a,b; Wu and Peng, 2012), correlation of vertically propagating ballistic waves (Chao and Peng, 2009), or deconvolution of borehole and surface records during strong motion (Nakata and Snieder, 2011) show changes during – or seconds after – strong ground motion that follow a linear trend on a logarithmic time axis, i.e., that are very fast in the beginning.

In fact, it is clear that the processes that occur in a laboratory-sized sample also occur in the field embedded in a larger environment, which leaves only the question of the upper bound of the time scale. Which process can act over time scales from 1 s to years (10^8 s)? A possible explanation is offered by the thermodynamic origin of the time dependence that is invoked in most models. The characteristic time τ of the thermodynamic transitions is related to the height of the energy barrier E_b by Arrhenius's law as $\tau \propto \exp(-E_b/k_B T)$, where k_B and T are Boltzmann's constant and temperature, respectively. For large barrier energy small changes of E_b can lead to large changes of the order of the characteristic time τ.

9.6 Summary

Noise-based seismic monitoring allows us to continuously track minute changes of the elastic material properties in the natural environment. Changes of the seismic velocity that lead to an apparent stretching of waveforms as well as structural changes that show up as a slight random waveform change can be inferred. Highest sensitivity is obtained by using coda waves that propagate for long times in the target medium and accumulate the effect of changes.

Systematic changes due to volcanic activity have been reported that are best explained by stress changes resulting from inflation or deflation of a magmatic reservoir. Volcanoes are also very sensitive to dynamic strain caused by propagating elastic waves originating from earthquakes. Such coseismic changes that systematically decrease the seismic velocity and introduce a de-correlation of waveforms are also observed in near-surface materials close to the epicentral area. Deeply rooted changes are reported less frequently as they are superimposed by the stronger near-surface effects. Frequently observed periodic changes are caused by environmental effects like hydrological changes or thermoelastic stress due to temperature differences. Man-made changes in geotechnical structures such as mines, dams, and injection sites can also be monitored.

Even though the observed changes in seismic velocity usually only amount to a fraction of a percent, their amplitude is much larger than expected from the properties of the rock-forming minerals, indicating that the sensitivity is related to the structure of the material. Micro-inhomogeneous materials such as rocks and granular media exhibit mesoscopic structural nonlinearity that leads to

- strong stress sensitivity
- dynamic softening
- slow dynamic recovery

which can now be observed in the field using the ambient seismic field. The mesoscopic nonlinearity strongly depends on the damage, i.e., the fracture density of the material, and therefore decreases quickly with depth and confining pressure.

Models for the mesoscopic nonlinearity are often phenomenological. Those attempts that describe observations on a basis of physical processes include the smallest spatial scale of material contacts and cracks. The temporal dynamics of recovery process are mostly linked to a variety of processes with different timescales.

Acknowledgments

We would like to thank two anonymous reviewers for their thoughtful comments that helped to improve this chapter.

References

Ardhuin, F., Gualtieri, L., and Stutzmann, E. 2018. Physics of ambient noise generation by ocean waves. Pages – of: Nakata, N., Gualtieri, L., and Fichtner, A. (eds.), *Seismic Ambient Noise*. Cambridge University Press, Cambridge, UK.

Barthel, E. 2008. Adhesive elastic contacts: JKR and more. *J. Phys. D. Appl. Phys.*, **41**(16), 163001.

Battaglia, J., Ferrazzini, V., Staudacher, T., Aki, K., and Cheminée, J.-L. 2005. Pre-eruptive migration of earthquakes at the Piton de la Fournaise volcano (Réunion Island). *Geophys. J. Int.*, **161**(2), 549–558.

Brenguier, F., Campillo, M., Hadziioannou, C., Shapiro, N. M., Nadeau, R. M., and Larose, E. 2008a. Postseismic relaxation along the San Andreas fault at Parkfield from continuous seismological observations. *Science*, **321**(5895), 1478–81.

Brenguier, F., Campillo, M., Takeda, T., Aoki, Y., Shapiro, N. M., Briand, X., Emoto, K., and Miyake, H. 2014. Mapping pressurized volcanic fluids from induced crustal seismic velocity drops. *Science (80-.).*, **345**(6192), 80–82.

Brenguier, F., Kowalski, P., Staudacher, T., Ferrazzini, V., Lauret, F., Boissier, P., Catherine, P., Lemarchand, A., Pequegnat, C., Meric, O., Pardo, C., Peltier, A., Tait, S., Shapiro, N. M., Campillo, M., and Di Muro, A. 2012. First results from the UnderVolc High Resolution Seismic and GPS Network deployed on Piton de la Fournaise volcano. *Seismol. Res. Lett.*, **83**(1), 97–102.

Brenguier, F., Shapiro, N. M., Campillo, M., Ferrazzini, V., Duputel, Z., Coutant, O., and Nercessian, A. 2008b. Towards forecasting volcanic eruptions using seismic noise. *Nat. Geosci.*, **1**(2), 126–130.

Budi-Santoso, A., and Lesage, P. 2016. Velocity variations associated with the large 2010 eruption of Merapi volcano, Java, retrieved from seismic multiplets and ambient noise cross-correlation. *Geophys. J. Int.*, **206**(1), 221–240.

Chao, K., and Peng, Z. 2009. Temporal changes of seismic velocity and anisotropy in the shallow crust induced by the 1999 October 22 M6.4 Chia-Yi, Taiwan earthquake. *Geophys. J. Int.*, **179**(3), 1800–1816.

Chaves, E. J., and Schwartz, S. Y. 2016. Monitoring transient changes within overpressured regions of subduction zones using ambient seismic noise. *Sci. Adv.*, **2**(1), e1501289.

Chen, J. H., Froment, B., Liu, Q. Y., and Campillo, M. 2010. Distribution of seismic wave speed changes associated with the 12 May 2008 Mw 7.9 Wenchuan earthquake. *Geophys. Res. Lett.*, **37**(18), L18302 1–4.

Clarke, D., Brenguier, F., Froger, J.-L., Shapiro, N. M., Peltier, A., and Staudacher, T. 2013. Timing of a large volcanic flank movement at Piton de la Fournaise Volcano using noise-based seismic monitoring and ground deformation measurements. *Geophys. J. Int.*, **195**(2), 1132–1140.

Colombi, A., Chaput, J., Brenguier, F., Hillers, G., Roux, P., and Campillo, M. 2014. On the temporal stability of the coda of ambient noise correlations. *Comptes Rendus Geosci.*, **346**(11–12), 307–316.

Dai, S., Wuttke, F., and Santamarina, J. C. 2013. Coda wave analysis to monitor processes in soils. *J. Geotech. Geoenvironmental Eng.*, 1504–1511.

Dales, P., Audet, P., and Olivier, G. 2017. Seismic interferometry using persistent noise sources for temporal subsurface monitoring. *Geophys. Res. Lett.*, 863–870.

Donaldson, C., Caudron, C., Green, R. G., Thelen, W. A., and White, R. S. 2017. Relative seismic velocity variations correlate with deformation at Kilauea volcano. *Sci. Adv.*, **3**, 1–12.

Fichtner, A., and Tsai, V. 2018. Theoretical foundations of noise interferometry. Pages – of: Nakata, N., Gualtieri, L., and Fichtner, A. (eds), *Seismic Ambient Noise*. Cambridge University Press, Cambridge, UK.

Field, E. H., Zeng, Y., Johnson, P. A., and Beresnev, I. A. 1998. Nonlinear sediment response during the 1994 Northridge earthquake: Observations and finite source simulations. *J. Geophys. Res.*, **103**(B11), 26869.

Froment, B., Campillo, M., Chen, J., and Liu, Q. 2013. Deformation at depth associated with the 12 May 2008 MW 7.9 Wenchuan earthquake from seismic ambient noise monitoring. *Geophys. Res. Lett.*, **40**(1), 78–82.

Froment, B., Campillo, M., Roux, P., Gouédard, P., Verdel, A., and Weaver, R. L. 2010. Estimation of the effect of nonisotropically distributed energy on the apparent arrival time in correlations. *Geophysics*, **75**(5), SA85.

Gassenmeier, M., Sens-Schönfelder, C., Delatre, M., and Korn, M. 2015. Monitoring of environmental influences on seismic velocity at the geological storage site for CO2 in Ketzin (Germany) with ambient seismic noise. *Geophys. J. Int.*, **200**(1), 524–533.

Gassenmeier, M., Sens-Schönfelder, C., Eulenfeld, T., Bartsch, M., Victor, P., Tilmann, F., and Korn, M. 2016. Field observations of seismic velocity changes caused by shaking-induced damage and healing due to mesoscopic nonlinearity. *Geophys. J. Int.*, **204**(3), 1490–1502.

Grêt, A., Snieder, R., and Özbay, U. 2006. Monitoring in situ stress changes in a mining environment with coda wave interferometry. *Geophys. J. Int.*, **167**(2), 504–508.

Gusev, V., and Tournat, V. 2005. Amplitude- and frequency-dependent nonlinearities in the presence of thermally-induced transitions in the Preisach model of acoustic hysteresis. *Phys. Rev. B - Condens. Matter Mater. Phys.*, **72**(5), 1–19.

Guyer, R., and Johnson, P. A. 2009. *Nonlinear Mesoscopic Elasticity: The Complex Behaviour of Rocks, Soil, Concrete*. Wiley.

Guyer, R., and Johnson, P. A. 1999. Nonlinear mesoscopic elasticity: evidence for a new class of materials. *Phys. Today*, 30–36.

Guyer, R., McCall, K., and Boitnott, G. 1995. Hysteresis, Discrete Memory, and Nonlinear Wave Propagation in Rock: A New Paradigm. *Phys. Rev. Lett.*, **74**(17), 3491–3494.

Guyer, R. A., McCall, K. R., Boitnott, G. N., Hilbert Jr., L. B., and Plona, T. J. 1997. Quantitative implementation of Preisach-Mayergoyz space to find static and dynamic elastic moduli in rock. *J. Geophys. Res.*, **102**, 5281–5293.

Guyer, R., TenCate, J., and Johnson, P. 1999. Hysteresis and the Dynamic Elasticity of Consolidated Granular Materials. *Phys. Rev. Lett.*, **82**(16), 3280–3283.

Hadziioannou, C., Larose, E., and Coutant, O. 2009. Stability of monitoring weak changes in multiply scattering media with ambient noise correlation: Laboratory experiments. *J. Acoust. Soc. Am.*, **125**, 3688.

Hillers, G., Ben-Zion, Y., Campillo, M., and Zigone, D. 2015a. Seasonal variations of seismic velocities in the San Jacinto fault area observed with ambient seismic noise. *Geophys. J. Int.*, **202**(2), 920–932.

Hillers, G., Campillo, M., and Ma, K. 2014. Seismic velocity variations at TCDP are controlled by MJO driven precipitation pattern and high fluid discharge properties. *Earth Planet. Sci. Lett.*, **391**, 121–127.

Hillers, G., Husen, S., Obermann, A., Planès, T., Larose, E., and Campillo, M. 2015b. Noise-based monitoring and imaging of aseismic transient deformation induced by the 2006 Basel reservoir stimulation. *Geophysics*, **80**(4), KS51–KS68.

Hobiger, M., Wegler, U., Shiomi, K., and Nakahara, H. 2012. Coseismic and postseismic elastic wave velocity variations caused by the 2008 Iwate-Miyagi Nairiku earthquake, Japan. *J. Geophys. Res.*, **117**(B9), 1–19.

Hobiger, M., Wegler, U., Shiomi, K., and Nakahara, H. 2013. Single-station cross-correlation analysis of ambient seismic noise: Application to stations in the surroundings of the 2008 Iwate-Miyagi Nairiku earthquake. *Geophys. J. Int.*, **0**, 1–11.

Hobiger, M., Wegler, U., Shiomi, K., and Nakahara, H. 2016. Coseismic and post-seismic velocity changes detected by passive image interferometry: Comparison of one great and five strong earthquakes in Japan. *Geophys. J. Int.*, **205**(2), 1053–1073.

Jaxybulatov, K., Shapiro, N. M., Koulakov, I., Mordret, A., Landes, M., and Sens-Schonfelder, C. 2014. A large magmatic sill complex beneath the Toba caldera. *Science*, **346**(6209), 617–619.

Jia, X., Brunet, T., and Laurent, J. 2011. Elastic weakening of a dense granular pack by acoustic fluidization: Slipping, compaction, and aging. *Phys. Rev. E - Stat. Nonlinear, Soft Matter Phys.*, **84**(2), 2–5.

Johnson, P., and Sutin, A. 2005. Slow dynamics and anomalous nonlinear fast dynamics in diverse solids. *J. Acoust. Soc. Am.*, **117**(1), 124–130.

Koulakov, I., Gordeev, E. I., Dobretsov, N. L., Vernikovsky, V. A., Senyukov, S., Jakovlev, A., and Jaxybulatov, K. 2013. Rapid changes in magma storage beneath the Klyuchevskoy group of volcanoes inferred from time-dependent seismic tomography. *J. Volcanol. Geotherm. Res.*, **263**, 75–91.

Lees, J. M. 2007. Seismic tomography of magmatic systems. *J. Volcanol. Geotherm. Res.*, **167**(1-4), 37–56.

Lesage, P., Reyes-Dávila, G., and Arámbula-Mendoza, R. 2014. Large tectonic earthquakes induce sharp temporary decreases in seismic velocity in Volcán de Colima, Mexico. *J. Geophys. Res. Solid Earth*, 4360–4376.

Li, Y.-G., and Vidale, J. 2001. Healing of the the 1992 M7.5 shallow fault zone from 1994-1998 Landers, California, earthquake. *Geophys. Res. Lett.*, **28**(15), 2999–3002.

Li, Y. G., Vidale, J. E., Day, S. M., Oglesby, D. D., and Cochran, E. 2003. Postseismic fault healing on the rupture zone of the 1999 M 7.1 Hector Mine, California, earthquake. *Bull. Seismol. Soc. Am.*, **93**(2), 854–869.

Lyakhovsky, V., Hamiel, Y., Ampuero, J. P., and Ben-Zion, Y. 2009. Non-linear damage rheology and wave resonance in rocks. *Geophys. J. Int.*, **178**(2), 910–920.

Massin, F., Ferrazzini, V., Bachèlery, P., Nercessian, A., Duputel, Z., and Staudacher, T. 2011. Structures and evolution of the plumbing system of Piton de la Fournaise volcano inferred from clustering of 2007 eruptive cycle seismicity. *J. Volcanol. Geotherm. Res.*, **202**(1–2), 96–106.

Meier, U., Shapiro, N. M., and Brenguier, F. 2010. Detecting seasonal variations in seismic velocities within Los Angeles basin from correlations of ambient seismic noise. *Geophys. J. Int.*, March, 985–996.

Mordret, A., Jolly, A., Duputel, Z., and Fournier, N. 2010. Monitoring of phreatic eruptions using interferometry on retrieved cross-correlation function from ambient seismic noise: results from Mt. Ruapehu, New Zealand. *J. Volcanol. Geotherm. Res.*, **191**(1–2), 46–59.

Mordret, A., Mikesell, T. D., Harig, C., Lipovsky, B. P., and Prieto, G. A. 2016. Monitoring southwest Greenlands ice sheet melt with ambient seismic noise. *Sci. Adv.*, **2**(5), e1501538–e1501538.

Nakata, N., and Snieder, R. 2011. Near-surface weakening in Japan after the 2011 Tohoku-Oki earthquake. *Geophys. Res. Lett.*, **38**(17), 1–5.

Nishimura, T., Tanaka, S., Yamawaki, T., Yamamoto, H., Sano, T., Sato, M., Nakahara, H., Uchida, N., Hori, S., and Sato, H. 2005. Temporal changes in seismic velocity of the crust around Iwate volcano, Japan, as inferred from analyses of repeated active seismic experiment data from 1998 to 2003. *Earth, Planets Sp.*, **57**(6), 491–505.

Nishimura, T., Uchida, N., Sato, H., Ohtake, M., Tanaka, S., and Hamaguchi, H. 2000. Temporal changes of the crustal structure associated with the M6.1 earthquake on September 3, 1998, and the volcanic activity of Mount Iwate, Japan. *Geophys. Res. Lett.*, **27**(2), 269–272.

Niu, F., Silver, P. G., Daley, T. M., Cheng, X., and Majer, E. L. 2008. Preseismic velocity changes observed from active source monitoring at the Parkfield SAFOD drill site. *Nature*, **454**(7201), 204–8.

Obermann, A., Froment, B., Campillo, M., Larose, E., Planès, T., Valette, B., Chen, J. H., and Liu, Q. Y. 2014. Seismic noise correlations to image structural and mechanical changes associated with the Mw 7.9 2008 Wenchuan earthquake. *Journal*, 1–14.

Obermann, A., Kraft, T., Larose, E., and Wiemer, S. 2015. Potential of ambient seismic noise techniques to monitor the St. Gallen geothermal site (Switzerland). *J. Geophys. Res. Solid Earth*, **120**, 1–16.

Obermann, A., Planès, T., Larose, E., and Campillo, M. 2013b. Imaging preeruptive and coeruptive structural and mechanical changes of a volcano with ambient seismic noise. *J. Geophys. Res. Solid Earth*, **118**(12), 6285–6294.

Obermann, A., Planès, T., Larose, E., Sens-Schönfelder, C., and Campillo, M. 2013a. Depth sensitivity of seismic coda waves to velocity perturbations in an elastic heterogeneous medium. *Geophys. J. Int.*, April, 372–382.

Olivier, G., Brenguier, F., Campillo, M., Lynch, R., and Roux, P. 2015b. Body-wave reconstruction from ambient seismic noise correlations in an underground mine. *Geophysics*, **80**(3), KS11–KS25.

Olivier, G., Brenguier, F., Campillo, M., Roux, P., Shapiro, N. M., and Lynch, R. 2015a. Investigation of coseismic and postseismic processes using in situ measurements of seismic velocity variations in an underground mine. *Geophys. Res. Lett.*, **42**(21), 9261–9269.

Olivier, G., Brenguier, F., de Wit, T., and Lynch, R. 2017. Monitoring the stability of tailings dam walls with ambient seismic noise. *Lead. Edge*, **36**(4), 350a1–350a6.

Ostrovsky, L. A., and Johnson, P. A. 2001. Dynamic nonlinear elasticity in geomaterials. *La Riv. del Nuovo Cimentó*, **24**(7), 1–46.

Pasqualini, D., Heitmann, K., TenCate, J. A., Habib, S., Higdon, D., and Johnson, P. A. 2007. Nonequilibrium and nonlinear dynamics in Berea and Fontainebleau sandstones: Low-strain regime. *J. Geophys. Res. Solid Earth*, **112**(1), 1–16.

Patanè, D., Barberi, G., Cocina, O., De Gori, P., and Chiarabba, C. 2006. Time-resolved seismic tomography detects magma intrusions at Mount Etna. *Science*, **313**(5788), 821–3.

Peltier, A., Bachèlery, P., and Staudacher, T. 2009. Magma transport and storage at Piton de La Fournaise (La Réunion) between 1972 and 2007: A review of geophysical and geochemical data. *J. Volcanol. Geotherm. Res.*, **184**(1–2), 93–108.

Peng, Z., and Ben-Zion, Y. 2006. Temporal changes of shallow seismic velocity around the Karadere-Düzce branch of the north Anatolian fault and strong ground motion. *Pure Appl. Geophys.*, **163**(2–3), 567–600.

Planès, T., Mooney, M. A., Rittgers, J. B. R., Parekh, M. L., Behm, M., and Snieder, R. 2016. Time-lapse monitoring of internal erosion in earthen dams and levees using ambient seismic noise. *Géotechnique*, **66**(4), 301–312.

Planès, T., Rittgers, J. B., Mooney, M. A., Kanning, W., and Draganov, D. 2017. Monitoring the tidal response of a sea levee with ambient seismic noise. *J. Appl. Geophys.*, **138**, 255–263.

Planès, T., and Larose, E. 2013. A review of ultrasonic Coda Wave Interferometry in concrete. *Cement and Concrete Research*, **53**, 248–255.

Poupinet, G., Ellsworth, W., and Frechet, J. 1984. Monitoring velocity variations in the crust using earthquake doublets: An application to the Calaveras Fault, California. *J. Geophys. Res.*, **89**(4), 5719–5731.

Ratdomopurbo, A., and Poupinet, G. 1995. Monitoring a temporal change of seismic velocity in a volcano: Application to the 1992 eruption of Mt. Merapi (Indonesia). *Geophys. Res. Lett.*, **22**(7), 775–778.

Richter, T., Sens-Schönfelder, C., Kind, R., and Asch, G. 2014. Comprehensive observation and modeling of earthquake and temperature-related seismic velocity changes in northern Chile with passive image interferometry. *J. Geophys. Res.*, **119**, 4747–4765.

Ritzwoller, M. H., and Feng, L. 2018. Overview of pre- and post-processing of ambient noise correlations. Pages – of: Nakata, N., Gualtieri, L., and Fichtner, A. (eds.), *Seismic Ambient Noise*. Cambridge University Press, Cambridge, UK.

Rivet, D., Brenguier, F., Clarke, D., Shapiro, N. M., and Peltier, A. 2014. Long-term dynamics of Piton de la Fournaise volcano from 13 years of seismic velocity change measurements and GPS observations. *J. Geophys. Res. Solid Earth*, **119**, 1–13.

Rivière, J., Pimienta, L., Scuderi, M., Candela, T., Shokouhi, P., Fortin, J., Schubnel, A., Marone, C., and Johnson, P. A. 2016. Frequency, pressure, and strain dependence of nonlinear elasticity in Berea Sandstone. *Geophys. Res. Lett.*, **43**(7), 3226–3236.

Rivière, J., Remillieux, M. C., Ohara, Y., Anderson, B. E., Haupert, S., Ulrich, T. J., and Johnson, P. A. 2014. Dynamic acousto-elasticity in a fatigue-cracked sample. *J. Nondestruct. Eval.*, **33**(2), 216–225.

Rubinstein, J. L., and Beroza, G. C. 2004. Evidence for widespread nonlinear strong ground motion in the MW 6.9 Loma Prieta earthquake. *Bull. Seismol. Soc. Am.*, **94**(5), 1595–1608.

Rubinstein, J. L., and Beroza, G. C. 2005. Depth constraints on nonlinear strong ground motion from the 2004 Parkfield earthquake. *Geophys. Res. Lett.*, **32**(14), 1–5.

Salvermoser, J., Hadziioannou, C., and Stähler, S. C. 2015. Structural monitoring of a highway bridge using passive noise recordings from street traffic. *J. Acoust. Soc. Am.*, **138**(6), 3864–3872.

Sawazaki, K., and Snieder, R. 2013. Time-lapse changes of P- and S-wave velocities and shear wave splitting in the first year after the 2011 Tohoku earthquake, Japan: shallow subsurface. *Geophys. J. Int.*, **193**(1), 238–251.

Sawazaki, K., Sato, H., Nakahara, H., and Nishimura, T. 2009. Time-lapse changes of seismic velocity in the shallow ground caused by strong ground motion Shock of the 2000 Western-Tottori earthquake, Japan, as revealed from coda deconvolution analysis. *Bull. Seismol. Soc. Am.*, **99**(1), 352–366.

Schaff, D. P., and Beroza, G. C. 2004. Coseismic and postseismic velocity changes measured by repeating earthquakes. *J. Geophys. Res. B Solid Earth*, **109**(10), B10302.

Sens-Schönfelder, C. 2008. Synchronizing seismic networks with ambient noise. *Geophys. J. Int.*, **174**(3), 966–970.

Sens-Schönfelder, C., and Larose, E. 2008. Temporal changes in the lunar soil from correlation of diffuse vibrations. *Phys. Rev. E*, **78**(4), 1–4.

Sens-Schönfelder, C., and Wegler, U. 2006. Passive image interferometry and seasonal variations of seismic velocities at Merapi Volcano, Indonesia. *Geophys. Res. Lett.*, **33**(21), 1–5.

Sens-Schönfelder, C., and Wegler, U. 2011. Passive image interferometry for monitoring crustal changes with ambient seismic noise. *Comptes Rendus Geosci.*, **343** (June), 639–651.

Sens-Schönfelder, C., Pomponi, E., and Peltier, A. 2014. Dynamics of Piton de la Fournaise Volcano observed by passive image interferometry with multiple references. *J. Volcanol. Geotherm. Res.*, **276** (February), 32–45.

Shapiro, N. 2018. Applications with surface waves extracted from ambient seismic noise. Pages — of: Nakata, N., Gualtieri, L., and Fichtner, A. (eds.), *Seismic Ambient Noise*. Cambridge University Press, Cambridge, UK.

Shapiro, N., and Campillo, M. 2004. Emergence of broadband Rayleigh waves from correlations of the ambient seismic noise. *Geophys. Res. Lett.*, **31** (April), 7614.

Snieder, R. 2006. The Theory of Coda Wave Interferometry. *Pure Appl. Geophys.*, **163**(2–3), 455–473.

Snieder, R., Duran, A., and Obermann, A. 2018. Locating velocity changes in elastic media with coda wave interferometry. Pages – of: Nakata, N., Gualtieri, L., and Fichtner, A. (eds.), *Seismic Ambient Noise*. Cambridge University Press, Cambridge, UK.

Snieder, R., Grêt, A., Douma, H., and Scales, J. 2002. Coda wave interferometry for estimating nonlinear behavior in seismic velocity. *Science*, **295**(5563), 2253–5.

Snieder, R., Sens-Schönfelder, C., and Wu, R. 2017. The time dependence of rock healing as a universal relaxation process, a tutorial. *Geophys. J. Int.*, **208**(2), 1–9.

Snieder, R., Van Wijk, K., Haney, M., and Calvert, R. 2008. Cancellation of spurious arrivals in Green's function extraction and the generalized optical theorem. *Phys. Rev. E - Stat. Nonlinear, Soft Matter Phys.*, **78**(3), 1–8.

Stähler, S. C., Sens-Schönfelder, C., and Niederleithinger, E. 2011. Monitoring stress changes in a concrete bridge with coda wave interferometry. *J. Acoust. Soc. Am.*, **129**(4), 1945–52.

Taira, T., Nayak, A., Brenguier, F., and Manga, M. 2017. Monitoring reservoir response to earthquakes and fluid extraction, Salton Sea geothermal field, California. *Sci. Adv.*, in press.

Takagi, R., Okada, T., Nakahara, H., Umino, N., and Hasegawa, A. 2012. Coseismic velocity change in and around the focal region of the 2008 Iwate-Miyagi Nairiku earthquake. *J. Geophys. Res.*, **117**(B6), B06315.

TenCate, J. A., Pasqualini, D., Habib, S., Heitmann, K., Higdon, D., and Johnson, P. A. 2004. Nonlinear and nonequilibrium dynamics in geomaterials. *Phys. Rev. Lett.*, **93**(6), 4–7.

Tournat, V., Zaitsev, V., Gusev, V., Nazarov, V., Béquin, P., and Castagnède, B. 2004. Probing weak forces in granular media through nonlinear dynamic dilatancy: Clapping contacts and polarization anisotropy. *Phys. Rev. Lett.*, **92**(9), 099903–1.

Tremblay, N., Larose, E., and Rossetto, V. 2010. Probing slow dynamics of consolidated granular multicomposite materials by diffuse acoustic wave spectroscopy. *J. Acoust. Soc. Am.*, **3**(127), 1–6.

Vakhnenko, O. O., Vakhnenko, V. O., Shankland, T. J., and TenCate, J. A. 2006. Soft-ratchet modeling of slow dynamics in the nonlinear resonant response of sedimentary rocks. *AIP Conf. Proc.*, **838**, 120–123.

Van Den Abeele, K. E. A., Johnson, P. A., and Sutin, A., 2000. Nonlinear Elastic Wave Spectroscopy (NEWS) techniques to discern material damage, Part I: Nonlinear Wave Modulation Spectroscopy (NWMS). *Res. Nondestruct. Eval.*, **12**(1), 17–30.

Vidale, J. E., and Li, Y. G. 2003. Damage to the shallow Landers fault from the nearby Hector Mine earthquake. *Nature*, **421**(6922), 524–526.

Wang, Q. Y., Brenguier, F., Campillo, M., Lecointre, A., Takeda, T., and Aoki, Y. 2017. Seasonal crustal seismic velocity changes throughout Japan. *J. Geophys. Res. Solid Earth*, **122**(10), 7987–8002.

Weaver, R. L., Hadziioannou, C., Larose, E., and Campillo, M. 2011. On the precision of noise correlation interferometry. *Geophys. J. Int.*, **185**(3), 1384–1392.

Wegler, U., and Sens-Schönfelder, C. 2007. Fault zone monitoring with passive image interferometry. *Geophys. J. Int.*, **168**(3), 1029–1033.

Wegler, U., Lühr, B.-G., Snieder, R., and Ratdomopurbo, A. 2006. Increase of shear wave velocity before the 1998 eruption of Merapi volcano (Indonesia). *Geophys. Res. Lett.*, **33** (May), 9303.

Wegler, U., Nakahara, H., Sens-Schönfelder, C., Korn, M., and Shiomi, K. 2009. Sudden drop of seismic velocity after the 2004 M w 6.6 mid-Niigata earthquake, Japan, observed with Passive Image Interferometry. *J. Geophys. Res.*, **114**(B06305), 1–11.

Wu, C., and Peng, Z. 2012. Long-term change of site response after the M_w 9.0 Tohoku earthquake in Japan. *Earth, Planets Sp.*, **64**(12), 1259–1266.

Wu, C., Peng, Z., and Assimaki, D. 2009b. Temporal changes in site response associated with the strong ground motion of the 2004 M w 6.6 Mid-Niigata earthquake sequences in Japan. *Bull. Seismol. Soc. Am.*, **99**(6), 3487–3895.

Wu, C., Peng, Z., and Ben-Zion, Y. 2009a. Non-linearity and temporal changes of fault zone site response associated with strong ground motion. *Geophys. J. Int.*, **176**(1), 265–278.

Yamawaki, T., Nishimura, T., and Hamaguchi, H. 2004. Temporal change of seismic structure around Iwate volcano inferred from waveform correlation analysis of similar earthquakes. *Geophys. Res. Lett.*, **31**(24), 1–4.

Zaitsev, V. Y., Gusev, V. E., Tournat, V., and Richard, P. 2014. Slow relaxation and aging phenomena at the nanoscale in granular materials. *Phys. Rev. Lett.*, **112** (March).

Zhan, Z., Tsai, V. C., and Clayton, R. W. 2013. Spurious velocity changes caused by temporal variations in ambient noise frequency content. *Geophys. J. Int.*, **194**(3), 1574–1581.

Zhang, Y., Planès, T., Larose, E., Obermann, A., Rospars, C., and Moreau, G. 2016. Diffuse ultrasound monitoring of stress and damage development on a 15-ton concrete beam. *J. Acoust. Soc. Am.*, **139**(4), 1691–1701.

10

Near-Surface Engineering

KOICHI HAYASHI

Abstract

Knowledge of S-wave velocity (V_S) structure to depths of several tens of meters to several kilometers is very important in many areas, particularly in geotechnical, earthquake, and environmental engineering. Over the years, V_S has typically been measured using downhole or crosshole seismic methods that require a borehole(s) since conventional surface seismic exploration methods, such as S-wave refraction and reflection, usually did not work well at sites with unconsolidated sediments. Seismic exploration methods using surface waves have greatly progressed over the last 20 years. Active surface wave methods, in which sledgehammers, weight drops, or shakers are used as energy sources, have rapidly gained popularity for near-surface site investigations. Active surface wave methods, however, have obvious limitations on penetration depth. Passive surface-wave methods, ambient-noise surface-wave methods, or microtremor array measurements (MAM), in which surface waves are extracted from ambient noise, are now recognized as a highly effective way to supplement the penetration depth of active methods. Seismic ambient noise mainly consists of surface waves, and the vertical component of ambient noise excites a significant Rayleigh wave component. The broadband nature (0.1 Hz to several tens of Hz) of these waves and their associated dispersion characteristics allow for the construction of a V_S model to depths of several meters to several kilometers. A large number of methods have been proposed and used to calculate the phase velocity from ambient noise, all of which require acquisition of data with multiple receivers, and an associated analysis of their phase velocities. Spatial auto-correlation (SPAC) is one of the oldest and the most commonly used methods since it relies on simple, yet robust, theory and processing techniques. Considering its simplicity, applicability, and stability, MAM will play an increasingly important role in engineering site investigations. This chapter briefly introduces active and passive surface wave methods and summarizes the fundamental theory of MAM.

302

It focuses on SPAC, reviews some recent developments, and introduces examples of the applications of the method for near-surface engineering site investigations.

10.1 Seismic Exploration Methods in Near-Surface Engineering Investigations

Seismic exploration methods have played an important role in civil engineering site investigations for many years (e.g., Yilmaz, 2015). For example, a method based on P-wave refraction is one of the most popular tools for geotechnical investigation (Zelt et al., 2013). The refraction method has been used prior to the construction of many tunnels and dams in order to estimate rock classification at the site. The addition of digital data acquisition and tomographic inversion (Hayashi and Saito, 1998) has greatly improved the method's applicability and reliability (Hayashi and Takahashi, 2001); however, most of the refraction applications are related to the fields of rock mechanics and not soil engineering. In rock mechanics, P-wave velocity (V_P) can be used for estimating rigidity of rocks. V_P, however, is not related to soil rigidity in soil mechanics due to the fact that V_P just indicates water velocity (1500 m/s) below the ground water table in unconsolidated soil, where S-wave velocity (V_S) is much lower than half of the water velocity ($<$ 750 m/s). Therefore, V_S is more useful in the soil engineering field for evaluating soil rigidity, with the P-wave refraction method used mainly in ground water surveys.

A refraction method using S-waves is a candidate for estimating soil rigidity, but such methods are not widely used. One reason is that the contrast of V_S in unconsolidated soil layers is generally small, and often high-velocity layers overlie low-velocity layers. As well, it is often difficult to acquire reliable data. In the S-wave refraction method, SH seismic motion must be used. P-waves and Love waves, however, are sometimes included and they complicate first arrival picking. Both of these characteristics make the refraction analysis of near-surface S-waves difficult.

Although S-wave reflection methods are another candidate, the method has not been widely used in soil engineering for many of the same reasons; the inclusion in data of direct waves, refracted waves, and surface waves contaminates the reflected waves and complicates reflection analysis. Therefore, it is difficult to obtain not only layered structures but also V_S through the reflection method in the unconsolidated soil ground.

Only downhole and crosshole seismic methods (Garofalo et al., 2016b) have been widely used in soil engineering in order to obtain reliable V_S models. However, these methods require a borehole(s) and can only provide one-dimensional (1D) structure. V_S tomography for the near-surface structure is still in an experimental phase and a method has not been widely used for the civil engineering

Table 10.1 *Widely used active and passive surface wave methods.*

	Method	Data acquisition	Array	Processing	Outcome	Remark
Active	SASW	Several sensors and a controlled source	1D	Cross-correlation	Phase velocity	
	MASW	Several tens of sensors and an impulsive source	1D	Phase shift and stack	Phase velocity	
Passive	Microtremor array measurements (MAM)	Spatially un-aliased array	1D or 2D	Spatial auto-correlation (SPAC)	Phase velocity	
			2D	FK	Phase velocity	
			1D	Tau-p transform	Phase velocity	Usually called ReMi
	Seismic interferometry	Spatially aliased array	1D or 2D	Cross-correlation	Group velocity	Still R & D phase
	HVSR (H/V)	Single 3 component sensor	-	No dispersion curve analysis	Horizontal to vertical spectral ratio	S-wave velocity is not obtained

investigations. Therefore, non-destructive methods that can provide a V_S model of the ground have been eagerly waited for many years in soil engineering.

10.2 Active and Passive Surface Wave Methods

During the past few decades, there has been significant development in the use of surface wave methods for V_S estimation (Garofalo et al., 2016a). Table 10.1 summarizes the active and passive surface wave methods that are currently the most widely used.

A spectral analysis of surface waves (SASW) has been used for the determination of 1D V_S profiles down to a depth of 100 m (Nazarian et al., 1983). SASW surveys rely on a controlled source such as a shaker or vibrator, from which practitioners calculate phase differences between the two receiver waveforms via cross-correlation.

Park et al. (1999) proposed the multichannel analysis of surface waves (MASW) method, whereby practitioners transform the multichannel surface wave data from the time-distance domain into the phase velocity-frequency domain, which they then use to determine phase velocities. MASW is superior to SASW for analyzing dispersion curves and distinguishing the fundamental mode Rayleigh wave from other higher modes and body waves. Xia et al. (1999) and Miller et al. (1999) applied MASW to shot gathers along a survey line and delineated pseudo two-dimensional (2D) V_S sections. Hayashi and Suzuki (2004) applied common mid-point (CMP) analysis as used in 2D seismic reflection surveying (Yilmaz, 2001) to MASW, and utilized common midpoint cross-correlation (CMPCC) analysis to increase the lateral resolution of the surface wave methods in heterogeneous environments. Nakata (2016) applied CMPCC to ambient-noise data after calculating cross-correlation to extract coherent Love waves and estimated near-surface 2D velocity model.

During the past few decades, researchers have made considerable progress toward the development of passive surface wave methods utilizing ambient noise or microtremors. The methods are typically called microtremor array measurements (MAM) in engineering since 2D arrays are usually used for calculating phase velocity from ambient noise. The receiver geometry is similar to the one discussed in section 7.2 (Shapiro, 2018), but typically smaller. Aki (1957) investigated the surface waves encoded in ambient noise and proposed a theory of spatial auto-correlation (SPAC). Okada (2003) developed a large-scale MAM based on SPAC in order to estimate deep V_S structure.

The frequency-wavenumber (FK) method (Capon, 1969) is another popular method for calculating the phase velocity from ambient noise, and for many years both SPAC and FK have been widely used. There is fairly general agreement that SPAC works with fewer receivers and more spatially irregular arrays compared with FK (Foti et al., 2018). As a result, in recent years SPAC has become more popular than the FK.

Building designers commonly use the average V_S, down to a depth of 30 m (V_{S30}), as a proxy for site response analysis and earthquake resistance in their building designs. In order to evaluate the V_{S30} quickly and inexpensively, Louie (2001) proposed the refraction microtremor method (ReMi), in which surface waves are recorded using a 1D linear array. This method requires ambient noise data to be transformed into the phase velocity-frequency domain using a $\tau - p$ transform.

Recently, researchers have been studying ambient noise in terms of seismic interferometry (Wapenaar, 2004). Tsai and Moschetti (2010) demonstrated that seismic interferometry (SI) in the time domain is equivalent to performing SPAC in the frequency domain. The main difference between SPAC and SI is the application methods. Conventional passive surface wave methods, such as SPAC, FK, and ReMi, require spatially un-aliased data to calculate phase velocity, and receivers must be deployed with relatively small spacing. Unlike conventional methods, SI for engineering applications is mainly used to calculate group velocity using spatially aliased data, with researchers obtaining and processing data from relatively sparse arrays. For scientific applications, SI is also used to calculate phase velocity at long periods (Lin et al., 2008). The group velocity calculation, however, is not straightforward compared with the phase velocity calculation, and generally requires a much longer record length. As a result, work on the method is ongoing, and it is not routinely used in production environments in near-surface engineering community.

Horizontal to vertical spectral ratio (HVSR or H/V) of ambient noise is another widely used seismic tool for site investigation (Nakamura, 1989). It is generally agreed that the ambient noise mainly consists of surface waves, and the HVSR is concerned with measuring the ellipticity of the Rayleigh waves. In addition, there is fairly general agreement that the HVSR relates to a natural period of S-waves at the site. For the reasons given above, the HVSR is widely used in investigations related to the seismic site response (Lermo and Chavez-Garcia, 1994). Unlike other surface wave methods, the HVSR provides neither phase nor group velocity information, so that estimating V_S only from the HVSR is generally difficult. The HVSR is often used in conjunction with the inversion of dispersion curve to supplement each other for both engineering applications (Suzuki and Yamanaka, 2010) and scientific applications (Lin et al., 2012; Shen and Ritzwoller, 2016).

Passive surface wave methods have several other advantages over active methods. These include deeper penetration, and a significant decrease in related engineering effort. Among various passive surface wave methods, the SPAC is currently most widely used in engineering site investigations, which motivates its focus below.

10.3 Data Acquisition

10.3.1 Array Shape

Passive surface wave methods usually use 2D arrays, and a variety of array shapes is employed in the data acquisition. Three typical arrays are illustrated in Figure 10.1. The circular array gives the most effective azimuthal averaging. The sparse nested or common-base triangles are the most efficient geometry, since these give sufficient azimuthal averaging in most cases. In areas of restricted access, L-shaped arrays are also useful. In some cases, useful SPAC data can be obtained with a linear array, and most simply a two-station array (Hayashi, 2008). However, it should be noted that irregular 1D arrays would require the omnidirectional propagation of ambient noise, and that 2D regular arrays are generally desirable.

10.3.2 Array Size and Penetration Depth

Figure 10.2 summarizes the relationship between array size and the maximum wavelength obtained from SPAC with various array shapes. The ratio of maximum

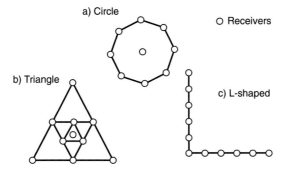

Figure 10.1. Three array geometries that have been typically used for microtremor array measurements: circular array using nine stations (a), nested triangles with 10 stations (b), and L-shaped array with 11 stations (c).

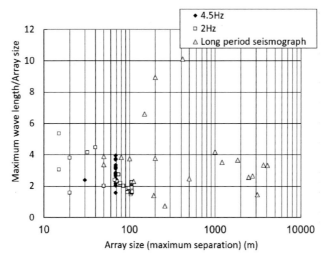

Figure 10.2. Relationship between array size (maximum receiver separation) and maximum wavelength obtained from SPAC with various array size, shapes, and sensors at many different sites mainly in Japan and western United States. Array size ranges from 15 to 4000 m, three different sensors (2 Hz and 4.5 Hz geophones and long-period seismometers) were used in the measurements.

wavelength to array size ranges from 0.7 to 10 and varies depending on the site. We can see that most data fall within the range of 2–4 wavelengths regardless of array size and shape. It is generally agreed that the penetration depth of surface wave methods is roughly one half to one quarter of maximum wavelength (e.g., Xia et al., 1999). We therefore conclude that the approximate penetration depth of the passive surface wave methods using SPAC is between 0.5 and 1 of the array size or maximum receiver separation (Foti et al., 2018).

10.3.3 Record Length

Clearly, longer data is better for statistical analysis in SPAC. However, the need for long data acquisition times decreases the convenience of the method, and recording microtremor data with appropriate data length is very important in practical surveys. Figure 10.3 shows an example of the relationship between data length and quality (Hayashi, 2008). Three hours of microtremor data were recorded with 2 Hz geophones in a triangular array with maximum receiver separation of 50 m at Pisa, Italy. Coherencies were calculated and compared with Bessel functions, and the root mean square (RMS) errors between the observed and theoretical coherencies are plotted against data length. It is clear that the RMS errors recorded after 10 minutes and after 3 hours are almost identical, which implies that a recording time of approximately 10 minutes is enough for phase velocity analysis using SPAC at this site. We performed similar experiments at many sites using different array sizes and arrived at the conclusion that 10–20 minutes of data is usually enough for arrays with a spacing of less than 100 m. The record length needs to be increased with the array size. Although data quality varies between sites, 30–40 minutes of data are required for the array size of several hundred meters, and approximately one hour of data is required for the array larger than one kilometer in general.

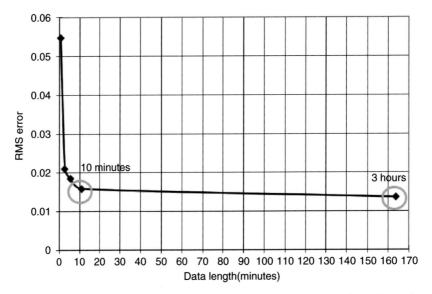

Figure 10.3. Example of the relationship between data length and quality obtained at Pisa, Italy. The array was a triangular with maximum receiver separation of 50 m. Vertical component of geophones with a natural frequency of 2 Hz were used in data acquisition. The root mean square (RMS) error between the observed and theoretical coherencies in a frequency range between 2 and 15 Hz are plotted against data length (from Hayashi, 2008).

10.4 Processing

This section summarizes various methods of phase velocity calculation used in active and passive surface wave methods. Although the cross-correlation or $\tau - p$ method is usually used in the processing of active data, these methods can be also applied to cross-correlations obtained from ambient noise in the context of seismic interferometry.

10.4.1 Cross-Correlation

The simplest method for obtaining phase velocity is thought to be the cross-correlation of two traces recorded at different positions. Let us start with two traces $f(t)$ and $g(t)$ obtained at two receivers with separation Δx. Waves are propagating parallel to a receiver array. Two traces are transformed into frequency domain by the Fourier transform and written as $F(\omega)$ and $G(\omega)$ using angular frequency ω:

$$F(\omega) = \frac{1}{2\pi} \int_{-\infty}^{+\infty} f(t)e^{-i\omega t} dt,$$

$$G(\omega) = \frac{1}{2\pi} \int_{-\infty}^{+\infty} g(t)e^{-i\omega t} dt. \tag{10.1}$$

The cross-correlation of the two traces $(CC_{fg}(\omega))$ can be defined in the frequency domain as

$$CC_{fg}(\omega) = F(\omega)\overline{G(\omega)} = A_f(\omega)A_g(\omega)e^{i\Delta\phi(\omega)}, \tag{10.2}$$

where $A_f(\omega)$ and $A_g(\omega)$ are amplitudes of $F(\omega)$ and $G(\omega)$, respectively, $\overline{G(\omega)}$ is the complex conjugate of $G(\omega)$, and $\Delta\phi(\omega)$ is the phase spectrum of the cross-correlation $CC_{fg}(\omega)$, equal to the phase difference of two traces. $\Delta\phi(\omega)$ can be simply calculated from the cross-correlation $CC_{fg}(\omega)$ as follows:

$$\Delta\phi(\omega) = \arctan\left(\frac{\text{Im}\left(CC_{fg}(\omega)\right)}{\text{Re}\left(CC_{fg}(\omega)\right)}\right), \tag{10.3}$$

and phase velocity $c(\omega)$ is directly related to the phase difference as

$$c(\omega) = \frac{\omega \Delta x}{\Delta\phi(\omega) + 2n\pi}, \tag{10.4}$$

where n is an integer.

10.4.2 $\tau - p$ Transform

The cross-correlation method in the previous section has serious limitations. These limitations are: (1) The receiver spacing must be chosen carefully in order to avoid phase wraparound. (2) The method cannot distinguish the fundamental mode of a dispersion curve from other modes or body waves. (3) The method cannot be applied to more than three traces. McMechan and Yedlin (1981) proposed a method that can calculate phase velocity directly from a multi-channel common shot gather, transforming time-domain data (time vs. distance) to the frequency domain (phase velocity vs. frequency) using the $\tau - p$ and Fourier transforms. This method has the additional advantage of allowing us to calculate the phase velocity directory from multi-channel (more than three traces) waveform data and to separate the fundamental mode of the phase velocity curve from higher modes and body waves visually. Park et al. (1999) also proposed a wavefield transformation, MASW, which calculates phase velocity directly from a multi-channel common shot gather (similar to the McMechan method). One significant difference is that McMechan and Yedlin (1981) calculate the apparent velocity (p) first, before transforming it to the frequency domain. On the contrary, Park et al. (1999) takes the opposite approach by transforming the gather into the frequency domain first and then calculating the phase velocity.

The method proposed by Park et al. (1999) can give a clear phase velocity curve even if the number of traces is limited and is summarized as follows:

1. Each observed trace ($f(x, t)$) is transformed into the frequency domain by the Fast Fourier Transform (FFT; $F(x, \omega)$).
2. The shot gather in the frequency domain is integrated over the spacing with respect to apparent phase velocities (c);

$$F(c, \omega) = \int_{-\infty}^{+\infty} \frac{F(x, \omega)}{|F(x, \omega)|} e^{i\omega x / c} dx. \tag{10.5}$$

3. The integration is repeated through all phase velocities (c) to be calculated.
4. The absolute value is calculated and plotted on phase velocity vs. angular frequency ($p(c, \omega) = |F(c, \omega)|$; Figure 10.4).
5. Finally, phase velocities are determined as the maximum amplitude at each frequency.

10.4.3 Spatial Auto-Correlation (SPAC)

In order to calculate the phase velocity of surface waves by either cross-correlation or MASW, the propagating direction has to be known. Microtremors, however, do not propagate specific direction, making it difficult to determine the propagation direction. For these data, Aki (1957) proposed the SPAC method in which microtremor data is statistically analyzed, enabling the extraction of phase velocity.

To summarize the SPAC method, let us start with two traces, $f(t)$ and $g(t)$, obtained at two receivers with separation Δx (Figure 10.5a). Waves are propagating parallel to a receiver array. Two traces are transformed into frequency

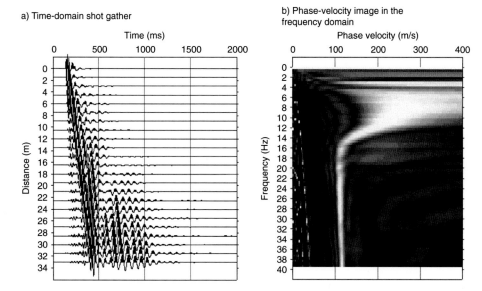

Figure 10.4. Example of phase velocity transformation. (a) Time-domain shot gather. (b) Phase-velocity image in frequency domain. Difference of brightness in (b) indicates difference of amplitude, and white and black indicate large and small amplitude respectively. Dispersion curve is defined as maximum amplitude (shown as white) in each frequency.

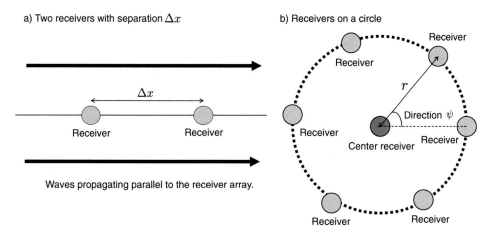

Figure 10.5. Spatial auto-correlation in 1D (a) and 2D (b). In the 1D wave propagation (a), waves propagate parallel to the receiver array and cross-correlation between two receivers goes to a cosine function. In the 2D array (b), cross-correlations are calculated between a receiver at the middle of the circle and receivers on the circle. The directional average of cosine functions goes to Bessel function.

domain by the Fourier transform and written as $F(\omega)$ and $G(\omega)$. Cross-correlation of the two traces $(CC_{fg}(\omega))$ can be defined as equation (10.2). The steps are as follows:

1. Substituting equation (10.2) into equation (10.4) gives

$$CC_{fg}(\omega) = A_f(\omega)A_g(\omega)e^{i\omega\Delta x/c(\omega)} \tag{10.6}$$

2. The complex coherency of two traces $COH_{fg}(\omega)$ is computed;

$$COH_{fg}(\omega) = \frac{CC_{fg}(\omega)}{A_f(\omega)A_g(\omega)} = e^{i\omega\Delta x/c(\omega)}, \tag{10.7}$$

and taking the real part of this gives

$$\mathrm{Re}\left(COH_{fg}(\omega)\right) = \cos\left(\frac{\omega\Delta x}{c(\omega)}\right). \tag{10.8}$$

3. SPAC for 2D array data is defined as the directional average of complex coherencies (equation (10.7)) giving

$$SPAC(r, \omega) = \frac{1}{2\pi}\int_0^{2\pi} COH(r, \psi, \omega)d\psi, \tag{10.9}$$

where r is the separation of two receivers (or radius of a circle), and ψ is the direction of two sensors. In essence, equation (10.9) calculates the complex coherencies for two sensors with separation r and angular separation ψ and averages these coherencies around a circle (Figure 10.5b).

4. The directional average of trigonometric functions reduces to a Bessel function as in Okada (2003), or

$$J_0(kr) = \frac{1}{2\pi}\int_0^{2\pi} \cos(kr, \psi)d\psi, \tag{10.10}$$

where k is the wavenumber. Using $k = \omega/c(\omega)$, equation (10.10) can be written as

$$J_0\left(\frac{\omega r}{c(\omega)}\right) = \frac{1}{2\pi}\int_0^{2\pi} \cos\left(\frac{\omega r}{c(\omega)}, \psi\right)d\psi, \tag{10.11}$$

5. Finally, equations (10.9) and (10.11) are combined to get

$$\mathrm{Re}\left(SPAC(r, \omega)\right) = J_0\left(\frac{\omega r}{c(\omega)}\right). \tag{10.12}$$

The left term of the equation (10.12) can be calculated from observed microtremor data, and resembles a Bessel function as shown in Figure 10.6a.

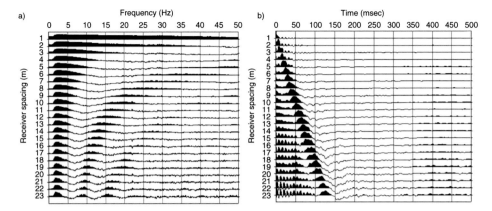

Figure 10.6. Example of SPAC in frequency domain (a) and time domain (b). Firstly, the SPAC was calculated in frequency domain from ambient noise data. (a) shows the real part of the SPAC and we can recognize that the SPAC appeared as Bessel functions and their shape was changing with receiver spacing. Secondly, inverse Fourier transform transformed the SPAC in frequency domain (a) to time domain (b). The SPAC in time domain (b) appears as a shot gather and it is equivalent to seismic interferometry (SI).

In reality, the phase velocity is calculated by theoretical relationship in equation (10.12) with the measured spatial auto-correlation, and tweaking the phase velocity $c(\omega)$ to minimize the error. The velocity that minimizes this error can be considered as the phase velocity at the angular frequency ω. The imaginary part of $SPAC(r, \omega)$ goes to zero in the directional average of ambient noise. It should be noted that the seismic interferometry (SI) is essentially the same as SPAC when we process surface waves. SPAC in the frequency domain (Figure 10.6a) is equivalent to SI in the time domain (Figure 10.6b).

The spatial auto-correlation defined in equation (10.12) can only be applied to isotropic arrays, such as circles or triangles, and cannot be applied to anisotropic arrays, such as lines or L-shapes, due to the required directional average. However, if we assume that ambient noise does not come from some specific direction, and comes from all directions equally, the directional average in equation (10.9) can be calculated even if arrays are anisotropic (Capon, 1973). Generally, the direction of ambient-noise propagation is not stable. Averaging a long enough time of ambient-noise data enables us to calculate the directional average of equation (10.9) correctly.

10.4.4 Inversion

A V_S model is generally obtained from phase velocity curves, which are first calculated through the waveform processing outlined in the previous sections by a nonlinear inversion. There are many nonlinear inversion methods, such as a least-squares method, a genetic algorithm, and simulated annealing (e.g., Menke, 2012). For the sake of simplicity, this section describes a traditional nonlinear, least-squares method in which the number and thickness of subsurface layers are fixed through the inversion process, and the layer-specific V_S is the only unknown. As an alternative, both V_S and thickness of each layer are often estimated in the inversion as mentioned in section 10.7. V_P and density can be related to the V_S linearly with empirical equations (e.g., Kitsunczaki et al., 1990; Ludwig et al., 1970) in each step of iteration. The 1D velocity calculation based on Monte Carlo is discussed in Chapter 7 (Shapiro, 2018).

For the inversion, the set of V_S values in each layer can be written as a 1D vector $\mathbf{x}$ as

$$\mathbf{x}^T = \left(V_{S_1}, V_{S_2}, \cdots V_{S_M}\right), \tag{10.13}$$

where V_{S_i} defines the V_S values for the ith layer. As an objective function, we minimize the L2 difference between the estimated and calculated phase velocity values:

$$\left[\sum_i^N \left(f_i^{obs} - f_i^{cal}(\mathbf{x})\right)^2\right]^{1/2}, \tag{10.14}$$

where N is the number of observed phase velocity data, f^{obs} observed phase velocities, and f^{cal} synthetic phase velocities computed from the given V_S model. A theoretical phase velocity curve can be calculated by the matrix method (e.g., Saito and Kabasawa, 1993). Now, a Jacobian matrix can be written as

$$\mathbf{a} = \begin{pmatrix} \frac{\partial f_1^{cal}(\mathbf{x})}{\partial V_{S_1}} & \frac{\partial f_1^{cal}(\mathbf{x})}{\partial V_{S_2}} & \cdots & \frac{\partial f_1^{cal}(\mathbf{x})}{\partial V_{S_M}} \\ \frac{\partial f_2^{cal}(\mathbf{x})}{\partial V_{S_1}} & \frac{\partial f_2^{cal}(\mathbf{x})}{\partial V_{S_2}} & \cdots & \frac{\partial f_2^{cal}(\mathbf{x})}{\partial V_{S_M}} \\ \vdots & \vdots & \vdots & \vdots \\ \frac{\partial f_N^{cal}(\mathbf{x})}{\partial V_{S_1}} & \frac{\partial f_N^{cal}(\mathbf{x})}{\partial V_{S_2}} & \cdots & \frac{\partial f_N^{cal}(\mathbf{x})}{\partial V_{S_M}} \end{pmatrix}. \tag{10.15}$$

The unknown vector $\mathbf{x}$ is in derivatives, and it makes the inversion nonlinear. One of the consequences is that an initial guess model must be constructed, which must predict the data sufficiently well in order to obtain convergence. In actual calculation, elements of the Jacobian matrix $\mathbf{a}$ are usually calculated numerically using a finite-difference method (Xia et al., 1999). A residual between observed and theoretical phase-velocities can be expressed as vector $\mathbf{y}$:

$$\mathbf{y}^T = \begin{pmatrix} f_1^{obs} - f_1^{cal}(\mathbf{x}) \\ f_2^{obs} - f_2^{cal}(\mathbf{x}) \\ \vdots \\ f_N^{obs} - f_N^{cal}(\mathbf{x}) \end{pmatrix}. \tag{10.16}$$

The correction vector $\Delta\mathbf{x}$ to a given guess for $f(\mathbf{x})$ can be calculated by the least-squares method, which is given by

$$\left(\mathbf{a}^T\mathbf{a} + \varepsilon\mathbf{I}\right)\Delta\mathbf{x} = \mathbf{a}^T\mathbf{y}, \tag{10.17}$$

where ε is a damping parameter stabilizing the inversion and $\mathbf{I}$ is an identity matrix. In the lth iteration, a new estimated model $\mathbf{x}^{l+1}$ is calculated as $\mathbf{x}^{l+1} = \mathbf{x}^l + \gamma\Delta\mathbf{x}$, where γ is a constant specifying the step length.

Generally, the nonlinear and non-unique nature of geophysical inverse problems requires the introduction of spatial regularization to converge to a physically meaningful result. In this case, it is common to regularize the variance in V_S between each layer as follows:

$$\left(\mathbf{a}^T\mathbf{a} + \alpha\mathbf{r_V}^T\mathbf{r_V} + \varepsilon\mathbf{I}\right)\Delta\mathbf{x} = \mathbf{a}^T\mathbf{y}, \tag{10.18}$$

where $\mathbf{r}_V$ is the difference of V_S between two successive layers given by

$$\mathbf{r_V} = \begin{pmatrix} 1 & -1 & 0 & \cdots & 0 & 0 \\ 0 & 1 & -1 & \cdots & 0 & 0 \\ \vdots & \vdots & \vdots & \vdots & \vdots & \vdots \\ 0 & 0 & 0 & \cdots & 1 & -1 \end{pmatrix} \begin{pmatrix} V_{S_1} \\ V_{S_2} \\ \vdots \\ V_{S_N} \end{pmatrix}, \tag{10.19}$$

and α is the weight of regularization, and the large α makes an inverted model smoother.

10.5 Higher Modes and Inversion Using Genetic Algorithm

Most active and passive surface wave methods assume that dispersion curves are dominated by the fundamental mode, and analysis including higher modes remains as an active research topic. The higher modes may, however, dominate in several types of velocity structures, such as in situations where a high-velocity layer overlays a low-velocity layer, or where high-velocity layers are embedded between low-velocity layers. These complex velocity models often appear in real engineering problems. This section introduces characteristics of data including higher modes and discusses the associated velocity models. We focus on active data for simplicity, although passive data show similar characteristics.

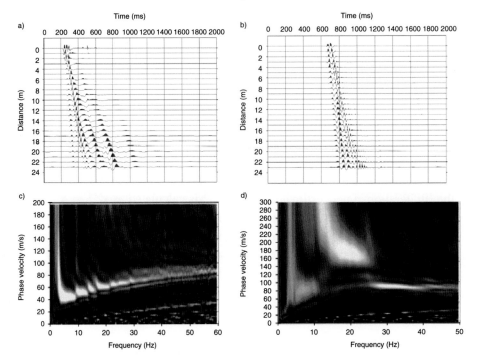

Figure 10.7. Observed waveform data (a and b) and their phase velocity images (c and d) for models Case-1 and 2. Data were obtained from active surface wave methods using a 10 kg sledgehammer and 4.5 Hz geophones (vertical component). Twenty-four geophones were deployed with 1 m spacings, and sources were placed 1 m away from the first receiver. The data (a and c) were recorded on an alluvium plain where a high-velocity thin layer overlays a low-velocity peat layer. The data (b and d) were recorded at a site where a high-velocity thin sand or gravel layer is embedded in low-velocity soft clay layers.

10.5.1 Example of Observed Waveform Data

Figure 10.7 shows examples of waveform data and their phase velocity images in the frequency domain. The waveform data in Figure 10.7a was recorded on an alluvium plain where a high-velocity thin layer overlays a low-velocity peat layer. In the phase velocity image (Figure 10.7c), above 5 Hz, phase velocity increases as frequency increases, but the dispersion curve is discontinuous at frequencies above 10 Hz. This type of dispersion curve cannot be explained by a fundamental mode of Rayleigh waves. In the following discussion, this type of curve and model will be referred to as Case 1.

The waveform data in Figure 10.7b was recorded at a site where a high-velocity thin sand or gravel layer is embedded in low-velocity soft clay layers. In the phase velocity image (Figure 10.7d), a dispersion curve is clearly discontinuous around 10 Hz and it is obvious that the velocities estimated for higher frequencies do not

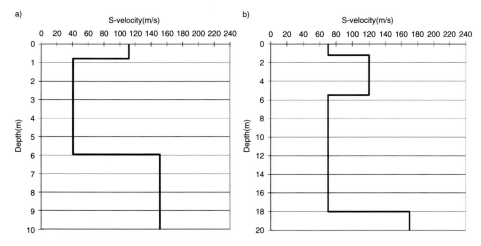

Figure 10.8. Estimated velocity models for Case-1 (a) and Case-2 (b) that generate significant higher modes.

correspond to the fundamental mode. This being said, there is a vague dispersion curve below a phase velocity of 100 m/sec, which may correspond to the fundamental mode of Rayleigh waves. This curve is difficult to delineate, however. In the following discussion, this type of curve and model will be referred to as Case 2.

V_S models that generate phase velocity images similar to those shown in Figure 10.7 are shown in Figure 10.8. Figure 10.9 shows theoretical waveforms (a and b) for the models shown in Figure 10.8 calculated by the discrete wave number method (DWM) (Bouchon and Aki, 1977), and phase velocity images (c and d) for these theoretical waveforms are calculated by the multi-channel analysis of surface waves (MASW) method. DWM can calculate the full waveform including body waves using the same source-receiver configuration, and we can consider these waveforms as idealized replicates of observed data in terms of phase velocity and amplitude. The theoretical phase velocity images (Figure 10.9c–d) look similar to the observed phase velocity images shown in Figure 10.7c–d.

Complex dispersion curves including higher modes cannot be explained by a fundamental mode of Rayleigh waves; however, we can explain these complex dispersion curves in terms of a Normal Mode Solution (NMS), including higher modes. To obtain this solution we solve the characteristic equation of Rayleigh waves (Saito and Kabasawa, 1993). Both phase velocity and relative amplitude are calculated for each mode of Rayleigh waves. The NMS only takes into account surface waves, and a source at infinite distance is assumed.

The character of the DWM ensures that all the relevant modes are included in Figure 10.9. Solid lines indicate dispersion curves computed via NMS, broken lines indicate relative amplitudes, and white circles indicate phase velocities that have a maximum amplitude at the given frequency. We see from Figure 10.9 that observed

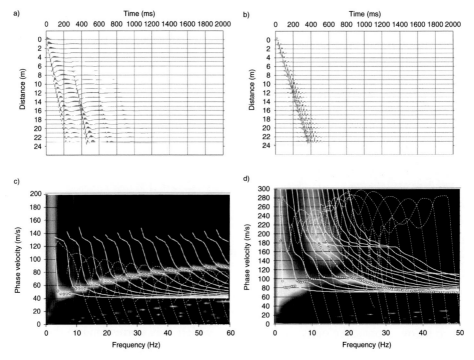

Figure 10.9. Theoretical waveform data (a and b) and their phase velocity images (c and d) for Case-1 (a and c) and Case-2 (b and d). Fundamental and higher-mode dispersion curves (solid lines) and their relative amplitude (broken lines) with a phase-velocity image of theoretical waveform data calculated by the DWM are shown in (c) and (d).

dispersion curves can be explained as phase velocities whose amplitude is maximum at each frequency, and discontinuity of dispersion curves is associated with transition from one mode to another mode. It also should be noted that the fundamental mode cannot be observed through the whole frequency range, and that the mode to which observed phase velocities belong may be difficult to recognize in actual observed data.

10.5.2 Inversion Using Genetic Algorithm

The conventional nonlinear, least-squares method described in previous section cannot be applied to the complex dispersion curves presented in this section, as the dispersion curves may be discontinuous. However, many researchers (e.g., Xia et al., 2000; Supranata et al., 2007) have proposed inversion methods that can include these higher modes. Here, we present a method that uses phase velocities with maximum amplitude at variable frequencies in order to estimate V_S models in such complex geological contexts.

In this method, maximum amplitude phase velocities are compared with observed phase velocities. The method uses a Normal Mode Solution (NMS) for the theoretical phase velocity calculation and assumes a source is located at infinite distance. It takes into account only surface waves and neglects body waves. The theoretical dispersion curve is defined by taking the phase velocity of the mode with maximum amplitude at a given frequency.

The discontinuous nature of the phase velocity curves requires that we forgo partial derivative calculation. In order to estimate velocity models without the calculation of partial derivatives, a genetic algorithm (Yamanaka and Ishida, 1995) is used in the inversion. The method is characterized as a global search method and can avoid local minima to which the inversion of dispersion curve may fall. One clear disadvantage of the genetic algorithm (GA) is that the method requires a large amount of forward modeling compared to the conventional iterative nonlinear, least-squares method. From this point of view, the NMS is much better than the full-waveform calculation for forward modeling of inversion, as it is computationally less time consuming. The agreement between the dispersion characteristics computed by both methods (Figure 10.9) allows us to proceed with confidence.

Dispersion curves shown in Figures 10.9c–d calculated by the discrete wave number method (DWM) are treated as observed data. Initial V_S models are created by a simple wavelength transformation (Xia et al., 1999) in which wavelength is calculated from phase velocity divided by frequency, then divided by three, and plotted at depth (Figures 10.10a–b). V_S models are represented as 15 thin layers, and the V_S of each layer is optimized by GA (Hayashi, 2008). The search area of V_S in GA is set to $\pm 50\%$ of the initial V_S. The number of higher modes in NMS is up to 15 modes. Spatial regularization similar to equation (10.20) was applied to converge to a physically meaningful result.

Figure 10.10 shows the velocity models (a and b) that have a minimum residual defined by equation (10.14) and a comparison of observed (calculated by DWM) and theoretical (calculated by NMS) dispersion curves (c and d). Although the obtained velocity models are smoother compared with true models, we can see that almost true velocity models are obtained, and the residual between observed and theoretical dispersion curves is much smaller than the initial models.

How to process surface wave data including higher modes is still an active research topic, and many different approaches have been proposed. The approach using GA is just one example of various processing approaches. Analyzing surface wave data in terms of guided waves or group velocity is another popular method (Ritzwoller and Levshin, 2002). Active and passive surface waves generally may include more or less higher modes depending on velocity structures. It is important to take higher modes into consideration when we process surface wave data, particularly for engineering applications.

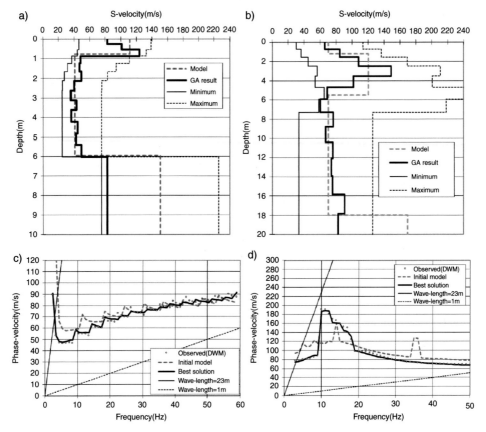

Figure 10.10. Comparison of true and resultant S-wave velocity models obtained by inversion (a and b) and comparison of observed (DWM) and theoretical (NMS) dispersion curves (c and d) for Case-1 (a and c) and Case-2 (b and d). (a) and (b) also show the search area (minimum and maximum) of genetic algorithm.

10.6 Application to Buried Channel Delineation

10.6.1 Near-Surface S-Wave Velocity Structure and Local Site Effect

Surface ground motion from earthquakes is highly dependent on subsurface geological structure. The local site effect may be defined as the effect of near-surface geological structure on surface ground motion during an earthquake. To estimate the local site effect, V_S to a depth of several tens of meters, such as the average V_S down to 30 m (AV_{S30}), is very popular worldwide. Table 10.2 summarizes the AV_{S30} and the site class mentioned in the international building code (FEMA, 2003).

Only a few attempts have so far been made at new observation methods for evaluating the local site effects in recent decades. The methods included the multi-channel analysis of surface waves (Park et al., 1999), the refraction

Table 10.2 *Site class definition (FEMA, 2003) based on average S-wave velocity to a depth of 30 m.*

Site class	Soil profile name	Soil shear wave velocity (Vs)	
		feet/sec	m/sec
A	Hard rock	$V_S > 5000$	$V_S > 1500$
B	Rock	$2500 < V_S \leq 5000$	$760 < V_S \leq 1500$
C	Very dense soil and soft rock	$1200 < V_S \leq 2500$	$360 < V_S \leq 760$
D	Stiff soil profile	$600 < V_S \leq 1200$	$180 < V_S \leq 360$
E	Soft soil profile	$V_S < 600$	$V_S < 180$

microtremor method (Louie et al., 2001), and the horizontal to vertical spectra ratio of ambient noise (Nakamura, 1989). However, a great deal of effort has been made in the collection of existing borehole data in the local site effects evaluation. What seems to be lacking, however, is the development of new non-destructive observation methods for near-surface velocity structures. For example, most existing borehole data has only blow counts (N-value) obtained by the standard penetration test (ASTM, 2011) and no V_S, which is the most important for the local site effects. Therefore, people have recently started to apply active and passive surface wave methods for estimating V_S structures down to a depth of several tens to several hundreds of meters from the surface. What follows is a description of a case study in the Saitama prefecture in Japan (Hayashi and Inazaki, 2005; Hayashi et al., 2006), where we intended to map a bedrock depth using the passive surface wave method.

10.6.2 Outline of Test Site

The test site is a 13.5×12 km rectangle in Saitama prefecture, Japan (Figure 10.11a). The site is placed in the Nakagawa Lowland area, and topography is almost flat. National Institute of Advanced Industrial Science and Technology (AIST) has collected existing borehole data in the site, and the approximate geological condition is already well known. A suspension PS-logging (Kitsunezaki, 1980) has been carried out in a borehole in the test site as well. The test site consists of buried channels and buried terraces (Figure 10.12a). The depth to bedrock (thickness of alluvium) is about 50 m in the channels and 15 m on the terraces. The main purpose of the passive surface wave method is the delineation of buried channels and terraces.

10.6.3 Data Acquisition and Analysis

Data acquisition was carried out at 218 points in the test site. Geophones that have the natural frequency of 2 Hz were used as receivers. Triangular arrays were

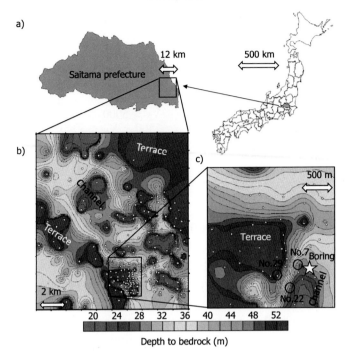

Figure 10.11. (a) The test site is a 13.5 × 12 km rectangle in Saitama prefecture, Japan. (b) Depth of the VS of 250 m/s in the three-dimensional VS model obtained through the survey. White circles indicate the points where passive surface wave method was carried out. The depth can be considered as the estimated depth of alluvium bottom (bedrock). (c) Closeup around the borehole. The white star indicates the location of the boring shown in Figure 10.13. Black circles (No.7, No.22, and No.29) indicate the sites mentioned in Figures 10.12 and 10.13. (A black-and-white version of this figure appears in some formats. For the color version, please refer to the plate section.)

used at four points and L-shaped arrays were used at 214 points. L-shaped arrays were deployed along road crossings. The triangular arrays consist of 10 receivers, and the L-shaped arrays consist of 11 receivers. Figures 10.1b–c schematically show the configuration of triangular and L-shaped arrays. All receivers were connected to the seismograph through a spread cable. The size of array was 40 to 80 m, sampling time was 2 ms, and about 10 minutes of ambient noise data were recorded. It took about one hour for the data acquisition at each point.

In the phase velocity analysis, the SPAC (Okada, 2003) described in section 10.4.3 was employed. In this survey, phase velocity curves (e.g., Figure 10.12b) were calculated in the frequency range between 2 and 10 Hz. Phase velocity curves can be obtained clearly in all observation points. A 1D inversion using a nonlinear, least-squares method similar to that presented in section 10.4.4 has been applied

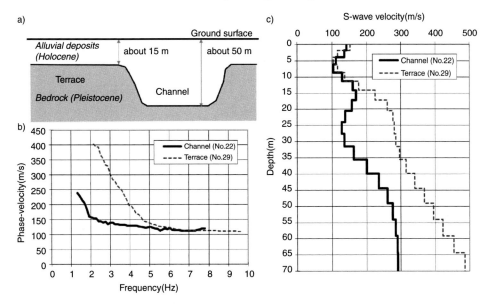

Figure 10.12. Schematic diagram of geology at the test site (a). The test site consists of buried channels and buried terraces. Typical phase velocity curves (b) and S-wave velocity profiles (c) on terrace (No.29) and in channel (No.22) obtained from passive surface waves method. The locations of the profiles (No.22 and No.29) are shown in Figure 10.11 (modified from Hayashi et al., 2006).

to the phase velocity curves, and 1D V_S structures down to the depth of 70 m were obtained (Figure 10.12c). In the inversion, we used the following relationship between V_P (km/s) and V_S (km/s) (Kitsunezaki et al., 1990):

$$V_P = 1.29 + 1.11 V_S, \tag{10.20}$$

and between density (ρ; g/cm^3) and V_P (km/s) as (Ludwig et al., 1970):

$$\rho = 1.2475 + 0.3999 V_P - 0.026 V_P^2. \tag{10.21}$$

The resultant 1D V_S profiles are horizontally interpolated into a three-dimensional (3D) V_S model with the minimum curvature method.

10.6.4 Survey Results

Figure 10.12b compares the typical phase velocity curves on a buried terrace and in buried channel. It is obvious that the phase velocity curves on terrace and in channel have large differences. It implies that the V_S structures also have a large difference between terrace and channel.

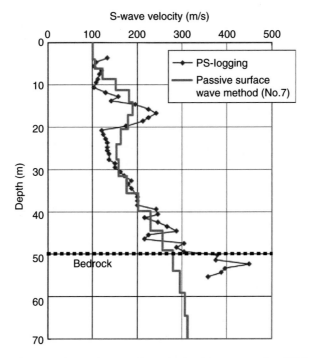

Figure 10.13. Comparison of V_S profiles obtained by suspension PS-logging and the passive surface wave method around the borehole (No.7). Thick broken line indicates the depth of bedrcok confirmed at the boring. Locations of the borehole and site No.7 are shown in Figure 10.11 (modified from Hayashi et al., 2006).

A nonlinear, least-squares inversion was applied to each dispersion curve and 1D V_S profiles were estimated. Figure 10.12c shows the typical V_S profiles on the buried terrace and in the buried channel. It is obvious that V_S beneath a depth of 10 m on terrace and in channel have large differences, Figure 10.13 shows the comparison of V_S profiles obtained by suspension PS-logging and the passive surface wave method at the borehole shown as a white star in Figure 10.11c. Although a high-velocity layer placed in the depth between 15 and 20 m is not clear, the velocity structure obtained through the passive surface wave method agrees with PS-logging very well.

The bottom of the alluvial layer is defined at the depth of 50 m in the borehole. V_S from the passive surface wave method at the depth of 50 m is approximately 250 m/s. Therefore, we assumed the V_S of 250 m/s is the boundary of alluvium and Pleistocene (bedrock). Figure 10.11 shows the depth of the V_S of 250 m/s in the 3D V_S model obtained through the passive surface wave method. The map shown in Figure 10.11 can be considered as the estimated depth of alluvium bottom (bedrock) obtained through the survey.

10.7 Application for Evaluating the Effect of Basin Geometry on Site Response

10.7.1 Basin Edge Effect

Observations from several recent severe earthquakes and subsequent research (e.g., Kawase, 1996; Hayashi et al., 2008) have revealed that 2D or 3D deep V_S structure (to a depth of several kilometers) has a large effect on intermediate- to long-period (0.5 to 5 s) ground motion in tectonic basins, such as the San Francisco Bay area. Such local site amplification due to the 2D or 3D structure is sometimes called the "Basin edge effect" (Kawase, 1996; Graves et al., 1998). Most studies on basin velocity structure rely on geological information, surface and borehole geophysical data, and observed earthquake records. In general, geophysical data and seismic stations are too sparsely distributed and much of the borehole data is too shallow to adequately characterize deep V_S structure. To establish more accurate basin velocity structure, there is a need for more closely spaced deep V_S measurements.

We measured V_S profiles at eleven sites in the east San Francisco Bay area using active and passive surface wave methods (Hayashi and Craig, 2017). The sites were placed around the Hayward fault and the Calaveras fault. The 30-year probabilities of magnitude 6.7 or greater earthquakes on the Hayward-Rogers Creek and Calaveras faults have been estimated at 32% and 25%, respectively (Field et al., 2015). These faults run through densely populated areas, and knowledge of a detailed 2D or 3D V_S structure along the faults is needed in order to estimate local site effects due to a potential earthquake.

10.7.2 Site of Investigation

Sites of investigation are shown in Figure 10.14. The Hayward and Calaveras faults are oriented northwest-southeast. Displacement along the faults is primarily strike slip. The two faults bound the East Bay Hills block, which has 50 to 500 m of topographic relief. As shown on the site map (Figure 10.14), seven sites are located on the west side of the Hayward fault (11, 57, 65, 66, 67, 68, and 84), two sites are located on the east side of the Calaveras fault (59 and 69), and two sites are located between the two faults (58 and 64).

10.7.3 Data Acquisition

Data acquisition methods included MASW, HVSR, a passive surface wave method using geophones in a linear array (Linear-MAM), and three-component, long-period accelerometers in a large array (Large-MAM).

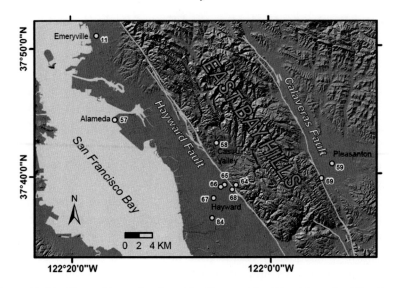

Figure 10.14. Sites of investigation. 11 - Emeryville, 57 - Alameda, 58 - Castro Valley, 59 - Pleasanton, 64 - CSU East Bay, 65 - Charles Ave, 66 - Huntwood Ave, 67 - Southgate Park, 68 - Cemetery, 69 - Alviso Adobe, 84 - Eden Shores Park (from Hayashi and Craig, 2017). (A black-and-white version of this figure appears in some formats. For the color version, please refer to the plate section.)

MASW surveys (Park et al., 1999; Xia et al., 1999) were conducted using 24 or 48 receivers, 4.5 Hz geophones, and a 1 or 2 m receiver interval. A 10 kg sledge-hammer was used as an energy source. Shot records with a sample rate of 1 ms and data length of 2.0 s were recorded using a 24- or 48-channel seismographic system. Ambient noise was also recorded for Linear-MAM with the geophone array used for the MASW survey described above. About 10 minutes of ambient noise data with a 2 ms sampling rate were recorded at each site.

We performed large-scale microtremor array measurements (Large-MAM) in which maximum separation ranged from 200 to 2300 m, depending on the site. We recorded microtremor data for an interval ranging from 10 minutes to 1 hour using a 10 ms sample rate, for a total of several hours of data acquisition per site. Seismographs were placed in relatively quiet locations such as parks and residential areas. The seismographs utilize three-component accelerometers and include a GPS clock to synchronize data between multiple seismographs. Recorded three-component ambient noise data was used for the HVSR analysis.

10.7.4 Data Processing

A $\tau - p$ transform (Park et al., 1999) described in section 10.4.2 was used to cal-culate dispersion curves from shot gathers obtained by MASW, and SPAC (Okada,

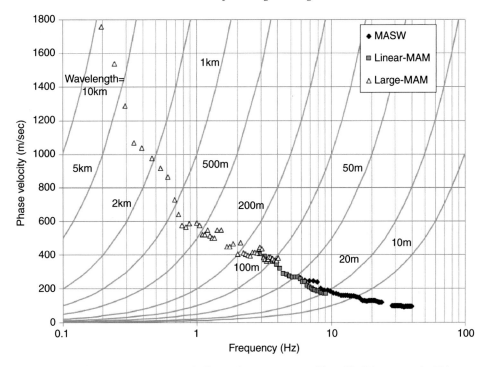

Figure 10.15. Comparison of dispersion curves at Site 59 (Pleasanton). Triangles indicate phase velocities obtained from Large-MAM. Squares indicate phase velocities obtained from Linear-MAM. Diamonds indicate phase velocities obtained from MASW. Solid lines indicate wavelength (from Hayashi and Craig, 2017).

2003) described in section 10.4.3 was used to calculate dispersion curves from ambient noise obtained by Linear- and Large-MAM.

Dispersion curves obtained by the three different methods are in excellent agreement in the frequency ranges where they overlap (Figure 10.15). Maximum wavelengths obtained using the Large-MAM, Linear-MAM, and MASW were about 10,000, 150, and 30 m, respectively. As a rule of thumb, the penetration depth of the surface wave method is about one-half to one-third of the maximum Rayleigh wave wavelength (Xia et al., 1999). The penetration depth of the Large-MAM method is much greater than that of conventional surface wave methods, such as MASW or ReMi (Linear-MAM). Phase velocities obtained from the three methods were combined to produce a single dispersion curve for each site.

An inversion scheme (Suzuki and Yamanaka, 2010) was applied to the observed dispersion curves to develop V_S profiles for the eleven sites. During the inversion, the observed data were the phase velocities of a dispersion curve, the horizontal to vertical spectral ratio (HVSR), and the peak frequency of the HVSR. The unknown parameters were layer thickness and V_S. A genetic algorithm (Yamanaka and Ishida, 1995), similar to that presented in section 10.5.2, was used for optimization. The search area for the inversion was determined based on initial velocity models created by a simple wavelength transformation in which wavelengths calculated from phase velocity and frequency pairs were divided by three and mapped as depth.

The theoretical phase velocity and HVSR were defined as an effective mode that was generated by calculating the weighted average of the fundamental mode and higher modes (up to the 5th mode) based on the medium response. The inversion was performed based on minimization of differences between the observed and the effective mode phase velocities and HVSR.

An example of observed and theoretical dispersion curves at Pleasanton (site 59) is shown in Figure 10.16, where the observed curve is shown in red and the yellow circles indicate the effective mode of theoretical phase velocities. The theoretical dispersion curve (effective mode) agrees reasonably well with the observed data.

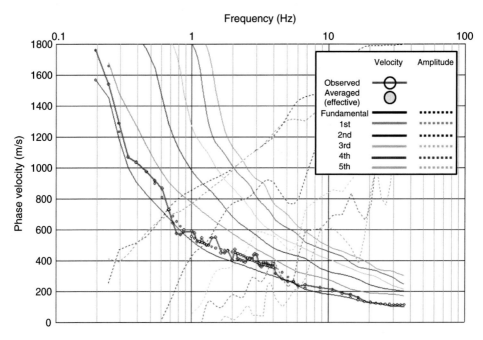

Figure 10.16. Comparison of observed and theoretical dispersion curves (Site 59: Pleasanton). Solid line with circles indicates observed dispersion curve. Solid and broken lines indicate fundamental and higher-mode theoretical dispersion curves and their relative amplitude (response of the medium). Grey circles (yellow circles in the color version) indicate the effective mode of theoretical phase velocities (from Hayashi and Craig, 2017). (A black-and-white version of this figure appears in some formats. For the color version, please refer to the plate section.)

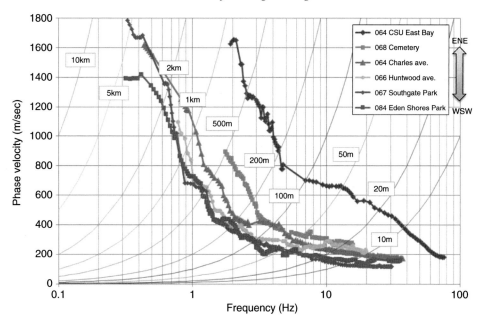

Figure 10.17. Comparison of dispersion curves at Hayward. Lines with markers indicate observed phase velocities obtained from active and passive surface wave methods. Solid lines indicate wavelength (from Hayashi and Craig, 2017). (A black-and-white version of this figure appears in some formats. For the color version, please refer to the plate section.)

10.7.5 Analysis Results

Figure 10.17 shows dispersion curves for six sites in Hayward. Phase velocities at Site 64 (CSU East Bay), located immediately to the east of the Hayward fault, are substantially higher than those at the other four sites. Phase velocities decrease from east to west, for example, from Site 68 (Cemetery) to Site 84 (Eden Shores Park) as the distance from the Hayward fault increases. Horizontal to vertical spectral ratio (HVSR) for six sites in Hayward is shown in Figure 10.18. There is no clear peak in a HVSR at Site 64 (CSU East Bay), located immediately to the east of the Hayward fault. In contrast, there are clear peaks in the HVSR of Site 68 (Cemetery) to 84 (Eden Shores Park). The peaks decrease from 1.5 Hz to 0.5 Hz as distance from the Hayward fault increases. Both changes of dispersion curves and HVSRs indicate that V_S decrease from east to west as distance from the Hayward Faults increases and the basin sediments evidently thicken.

A comparison of V_S profiles at Hayward is shown in Figure 10.19. A very thin near-surface layer (about 4 m thick) with V_S less than 300 m/s is present at Site 64 (CSU East Bay). At five other sites, a shallow stiff sediment layer with V_S more than 300 m/s was determined at a depth ranging from 8 to 53 m, depending on the

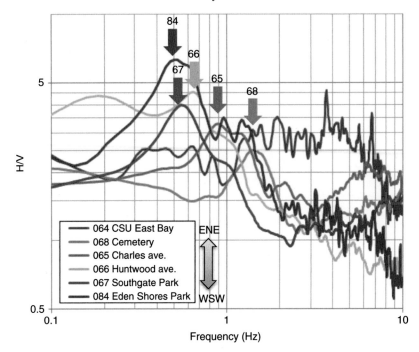

Figure 10.18. Comparison of horizontal to vertical spectral ratio (HVSR) obtained from three-component ambient noise data at Hayward. Solid lines indicate observed HVSR and down arrows indicate the peaks of HVSR. There are clear peaks in the HVSR of Site 68 (Cemetery) to 84 (Eden Shores Park). The peaks decrease from 1.5 Hz to 0.5 Hz as distance from the Hayward fault increases (from Hayashi and Craig, 2017). (A black-and-white version of this figure appears in some formats. For the color version, please refer to the plate section.)

site. An intermediate velocity layer with V_S greater than 700 m/s was calculated for a depth range from 40 to 250 m. This layer is shallow at Site 64 (CSU East Bay) and gets deeper from east (Site 68) to west (Site 84) as distance from the Hayward Fault to each site increases. A velocity layer with V_S greater than 1200 m/s was detected at all sites. The depth to the layer is about 160 m at Site 64 and 400 to 700 m at other sites. A velocity layer with V_S greater than 1500 m/s was determined at a depth of about 200 m at Site 64 (CSU East Bay), 800 m at Site 65 (Charles Ave.), and greater than 1400 m at Sites 67 (Southgate Park) and 84 (Eden Shores Park).

A schematic V_S section based on the V_S profiles obtained by this study is shown in Figure 10.20. V_S profiles show significant differences among the sites. On the west side of the Hayward fault and the east side of the Calaveras fault there is a low-velocity layer at the surface, with V_S less than 700 m/s to a depth of 100 to 300 m. A thick intermediate-velocity layer with V_S ranging from 700 to 1500 m/s lies beneath the low-velocity layer. Bedrock with V_S greater than 1500 m/s was measured at a depth of approximately 1700 m. Between the Hayward fault and the

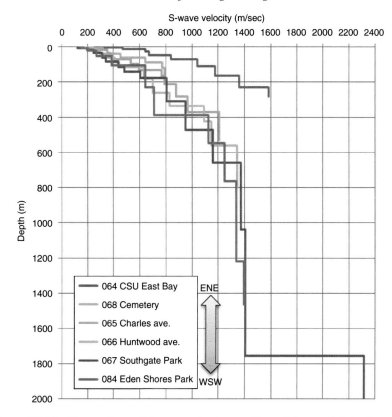

Figure 10.19. Comparison of S-wave velocity profiles at Hayward. It is clear that VS decrease from east to west as distance from the Hayward fault increases and the basin sediments evidently thicken (from Hayashi and Craig, 2017). (A black-and-white version of this figure appears in some formats. For the color version, please refer to the plate section.)

Calaveras fault, thicknesses of the low-velocity layer and the intermediate-velocity layer are less than 50 m and 250 m, respectively. Depth to bedrock is less than 300 m at these sites. As we cross the Hayward fault from east to west, depths to an intermediate-velocity layer with V_S greater than 700 m/s and a layer with V_S greater than 1200 m/s increase about 40 and 300 m respectively between Site 64 and Site 68, which are only 800 m apart. Depth to a bedrock with V_S greater than 1500 m/s increases by at least 500 m as well.

10.7.6 Two-Dimensional Amplification Across the Hayward Fault

To evaluate the effect of a significant change of bedrock depth on surface ground motion due to an earthquake, a representative V_S cross section perpendicular to the

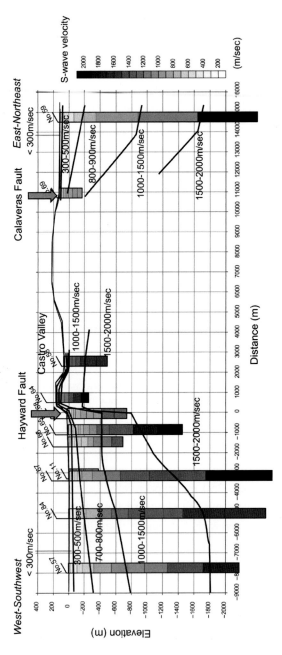

Figure 10.20. Schematic V_S section based on the V_S profiles obtained in the study. The horizontal axis indicates distance from the Hayward fault to each investigation site. Down arrows indicate the locations of the Hayward and Calaveras faults (from Hayashi and Craig, 2017).

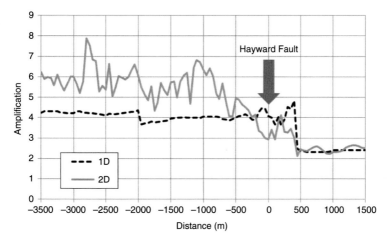

Figure 10.21. Comparison of maximum amplification in a frequency range of 0 to 5 Hz. The horizontal axis indicates distance from the Hayward fault and the vertical axis indicates amplification that is surface ground motion divided by the initiated wave. Down arrow indicates the location of the Hayward fault (from Hayashi and Craig, 2017).

Hayward fault was constructed and theoretical amplification of SH waves including 2D structure was calculated using a viscoelastic finite-difference method (Levander, 1988; Robertsson et al., 1994). A plane SH wave was initiated at the bottom of the model, and response at the ground surface was recorded. The surface ground motion was divided by the initiated wave and examined in frequency domain as amplification. To evaluate the effect of 2D structure, 1D amplification was also calculated by a reflectivity method (Mullar, 1985).

A comparison of maximum amplification in a frequency range of 0 to 5 Hz is shown in Figure 10.21. On the west side of the Hayward fault, 2D amplification at a frequency of 1 Hz is approximately six times that of incident waves. The 2D amplification is particularly large (seven to eight times) at distances of -2800 and -1000. In contrast, amplification is about four times in the 1D calculation. The calculation results imply that the low-frequency (0.5 to 5 Hz) component of ground motion can be locally amplified on the west side of the Hayward fault because of the effect of a lateral change in the depth of basement.

10.8 Conclusions

This chapter introduced ambient-noise surface wave methods for near-surface engineering, summarized the fundamental theory, reviewed some recent developments, and introduced examples of the applications for near-surface engineering site investigations. Most conventional seismic exploration methods utilize deterministic wave fields, and evaluating data quality is generally straightforward. Unlike the conventional methods, passive surface wave methods utilize stochastic wave fields and evaluating the quality or reliability of data is not intuitive. Although many processing software packages are available, understanding the fundamental theory of the methods described in the chapter would be valuable for evaluating data quality. Preliminary processing in the field is also required to evaluate the data quality since raw data obtained from these methods looks noisy. Applicability and reliability of the method depend very much on site conditions, and it is dangerous to rely on only one method. Combining several different methods, for example active and passive methods, is generally recommended. Storing field examples in a database and sharing the database and empirical knowledge are also important.

Acknowledgments

I would like to thank all the people of OYO Corporation and Geometrics Inc. for giving me this opportunity to write the manuscript. I wish to express my gratitude to Dr. Tomio Inazaki of Public Works Research Institute for his support of field work and the development of surface wave methods. A very special thanks to all the people of Mony Exploration Corporation and Japan Measure Survey for their support of field work. I would like to thank Prof. Mitchell Craig of California State University, East Bay and his students for their support of field work and the development of surface wave methods. I would like to thank the editors for giving me this opportunity to write the chapter. Comments from two anonymous reviewers greatly improved the manuscript. I also appreciate the effort of Dennis Wilkison and Sue Wilkison in manuscript preparation.

References

Aki, K. 1957. Space and time spectra of stationary stochastic waves, with special reference to microtremors. *Bull. Earthq. Res. Inst.*, **35**, 415–456.

ASTM. 2011. Standard test method for standard penetration test (SPT) and split-barrel sampling of soils. *ASTM International*, D1586–11.

Bouchon, M., and Aki, K. 1977. Discrete wave number representation of elastic wave fields in three-space dimensions. *J. Geophys. Res.*, **84**, 3609–3614.

Capon, J. 1969. High-resolution frequency-wavenumber spectrum analysis. *Proc. of the IEEE*, **57**, 1408–1418.

Capon, J. 1973. Signal processing and frequency-wavenumber spectrum analysis for a large aperture seismic array. *Methods in Computational Physics*, **13**.

FEMA. 2003. NEHRP recommended provisions for seismic regulations for new buildings and other structures. *FEMA*.

Field, E. H., Biasi, G. P., Bird, P., Dawson, T. E., Felzer, K. R., Jackson, D. D., Johnson, K. M., Jordan, T. H., Madden, C., Michael, A. J., Milner, K. R., Page, M. T., Parsons, T., Powers, P. M., Shaw, B. E., Thatcher, W. R., Weldon, R. J., and Zeng, Y. 2015.

Long-term time-dependent probabilities for the third uniform California earthquake rupture forecast (UCERF3). *Bull. Seismol. Soc. Am.*, **111**, 511–543.

Foti, S., Hollender, S., Garofalo, F., Albarello, D., Asten, M., Bard, P.-Y., Comina, C., Cornou, C., Cox, B., Giuilio, G. D., Forbriger, T., Hayashi, K., Lunedei, E., Martin, A., Mercerat, D., Ohrnberger, M., Poggi, V., Renalier, F., Sicilia, D., and Socco, L. V. 2018. Guidelines for the good practice of surface wave analysis: a product of the InterPACIFIC project. *Bull. Earthquake Eng.*, **16**, 2367–2420.

Garofalo, F., Foti, S., Hollender, F., Bard, P. Y., Cornou, C., Cox, B. R., Dechamp, A., Ohrnberger, M., Perron, V., Sicilia, D., Teague, D., and Vergniault, C. 2016b. InterPACIFIC project: comparison of invasive and non-invasive methods for seismic site characterization. Part II: Intra-comparison between surface-wave and borehole methods. *Soil Dyn. Earthquake Eng.*, **82**, 241–254.

Garofalo, F., Foti, S., Hollender, F., Bard, P. Y., Cornou, C., Cox, B. R., Ohrnberger, M., Sicilia, D., Asten, M., Giulio, G. D., Forbriger, T., Guillier, B., Hayashi, K., Martin, A., Matsushima, S., Mercerat, D., Poggi, V., and Yamanaka, H. 2016a. InterPACIFIC project: comparison of invasive and non-invasive methods for seismic site characterization. Part I: Intra-comparison of surface wave methods. *Soil Dyn. Earthquake Eng.*, **82**, 222–240.

Graves, R. W., Pitarka, A., and Somerville, P. G. 1998. Ground-motion amplification in the Santa Monica area: Effects of shallow basin-edge structure. *Bull. Seismol. Soc. Am.*, **88**, 1224–1242.

Hayashi, K. 2008. *Development of surface-wave methods and its application to site investigations*. Ph.D. thesis, Kyoto University.

Hayashi, K., and Craig, M. 2017. S-wave velocity measurement and the effect of basin geometry on site response, east San Francisco Bay area, California, USA. *Phys. Chem. Earth.*, **98**, 49–61.

Hayashi, K., and Inazaki, T. 2005. Buried channel delineation using microtremor array measurements. *SEG Expanded Abstracts*, 1137–1140.

Hayashi, K., and Saito, H. 1998. High resolution seismic refraction method - Development and application. *Butsuri-Tansa (in Japanese)*, **51**, 471–491.

Hayashi, K., and Takahashi, T. 2001. High resolution seismic refraction method using surface and borehole data for site characterization of rocks. *Int. J. Rock Mech. Min. Sci.*, **38**, 807–813.

Hayashi, K., and Suzuki, H. 2004. CMP cross-correlation analysis of multi-channel surface-wave data. *Exploration Geophysics*, **35**, 7–13.

Hayashi, K., Hirade, T., Iiba, M., Inazaki, T., and Takahashi, H. 2008. Site investigation by surface-wave method and micro-tremor array measurements at central Anamizu, Ishikawa prefecture. *Butsuri-Tansa (in Japanese)*, **61**, 483–498.

Hayashi, K., Inazaki, T., and Suzuki, H. 2006. Buried incised channels delineation using microtremor array measurement at Soka and Misato cities in Saitama prefecture. *Bull. Geol. Sur. Japan (in Japanese)*, **57**, 309–325.

Kawase, H. 1996. Cause of the damage belt in Kobe: "The basin-edge effect," constructive interference of the direct S-wave with the basin-induced diffracted/Rayleigh waves. *Seismol. Res. Lett.*, **67**(5), 25–34.

Kitsunezaki, C. 1980. A new method for shear-wave logging. *Geophysics*, **45**, 1489–1506.

Kitsunezaki, C., Goto, N., Kobayashi, Y., Ikawa, T., Horike, M., Saito, T., Kurota, T., Yamane, K., and Okuzumi, K. 1990. Estimation of P- and S-wave velocities in deep soil deposits for evaluating ground vibrations in earthquake. *Sizen-saigai-kagaku (in Japanese)*, **9**, 1–17.

Lermo, J., and Chavez-Garcia, F. J. 1994. Site effect evaluation at Mexico City: dominant period and relative amplification from strong motion and microtremor records. *Soil Dyn. Earthquake Eng.*, **13**, 413–423.

Levander, A. R. 1988. Fourth-order finite-difference P-SV seismograms. *Geophysics*, **53**(11), 1425–1436.

Lin, F.-C., Moschetti, M. P., and Ritzwoller, M. H. 2008. Surface wave tomography of the western Uunited States from ambient seismic noise: Rayleigh and Love wave phase velocity maps. *Geophys. J. Int.*, **173**, 281–298.

Lin, F.-C., Schmandt, B., and Tsai, V. 2012. Joint inversion of Rayleigh wave phase velocity and ellipticity using USArray: Constraining velocity and density structure in the upper crust. *Geophys. Res. Lett.*, **39**, L12303.

Louie, J. N. 2001. Faster, better: Shear-wave velocity to 100 meters depth from refraction microtremor arrays. *Bull. Seismol. Soc. Am.*, **91**(2), 347–364.

Louie, J. N., Chavez-Perez, S., Henrys, S., and Bannister, S. 2001. Multimode migration of scattered and converted waves for structure of the Hikurangi Slab Interface, New Zealand.

Ludwig, W. J., Nafe, J. E., and Drake, C. L. 1970. Seismic refraction. *In the Sea*, **4**(1), 53–84.

McMechan, G. A., and Yedlin, M. J. 1981. Analysis of dispersive waves by wave field transformation. *Geophysics*, **46**, 869–874.

Menke, W. 2012. *Geophysical data analysis: discrete inverse theory*. Vol. 3. 225 Wyman Street, Waltham, MA 02451, USA: Academic press.

Miller, R. D., Xia, J., Park, C. B., and Ivanov, J. M. 1999. Multichannel analysis of surface waves to map bedrock. *The Leading Edge*, **18**, 1392–1396.

Mullar, G. 1985. The reflectivity method: a tutorial. *J. Geophys.*, **58**, 153–174.

Nakamura, Y. 1989. A method for dynamic characteristics estimation of subsurface using microtremors on the ground surface. *Quarterly Report of Railway Technical Research Institute*, **30**, 25–33.

Nakata, N. 2016. Near-surface S-wave velocities estimated from traffic-induced Love waves using seismic interferometry with double beamforming. *Interpretation*, **4**, SQ23–SQ31.

Nazarian, S., Stokoe, K. H., and Hudson, W. R. 1983. Use of spectral analysis of surface waves method for determination of moduli and thickness of pavement system. *Transp. Res. Rec.*, **930**, 38–45.

Okada, H. 2003. *The microtremor survey method*. Society of Exploration Geophysicists.

Park, C. B., Miller, R. D., and Xia, J. 1999. Multichannel analysis of surface waves. *Geophysics*, **64**(3), 800–808.

Ritzwoller, M. H., and Levshin, A. L. 2002. Estimating shallow shear velocities with marine multi-component seismic data. *Geophysics*, **67**, 1991–2004.

Robertsson, J. O. A., Blanch, J. O., and Symes, W. W. 1994. Viscoelastic finite-difference modeling. *Geophysics*, **59**, 1444–1456.

Saito, M., and Kabasawa, H. 1993. Computations of reflectivity and surface wave dispersion curves for layered media II. Rayleigh Wave Calculations. *BUTSURI-TANSA*, **46**(4), 283–298.

Shapiro, N. 2018. Applications with surface waves extracted from ambient seismic noise. Pages — of: Nakata, N., Gualtieri, L., and Fichtner, A. (eds.), *Seismic Ambient Noise*. Cambridge University Press, Cambridge, UK.

Shen, W., and Ritzwoller, M. H. 2016. Crustal and uppermost mantle structure beneath the United States. *J. Geophys. Res.*, **121**, 4306–4342.

Supranata, Y. E., Kalinski, M. E., and Ye, Q. 2007. Improving the uniqueness of surface wave inversion using multiple-mode dispersion data. *Int. J. Geomech.*, Sept., 333–343.

Suzuki, H., and Yamanaka, H. 2010. Joint inversion using earthquake ground motion records and microtremor survey data to S-wave profile of deep sedimentary layers. *Butsuri-Tansa (in Japanese)*, **65**, 215–227.

Tsai, V. C., and Moschetti, M. P. 2010. An explicit relationship between time-domain noise correlation and spatial autocorrelation (SPAC) results. *Geophys. J. Int.*, **182**, 454–460.

Wapenaar, K. 2004. Retrieving the elastodynamic Green's function of an arbitrary inhomogeneous medium by cross correlation. *Phys. Rev. Lett.*, **93**, 254301.

Xia, J., Miller, R. D., and Park, C. B. 1999. Configuration of near-surface shear-wave velocity by inverting surface wave. *Proceedings of the Symposium on the Application of Geophysics to Engineering and Environmental Problems*, 95–104.

———. 2000. Advantages of calculating shear-wave velocity from surface waves with higher modes. *SEG Expanded Abstracts*, 1295–1298.

Yamanaka, H., and Ishida, H. 1995. Phase velocity inversion using Genetic Algorithms. *J. Struct. Constr. Eng.*, **468**, 9–17.

Yilmaz, O. 2015. *Engineering Seismology with Applications to Geotechnical Engineering.* Society of Exploration Geophysicists.

———. 2001. *Seismic Data Analysis.* 2nd edn. Investigations in Geophysics, vol. 10. Tulsa, USA: Society of Exploration Geophysicists.

Zelt, C. A., Haines, S., Powers, M. H., Sheehan, J., Rohdewald, S., Link, C., Hayashi, K., Zhao, D., Zhou, H., Burton, B. L., Petersen, U. K., Bonal, N. D., and Doll, W. E. 2013. Blind test of methods for obtaining 2D near-surface seismic velocity models from first-arrival traveltimes. *J. Environ. Eng. Geophys.*, **18**, 183–194.

Epilogue

Ambient Noise Seismology – The *Status Quo*

Seismic ambient noise has been the subject of rigorous scientific studies at least since the pioneering work of Timoteo Bertelli in the second half of the nineteenth century. Not many physical phenomena have been studied for so long and are still *en vogue* today.

It is not an exaggeration to write that work on ambient noise during the past decade has revolutionized seismology, a field that some might have thought to be in its asymptotic stage. While ambient noise seismology continues to be an exciting and dynamic topic, we think it is the right time for a book that summarizes what has been achieved so far.

Studies on ambient noise have built a bridge between solid-Earth geophysics and the ocean and atmospheric sciences. Hints of the bridge have certainly been visible in the fog for a long time, but it has become concrete only recently. Observations of seismic noise can be used to characterize the ocean wave state, providing information that complements observations from satellites and buoys. Seismic noise has emerged as a powerful tool to track extreme weather conditions such as tropical cyclones.

Ambient noise seismology rests on and has motivated many beautiful theories. They relate ocean waves to ambient noise at periods of several seconds to nearly one hour, or explain the emergence of deterministic signals from spatial or temporal noise correlations. Early versions of many of these theories have been known for a long time as well, but their practical significance has only recently become obvious for the majority of seismologists.

Certainly, ambient noise seismology owes much of its current prominence to the emergence of Green's function approximations from noise correlations, and to the simplicity of this operation when using modern processing tools and computational resources. Green's function retrieval from noise has liberated seismologists, not completely but to a large extent, from the corset of energetic, transient sources such as earthquakes and explosions. The result is unprecedented data coverage and resolution in tomographic images of the Earth's interior. When averaged over sufficiently long times, the omnipresent ambient field opens a new window into the time-variable Earth, enabling the continuous monitoring of active systems, such as volcanoes, fault zones, and reservoirs.

The literal transformation of noise into signal that has marked seismological research during the past decade is strongly correlated with a boost of creativity in seismic signal processing. Rapid advances in our knowledge of ambient noise sources and Earth structure would be unthinkable without ingenious processing methods that suppress transient signals, or stabilize and accelerate Green's function retrieval.

Some Potential Future Developments

While predicting the future is technically not possible, trying to do so is definitely a lot of fun (especially when looking back to it years later). This is enough motivation for us to write a few lines on where we think the field of ambient noise seismology might go, either to overcome remaining limitations or to harness emerging potentials.

Interdisciplinary studies on the energy transfer between the atmosphere/ocean system and the solid Earth are clearly in their infancy but already foreshadow a colorful palette of new applications. The characterization of weather and climate phenomena using historical seismic data from the pre-satellite era is already within reach. When performed systematically, the analysis of seismic data could add independent information to ocean wave and climate models, and advance our knowledge on the long-time effects of climate change.

A key ingredient in achieving these goals is models of ambient noise sources with significantly improved resolution, in time, frequency, and space. This requirement will raise questions concerning (1) the configuration and type of seismic stations for noise source studies (traditional continental seismic stations versus stations on the ocean floor or floating in the ocean), and (2) the physics of ambient noise propagation from the source, across the ocean-continent boundary, and through a complex continental lithosphere. More specific problems in this context include the excitation of Love and Lg waves, and of high-frequency seismic radiation above 1 Hz.

More detailed models of seismic noise sources are also likely to enable a new class of interferometric techniques. Traditional interferometry critically relies on Green's function retrieval, which involves a leap of faith related to the unfulfilled requirement of wavefield equipartitioning. Interferometry without Green's function retrieval has the potential to improve tomographic resolution by reducing systematic modeling errors and by exploiting correlation waveforms that would otherwise be classified as unphysical. Regardless of the type of interferometry, much work remains to be done on the reliability of noise correlation waveforms, beyond the phase of the fundamental-mode surface waves. This has the potential to produce superior spatio-temporal resolution of Earth structure, possibly including seismic attenuation, with likely consequences for our understanding of the Earth's dynamics.

Emerging technologies such as deep learning, distributed acoustic sensing, or rotational ground motion measurements are likely to affect the future trajectory of ambient noise seismology. Whether they truly advance our understanding of system Earth remains to be seen.

These are just some of the possible new directions and developments we envision for the next years to come. Yet, from this short list it is clearly evident that, despite the impressive improvements of the last decades, we cannot conclude that the field of seismic ambient noise has expressed its full potential, and very exciting times are indeed expected to lie ahead of us.

                                                    Nori, Lucia & Andreas

Index

Printed in the United States
by Baker & Taylor Publisher Services